Pinpoint Math

Teacher's Guide
Level E

Volumes 1–6

Wright Group

Photo Credits
©iStock International Inc., cover.

Acknowledgements
Content Consultant:

Linda Proudfit, Ph.D.

After earning a B.A. and M.A in Mathematics from the University of Northern Iowa, Linda Proudfit taught junior- and senior-high mathematics in Iowa. Following this, she earned a Ph.D. in Mathematics Education from Indiana University. She currently is Coordinator of Elementary Education and Professor of Mathematics Education at Governors State University in University Park, IL.

Dr. Proudfit has made numerous presentations at professional meetings at the local, state, and national levels. Her main research interests are problem solving and algebraic thinking.

www.WrightGroup.com

Copyright © 2009 by Wright Group/McGraw-Hill.

All rights reserved. Except as permitted under the United States Copyright Act, no part of this publication may be reproduced or distributed in any form or by any means, or stored in a database or retrieval system, without the prior written permission from the publisher, unless otherwise indicated.

Permission is granted to reproduce the material contained on pages 361–402 on the condition that such material be reproduced only for classroom use; be provided to students, teachers, or families without charge; and be used solely in conjunction with *Pinpoint Math*.

Printed in USA.

Send all inquiries to:
Wright Group/McGraw-Hill
P.O. Box 812960
Chicago, IL 60681

ISBN 978-1-40-4568259
MHID 1-40-4568255

2 3 4 5 6 7 8 9 WDQ 13 12 11 10

Contents

Tutorial Guide x
Overview of *Pinpoint Math*™ xx

Volume 1: Number Sense and Place Value

Topic 1 Place Value through 100

Topic 1 Objective Chart 1
Topic 1 Introduction 2
Lesson 1-1 Tens and Ones 3–5
Lesson 1-2 Numbers to 100 6–8
Lesson 1-3 Compare and Order to 100 9–11
Lesson 1-4 Equivalent Forms 12–14
Topic 1 Summary 15
Topic 1 Mixed Review 16

Topic 2 Place Value through 1,000

Topic 2 Objective Chart 17
Topic 2 Introduction 18
Lesson 2-1 Numbers to 1,000 19–21
Lesson 2-2 Write Numbers to 1,000 22–24
Lesson 2-3 Compare and Order to 1,000 25–27
Topic 2 Summary 28
Topic 2 Mixed Review 29

Topic 3 Place Value beyond 1,000

Topic 3 Objective Chart 30
Topic 3 Introduction 31
Lesson 3-1 Place Value to 10,000 32–34
Lesson 3-2 Expanded Notation with Zeros 35–37
Lesson 3-3 Numbers in the Millions 38–40
Lesson 3-4 Round Numbers through Millions 41–43
Topic 3 Summary 44
Topic 3 Mixed Review 45

Contents

Volume 2: Basic Facts

Topic 4 Addition and Subtraction Facts
Topic 4 Objective Chart 46
Topic 4 Introduction 47
Lesson 4-1 Add or Subtract to Solve Problems 48–50
Lesson 4-2 Properties of Addition 51–53
Lesson 4-3 Addition Strategies 54–56
Lesson 4-4 Add Three 1-Digit Numbers 57–59
Lesson 4-5 Relate Addition and Subtraction 60–62
Lesson 4-6 Subtraction Strategies 63–65
Lesson 4-7 Fact Families 66–68
Topic 4 Summary 69
Topic 4 Mixed Review 70

Topic 5 Multiplication Facts
Topic 5 Objective Chart 71
Topic 5 Introduction 72
Lesson 5-1 Meaning of Multiplication 73–75
Lesson 5-2 Multiply by 2, 5, and 10 76–78
Lesson 5-3 Properties of Multiplication 79–81
Lesson 5-4 Multiplication Strategies 82–84
Lesson 5-5 Basic Multiplication Facts 85–87
Topic 5 Summary 88
Topic 5 Mixed Review 89

Topic 6 Division Facts
Topic 6 Objective Chart 90
Topic 6 Introduction 91
Lesson 6-1 Meaning of Division 92–94
Lesson 6-2 Properties of Zero and One 95–97
Lesson 6-3 Divide by 2, 3, 4, or 5 98–100
Lesson 6-4 Divide by 6, 7, 8, or 9 101–103
Lesson 6-5 Relate Multiplication and Division 104–106

Topic 6 Summary 107
Topic 6 Mixed Review 108

Volume 3: Add and Subtract
Topic 7 Add or Subtract 1- and 2-Digit Numbers
Topic 7 Objective Chart 109
Topic 7 Introduction 110
Lesson 7-1 1-Digit and 2-Digit Numbers 111–113
Lesson 7-2 Add 2-Digit Numbers 114–116
Lesson 7-3 Subtract 2-Digit Numbers 117–119
Topic 7 Summary 120
Topic 7 Mixed Review 121

Topic 8 Add or Subtract Multidigit Numbers
Topic 8 Objective Chart 122
Topic 8 Introduction 123
Lesson 8-1 Add or Subtract Mentally 124–126
Lesson 8-2 Add and Subtract with Estimation 127–129
Lesson 8-3 Add and Subtract 3-Digit Numbers 130–132
Lesson 8-4 Whole Numbers to 10,000 133–135
Lesson 8-5 Multidigit Numbers 136–138
Topic 8 Summary 139
Topic 8 Mixed Review 140

Volume 4: Multiply and Divide
Topic 9 Use Multiplication to Compute
Topic 9 Objective Chart 141
Topic 9 Introduction 142
Lesson 9-1 Multiply by Multiples of 10 143–145
Lesson 9-2 Estimate Products 146–148
Lesson 9-3 Multiply: Four Digits by One Digit 149–151
Lesson 9-4 Choose a Method and Multiply 152–154

Contents

Topic 9 Summary 155
Topic 9 Mixed Review 156

Topic 10 Use Division to Compute
Topic 10 Objective Chart 157
Topic 10 Introduction 158
Lesson 10-1 Divide Multiples of 10 159–161
Lesson 10-2 Estimate Quotients 162–164
Lesson 10-3 Divide by 1-Digit Numbers 165–167
Lesson 10-4 Choose a Method for Division 168–170
Lesson 10-5 Multiplication and Division 171–173
Topic 10 Summary 174
Topic 10 Mixed Review 175

Topic 11 Expressions and Equations
Topic 11 Objective Charts 176
Topic 11 Introduction 177
Lesson 11-1 Write Expressions for Patterns 178–180
Lesson 11-2 Write Expressions 181–183
Lesson 11-3 Write Equations with Unknowns 184–186
Lesson 11-4 Solve Equations with Unknowns 187–189
Topic 11 Summary 190
Topic 11 Mixed Review 191

Volume 5: Data, Geometry, and Measurement
Topic 12 Graphing
Topic 12 Objective Chart 192
Topic 12 Introduction 193
Lesson 12-1 Compare Data with Graphs 194–196
Lesson 12-2 Record Data 197–199
Lesson 12-3 Mean, Median, and Mode 200–202
Lesson 12-4 Show Data in More Than One Way 203–205

Lesson 12-5 Graph Ordered Pairs 206–208

Topic 12 Summary...209

Topic 12 Mixed Review 210

Topic 13 Basic Geometric Figures

Topic 13 Objective Chart 211

Topic 13 Introduction..212

Lesson 13-1 Angles and Lines 213–215

Lesson 13-2 Types of Polygons... 216–218

Lesson 13-3 Triangles.. 219–221

Lesson 13-4 Quadrilaterals 222–224

Lesson 13-5 Circles 225–227

Topic 13 Summary.. 228

Topic 13 Mixed Review..229

Topic 14 Measurement Conversion

Topic 14 Objective Chart 230

Topic 14 Introduction..231

Lesson 14-1 U.S. Customary Units... 232–234

Lesson 14-2 Basic Metric Prefixes 235–237

Lesson 14-3 Use the Metric System 238–240

Lesson 14-4 Factors in Unit Conversions 241–243

Lesson 14-5 Convert Units within a System................................. 244–246

Topic 14 Summary 247

Topic 14 Mixed Review 248

Topic 15 Measure Geometric Figures

Topic 15 Objective Charts 249

Topic 15 Introduction... 250

Lesson 15-1 Length 251–253

Lesson 15-2 Perimeter 254–256

Lesson 15-3 Area 257–259

Contents

Lesson 15-4 Area of Rectangles...................................... 260–262
Lesson 15-5 Volume 263–265
Topic 15 Summary.. 266
Topic 15 Mixed Review 267

Volume 6: Fractions and Decimals
Topic 16 Meaning of Fractions
Topic 16 Objective Chart 268
Topic 16 Introduction 269
Lesson 16-1 Basics of Fractions 270–272
Lesson 16-2 Unit Fractions 273–275
Lesson 16-3 Parts and the Whole 276–278
Lesson 16-4 Rename Values Greater Than 1.......................... 279–281
Topic 16 Summary 282
Topic 16 Mixed Review 283

Topic 17 Equivalence of Fractions
Topic 17 Objective Chart 284
Topic 17 Introduction.. 285
Lesson 17-1 Compare Fractions 286–288
Lesson 17-2 Factor Whole Numbers................................... 289–291
Lesson 17-3 Equivalent Fractions................................... 292–294
Lesson 17-4 Fractions in Lowest Terms 295–297
Lesson 17-5 Interpretations of Fractions........................... 298–300
Topic 17 Summary... 301
Topic 17 Mixed Review 302

Topic 18 Addition and Subtraction of Fractions
Topic 18 Objective Chart 303
Topic 18 Introduction 304
Lesson 18-1 Add and Subtract Fractions............................. 305–307
Lesson 18-2 Unlike Denominators: Add............................... 308–310
Lesson 18-3 Unlike Denominators: Subtract 311–313

Lesson 18-4 Add Mixed Numbers..314–316

Lesson 18-5 Subtract Mixed Numbers.......................................317–319

Topic 18 Summary..320

Topic 18 Mixed Review...321

Topic 19 Decimals and Money

Topic 19 Objective Chart..322

Topic 19 Introduction..323

Lesson 19-1 Money...324–326

Lesson 19-2 Tenths and Hundredths.....................................327–329

Lesson 19-3 Compare and Order Decimals............................330–332

Lesson 19-4 Use Coins and Bills..333–335

Lesson 19-5 Decimal Notation for Money.............................336–338

Lesson 19-6 Money in Fractions and Decimals......................339–341

Lesson 19-7 Unit Costs...342–344

Topic 19 Summary..345

Topic 19 Mixed Review...346

Topic 20 Decimal Operations and Comparisons

Topic 20 Objective Charts..347

Topic 20 Introduction..348

Lesson 20-1 Add and Subtract Decimals...............................349–351

Lesson 20-2 Decimal and Fraction Equivalents......................352–354

Lesson 20-3 Use a Number Line..355–357

Topic 20 Summary..358

Topic 20 Mixed Review...359

Teaching Aid Masters.. 360

Student Action Plan...383

Glossary..403

Index..412

Tutorial Guide

Each of the standards listed below has at least one animated tutorial for students to use with the lesson that matches the objective. If you are using the electronic components of *Pinpoint Math*, you will find a complete listing of Tutorial codes and titles when you access them either online or via CD-ROM.

Level E

Standards by Topic   Tutorial Codes

Volume 1 Number Sense and Place Value	
Topic 1 Place Value through 100	
1.1 Count and group objects in ones and tens.	1a Modeling Numbers with Base Ten Blocks
1.2 Count, read, and write whole numbers to 100.	1b Working with Whole Numbers, 0 to 100
1.3 Compare and order whole numbers to 100 using the symbols for less than, equal to, or greater than (<, =, >).	1c Ordering Whole Numbers, Example A
1.4 Represent equivalent forms of the same number through the use of physical models, diagrams, and number expressions.	1d Using Equivalent Forms to Represent Numbers, Example A
1.4 Represent equivalent forms of the same number through the use of physical models, diagrams, and number expressions.	1e Using Equivalent Forms to Represent Numbers, Example B
Topic 2 Place Value through 1,000	
2.1 Count, read, and write whole numbers to 1,000 and identify the place value for each digit.	2a Identifying the Place Value of Each Digit in a Whole Number, Example A
2.1 Count, read, and write whole numbers to 1,000 and identify the place value for each digit.	2b Ordering Whole Numbers, Example A
2.1 Count, read, and write whole numbers to 1,000 and identify the place value for each digit.	2c Working with Whole Numbers, 0 to 100
2.2 Use words, models, and expanded forms to represent numbers to 1,000.	2d Modeling Numbers with Base Ten Blocks
2.3 Order and compare whole numbers to 1,000 by using the symbols <, =, >.	2e Comparing Whole Numbers
Topic 3 Place Value beyond 1,000	
3.1 Identify the place value for each digit in numbers to 10,000.	3a Identifying the Place Value of Each Digit in a Whole Number, Example B
3.2 Use expanded notation to represent four-digit numbers.	3b Writing Numbers in Expanded Notation
3.3 Read and write whole numbers in the millions.	3c Writing Numbers in Word Form
3.4 Round whole numbers through the millions to the nearest ten, hundred, thousand, ten thousand, or hundred thousand.	3d Rounding

Standards by Topic   Tutorial Codes

Volume 2 Basic Facts	
Topic 4 Addition and Subtraction Facts	
4.1 Solve problems by choosing the operation and show the meanings of addition and subtraction.	4a Choosing the Operation to Solve Addition and Subtraction Problems

x

Level E

4.1 Solve problems by choosing the operation and show the meanings of addition and subtraction.	4b Using Addition Fact Strategies	**4.4** Find the sum of three 1-digit numbers.	4i Finding the Sum of Three Numbers
4.1 Solve problems by choosing the operation and show the meanings of addition and subtraction.	4c Using Subtraction Fact Strategies	**4.5** Illustrate the relationship between addition and subtraction.	4f Using Fact Families to Add and Subtract
4.1 Solve problems by choosing the operation and show the meanings of addition and subtraction.	4d Choosing the Operation to Solve Addition and Subtraction Word Problems	**4.5** Illustrate the relationship between addition and subtraction.	4c Using Subtraction Fact Strategies
4.1 Solve problems by choosing the operation and show the meanings of addition and subtraction.	4e Using Addition to Solve Word Problems	**4.6** Use different thinking strategies to subtract numbers.	4f Using Fact Families to Add and Subtract
		4.6 Use different thinking strategies to subtract numbers.	4c Using Subtraction Fact Strategies
		4.7 Write addition and subtraction fact families	4f Using Fact Families to Add and Subtract
		Topic 5 Multiplication Facts	
4.2 Apply various addition properties including the commutative, associative, and identity properties.	4f Using Fact Families to Add and Subtract	**5.1** Use repeated addition, arrays, and counting by multiples to do multiplication.	5a Understanding Multiplication, Example A
4.2 Apply various addition properties including the commutative, associative, and identity properties.	4g The Commutative and Associative Properties of Addition	**5.1** Use repeated addition, arrays, and counting by multiples to do multiplication.	5b Understanding Multiplication, Example B
		5.3 Recognize and use the basic properties of multiplication.	5c Using the Commutative and Associative Properties of Multiplication
4.3 Use different thinking strategies to add numbers.	4h Understanding Compatible Numbers	**5.3** Recognize and use the basic properties of multiplication.	5a Understanding Multiplication, Example A
4.3 Use different thinking strategies to add numbers.	4i Finding the Sum of Three Numbers	**5.3** Recognize and use the basic properties of multiplication.	5b Understanding Multiplication, Example B
4.3 Use different thinking strategies to add numbers.	4j Using Addition Fact Strategies	**5.3** Recognize and use the basic properties of multiplication.	5d Using Patterns to Solve Word Problems
4.3 Use different thinking strategies to add numbers.	4k Choosing the Operation to Solve Addition and Subtraction Word Problems	**5.4** Memorize to automaticity the multiplication table for numbers 1 - 10.	5e Using Fact Families to Multiply and Divide

Level E

Tutorial Guide

Topic 6 Division Concepts	
6.1 Use models and multiplication to understand division.	6a Understanding Division
6.1 Use models and multiplication to understand division.	6b Modeling Division
6.3 Find the basic facts involving division by 2, 3, 4, or 5.	6c Understanding Patterns in Division
6.4 Find the basic facts involving division by 6, 7, 8, or 9.	6c Understanding Patterns in Division
6.5 Write multiplication and division fact families.	6d Using Fact Families to Multiply and Divide

Standards by Topic   Tutorial Codes

Volume 3 Add or Subtract	
Topic 7 Add or Subtract 1- and 2-Digit Numbers	
7.1 Solve addition problems with one- and two-digit numbers.	7a Using Models to Solve Word Problems
7.2 Add two 2-digit numbers with and without regrouping.	7b Using the Partial Sums Algorithm
7.2 Add two 2-digit numbers with and without regrouping.	7c Solving Word Problems, Example A
7.2 Add two 2-digit numbers with and without regrouping.	7a Using Models to Solve Word Problems
7.3 Subtract two 2-digit numbers with and without regrouping.	7d Using the Same-Change Rule to Subtract
Topic 8 Add or Subtract Multidigit Numbers	
8.2 Estimate sums and differences of 2-digit numbers.	8a Using the Partial Sums Algorithm
8.2 Estimate sums and differences of 2-digit numbers.	8b Using the Same-Change Rule to Subtract
8.3 Find the sum or difference of two whole numbers up to three digits long.	8c Using the Standard Addition Algorithm, Example A
8.3 Find the sum or difference of two whole numbers up to three digits long.	8d Using the Standard Subtraction Algorithm, Example A
8.4 Find the sum or difference of two whole numbers between 0 and 10,000.	8e Using the Standard Addition Algorithm, Example B
8.4 Find the sum or difference of two whole numbers between 0 and 10,000.	8f Using the Standard Subtraction Algorithm, Example B
8.5 Demonstrate an understanding of, and the ability to use, standard algorithms for the addition and subtraction of multidigit numbers.	8c Using the Standard Addition Algorithm, Example A
8.5 Demonstrate an understanding of, and the ability to use, standard algorithms for the addition and subtraction of multidigit numbers.	8d Using the Standard Subtraction Algorithm, Example A
8.5 Demonstrate an understanding of, and the ability to use, standard algorithms for the addition and subtraction of multidigit numbers.	8e Using the Standard Addition Algorithm, Example B
8.5 Demonstrate an understanding of, and the ability to use, standard algorithms for the addition and subtraction of multidigit numbers.	8f Using the Standard Subtraction Algorithm, Example B

Standards by Topic   Tutorial Codes

Volume 4 Multiply and Divide

Topic 9 Use Multiplication to Compute

9.1 Multiply multiples of 10.	9a Using the Partial-Products Method
9.1 Multiply multiples of 10.	9b Using Multiples of 10, 100, and 1,000 to Multiply and Divide
9.2 Estimate products by rounding factors and using mental math techniques.	9a Using the Partial-Products Method
9.2 Estimate products by rounding factors and using mental math techniques.	9c Using the Standard Multiplication Algorithm
9.3 Solve simple problems involving multiplication of multidigit numbers by one-digit numbers.	9d Solving Word Problems, Example B
9.3 Solve simple problems involving multiplication of multidigit numbers by one-digit numbers.	9a Using the Partial-Products Method
9.4 Solve problems by using estimation, mental math, or pencil and paper.	9e Choosing a Method to Solve Multiplication and Division Word Problems

Topic 10 Use Division to Compute

10.1 Solve problems dividing multiples of 10, 100, and 1,000.	10a Using Multiples of 10, 100, and 1,000 to Multiply and Divide
10.2 Estimate quotients by rounding numbers and using mental math techniques.	10b Estimating Quotients by Rounding Numbers
10.3 Solve simple problems involving division of multidigit numbers by 1-digit numbers with and without remainders.	10c Modeling Division
10.3 Solve simple problems involving division of multidigit numbers by 1-digit numbers with and without remainders.	10d Solving Word Problems, Example C
10.3 Solve simple problems involving division of multidigit numbers by 1-digit numbers with and without remainders.	10e Using the Standard Long Division Algorithm, Example A
10.3 Solve simple problems involving division of multidigit numbers by 1-digit numbers with and without remainders.	10f Using a Standard Long Division Algorithm, Example B
10.3 Solve simple problems involving division of multidigit numbers by 1-digit numbers with and without remainders.	10g Using a Standard Long Division Algorithm, Example C
10.4 Solve division problems by choosing between using estimation, mental math, or pencil and paper to find the quotients.	10h Choosing a Method to Solve Multiplication and Division Word Problems
10.5 Solve problems by multiplying or dividing.	10i Solving Word Problems, Example B
10.5 Solve problems by multiplying or dividing.	10d Solving Word Problems, ExampleC
10.5 Solve problems by multiplying or dividing.	10j Using Multiplication to Check Division

Topic 11 Expressions and Equations

11.1 Record the rule for a pattern as an expression.	11a Using Patterns to Solve Word Problems
11.2 Write expressions for situations that include an unknown quantity.	11b Writing Expressions

Tutorial Guide

11.3 Write equations for word problems that include an unknown quantity.	11b Writing Equations
11.4 Write and solve simple equations for word problems that include an unknown quantity.	11d Solving Equations, Example A
11.4 Write and solve simple equations for word problems that include an unknown quantity.	11e Solving Equations, Example B

Standards by Topic   **Tutorial Codes**

Volume 5 Data, Geometry, and Measurement	
Topic 12 Meaning of Fractions	
12.1 Represent and compare data by using pictures, bar graphs, tally charts, and picture graphs.	12a Using a Tally Chart to Make a Bar Graph
12.1 Represent and compare data by using pictures, bar graphs, tally charts, and picture graphs.	12b Using a Data Table to Make a Bar Graph
12.2 Record numerical data in systematic ways, keeping track of what has been counted.	12a Using a Tally Chart to Make a Bar Graph
12.2 Record numerical data in systematic ways, keeping track of what has been counted.	12b Using a Data Table to Make a Bar Graph
12.3 Find the mean, median, and mode of a set of data.	12c Find the Mean, Median, and Mode of a Data Set
12.4 Represent the same data in more than one way.	12a Using a Tally Chart to Make a Bar Graph
12.4 Represent the same data in more than one way.	12b Using a Data Table to Make a Bar Graph

12.5 Use two-dimensional coordinate graphs to represent points and graph lines and simple figures.	12d Graphing Ordered Pairs
Topic 13 Basic Geometric Figures	
13.1 Draw, measure, and classify different types of angles and lines.	13a Measuring an Angle
13.1 Draw, measure, and classify different types of angles and lines.	13b Drawing an Angle
13.1 Draw, measure, and classify different types of angles and lines.	13c Identifying and Drawing Parallel Lines
13.1 Draw, measure, and classify different types of angles and lines.	13d Identifying and Drawing Perpendicular Lines
13.2 Define *polygon* and classify the different types of polygons.	13e Classifying Polygons
13.3 Explore, compare, and classify different types of triangles.	13f Sorting and Classifying Triangles
13.3 Explore, compare, and classify different types of triangles.	13g Finding Angle Measures in Triangles
13.4 Explore, compare, and classify different types of quadrilaterals.	13h Finding Angle Measures in Quadrilaterals
13.5 Explore circles and define their parts.	13i Finding the Circumference of a Circle, Given a Radius

Level E

Topic 14 Measurement Conversion	
14.1 Explore the basic units of measure in the United States.	14a Converting Units of Capacity
14.2 Explore the basic metric prefixes and what they mean.	14b Using the Metric System to Measure Length
14.2 Explore the basic metric prefixes and what they mean.	14c Using the Metric System to Measure Mass
14.3 Explore the basic metric units and their relationships.	14b Using the Metric System to Measure Length
14.3 Explore the basic metric units and their relationships.	14c Using the Metric System to Measure Mass
14.4 Express simple unit conversions in symbolic form.	14b Using the Metric System to Measure Length
14.4 Express simple unit conversions in symbolic form.	14d Converting Units of Length
14.4 Express simple unit conversions in symbolic form.	14e Converting Units of Time
14.4 Express simple unit conversions in symbolic form.	14a Converting Units of Capacity
14.5 Carry out simple unit conversions within a system of measurement.	14b Using the Metric System to Measure Length
14.5 Carry out simple unit conversions within a system of measurement.	14d Converting Units of Length
14.5 Carry out simple unit conversions within a system of measurement.	14e Converting Units of Time

Topic 15 Measure Geometric Figures	
15.1 Measure the length of an object to the nearest inch or centimeter.	15a Measuring Length to the Nearest Unit
15.2 Find the perimeter of a polygon with integer sides.	15b Finding Perimeter
15.3 Estimate or determine the area of figures by covering them with squares.	15c Finding Area
15.4 Measure the area of rectangular shapes by using appropriate units.	15c Finding Area
15.5 Estimate or determine the volume of solid figures by counting the number of cubes that would fill them.	15d Finding Volume

Standards by Topic   Tutorial Codes

Volume 6 Fractions and Decimals	
Topic 16 Meaning of Fractions	
16.1 Understand that fractions may refer to parts of a set or parts of a whole.	16a Relating Fractions to Models
16.1 Understand that fractions may refer to parts of a set or parts of a whole.	16b Modeling Fractions
16.2 Recognize, name, and compare unit fractions from $\frac{1}{12}$ to $\frac{1}{2}$.	16c Recognizing and Using Unit Fractions, Example A
16.2 Recognize, name, and compare unit fractions from $\frac{1}{12}$ to $\frac{1}{2}$.	16d Recognizing and Using Unit Fractions, Example B

Tutorial Guide

16.2 Recognize, name, and compare unit fractions from $\frac{1}{12}$ to $\frac{1}{2}$.	16e Comparing Unit Fractions	17.3 Multiply and divide by forms of 1 to write equivalent fractions.	17h Finding Equivalent Fractions, Exmaple A
16.2 Recognize, name, and compare unit fractions from $\frac{1}{12}$ to $\frac{1}{2}$.	16f Comparing Fractions Using the Symbols <, =, >, Example A	17.3 Multiply and divide by forms of 1 to write equivalent fractions.	17d Finding Equivalent Fractions, Example B
16.3 Know that when all fractional parts are included, such as four fourths, the result is equal to the whole and to one.	16g Modeling a Whole	17.3 Multiply and divide by forms of 1 to write equivalent fractions.	17i Finding Equivalent Fractions, Example C
16.4 Define and manipulate improper fractions and mixed numbers.	16h Changing a Mixed Number into an Improper Fraction	17.4 Rewrite fractions in lowest terms.	17j Simplifying Fractions, Example A
16.4 Define and manipulate improper fractions and mixed numbers.	16i Ordering Fractions and Mixed Numbers	17.4 Rewrite fractions in lowest terms.	17k Simplifying Fractions, Example B
Topic 17 Equivalence of Fractions		17.5 Explain different interpretations of fractions, such as parts of a whole, parts of a set, and division of whole numbers by whole numbers.	17l Modeling Fractions
17.1 Compare fractions represented by drawings, number lines, or concrete materials.	17a Comparing Unit Fractions		
17.1 Compare fractions represented by drawings, number lines, or concrete materials.	17b Comparing Fractions using the Symbols <, =, >, Example A	17.5 Explain different interpretations of fractions, such as parts of a whole, parts of a set, and division of whole numbers by whole numbers.	17m Recognizing and Using Unit Fractions, Example A
17.1 Compare fractions represented by drawings, number lines, or concrete materials.	17c Comparing Fractions using the Symbols <, =, >, Example B	17.5 Explain different interpretations of fractions, such as parts of a whole, parts of a set, and division of whole numbers by whole numbers.	17n Recognizing and Using Unit Fractions, Example B
17.1 Compare fractions represented by drawings, number lines, or concrete materials.	17d Finding Equivalent Fractions, Example B	17.5 Explain different interpretations of fractions, such as parts of a whole, parts of a set, and division of whole numbers by whole numbers.	17o Using Fractions to Compare Data
17.2 Find the factors of whole numbers.	17e Finding Factors of Numbers, Example A		
17.2 Find the factors of whole numbers.	17f Finding Factors of Numbers, Example B		
17.2 Find the factors of whole numbers.	17g Finding the Prime Factorization of a Number		

Topic 18 Addition and Subtraction of Fractions		Topic 19 Decimals and Money	
18.1 Add and subtract fractions with like denominators.	18a Adding and Subtracting Fractions with Common Denominators, Example A	**19.1** Know, understand, and use the dollar and cent signs for money.	19a Finding the Value of Coins
18.1 Add and subtract fractions with like denominators.	18b Adding and Subtracting Fractions with Common Denominators, Example B	**19.1** Know, understand, and use the dollar and cent signs for money.	19b Representing Money as Decimals and Fractions, Example B
18.1 Add and subtract fractions with like denominators.	18c Adding and Subtracting Fractions with Common Denominators, Example C	**19.2** Write tenths and hundredths in decimal notation.	19b Representing Money as Decimals and Fractions, Example B
18.2 Add fractions having unlike denominators and change answers to mixed numbers in lowest terms when appropriate.	18d Adding Fractions (with Like and Unlike Denominators), Example A	**19.3** Order and compare whole numbers and decimals to two decimal places.	19c Comparing Decimals Using the Symbols <, =, >, Example A
18.2 Add fractions having unlike denominators and change answers to mixed numbers in lowest terms when appropriate.	18e Adding Fractions (with Like and Unlike Denominators), Example A	**19.3** Order and compare whole numbers and decimals to two decimal places.	19d Comparing Decimals Using the Symbols <, =, >, Example A
18.3 Subtract and simplify fractions with unlike denominators.	18f Subtracting Fractions (with Unlike Denominators), Example A	**19.3** Order and compare whole numbers and decimals to two decimal places.	19g Ordering Decimals
18.3 Subtract and simplify fractions with unlike denominators.	18g Subtracting Fractions (with Unlike Denominators), Example B	**19.4** Count combinations of coins or bills.	19a Finding the Value of Coins
18.4 Add mixed numbers with like and unlike denominators.	18c Adding and Subtracting Fractions with Common Denominators, Example C	**19.5** Know and use the decimal notation and the dollar and cent symbols for money.	19a Finding the Value of Coins
18.4 Add mixed numbers with like and unlike denominators.	18h Solving Word Problems Involving Fractions, Example A	**19.6** Know and use the decimal notation and the dollar and cent symbols for money.	19b Representing Money as Decimals and Fractions, Example B
18.5 Subtract mixed numbers with like and unlike denominators.	18g Subtracting Fractions (with Unlike Denominators), Example B	**19.6** Know and understand that fractions and decimals are two different representations of the same concept.	19f Using Models to Find Equivalent Fractions and Decimals

Tutorial Guide

19.6 Know and understand that fractions and decimals are two different representations of the same concept.	19g Using Bills and Coins to Solve Problems, Example A	**Topic 20 Decimal Operations and Comparisons**	
		20.1 Add and subtract simple decimals.	20a Adding Decimals, Example A
		20.1 Add and subtract simple decimals.	20b Adding Decimals, Example B
19.6 Know and understand that fractions and decimals are two different representations of the same concept.	19h Using Bills and Coins to Solve Problems, Example B	**20.1** Add and subtract simple decimals.	20c Subtracting Decimals
		20.2 Find basic fraction and decimal equivalents.	20d Using Models to Find Equivalent Fractions and Decimals
19.6 Know and understand that fractions and decimals are two different representations of the same concept.	19i Representing Money as Decimals and Fractions, Example A	**20.3** Identify and represent on a number line fractions, decimals, and mixed numbers.	20e Graphing Positive Fractions
		20.3 Identify and represent on a number line fractions, decimals, and mixed numbers.	20f Graphing Position Decimals
19.6 Know and understand that fractions and decimals are two different representations of the same concept.	19b Representing Money as Decimals and Fractions, Example B		
19.6 Know and understand that fractions and decimals are two different representations of the same concept.	19j Representing Money as Decimals and Fractions, Example C		
19.7 Determine the unit cost when given the total cost and number of units.	19k Using Rates to Find the Unit Cost		
19.7 Determine the unit cost when given the total cost and number of units.	19l Comparing Unit Costs		

Pinpoint Math

A Three-Part Intervention Solution

Pinpoint Math's complete intervention solution incorporates three essential elements necessary to improve mathematics performance among struggling students.

Diagnostic Assessment

Progress Monitoring

Pinpoint Math
A COMPLETE INTERVENTION SOLUTION

Targeted Instruction

Program Organization

Topics for each *Pinpoint Math* grade level are divided into a six-volume set of student books. The Teacher's Guide and assessment materials are organized by volume.

Pinpoint Math provides the option to use a comprehensive, learning-based assessment management system for mathematics. Each diagnostic test, whether administered via paper and pencil or electronically, provides actionable data on student achievement to help teachers target instruction and measure student progress.

Core Components

Student Booklets

Student Booklets
6 per level

Teacher's Guides

Teacher's Guide
1 per level

Assessment

Assessment Resources
1 per level

Tutorials

Student Tutorials CD-ROM
1 per level

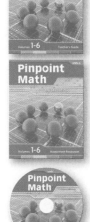

THE CORE COMPONENTS ARE THE SAME FOR EACH LEVEL, A–G

Optional Components

Online Subscription

A Online assessments

B Online student tutorials

C Computer-generated individual student action plans based on assessment results

D Computer-generated reports summarizing assessment results

Manipulatives

- One Manipulative Kit for use with Levels A–C
- One Manipulative Kit for use with Levels D–G

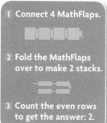

Program Organization

	LEVEL A	LEVEL B	LEVEL C
VOLUME 1	**Volume 1: Understand Numbers** TOPIC 1 Explore One to Ten TOPIC 2 Explore Eleven to Twenty	**Volume 1: Data and Number Concepts** TOPIC 1 Understand Numbers to 12 TOPIC 2 Use Numbers to 20 TOPIC 3 Explore Larger Numbers	**Volume 1: Number Sense** TOPIC 1 Explore Numbers TOPIC 2 Understand Numbers to 100
VOLUME 2	**Volume 2: Position, Location, and Comparison** TOPIC 3 Position and Location TOPIC 4 Compare Numbers and Quantities	**Volume 2: Understand Addition and Addition Facts** TOPIC 4 Understand Addition TOPIC 5 Add TOPIC 6 Explore Greater Sums	**Volume 2: Understand Place Value** TOPIC 3 Numbers to 100 TOPIC 4 Numbers to 1,000
VOLUME 3	**Volume 3: Sorting, Classifying, and Patterns** TOPIC 5 Sorting and Classifying TOPIC 6 Patterns	**Volume 3: Understand Subtraction and Subtraction Facts** TOPIC 7 Understand Subtraction TOPIC 8 Subtract TOPIC 9 Subtraction with Greater Numbers	**Volume 3: Basic Addition and Subtraction Facts** TOPIC 5 Focus on Addition Facts TOPIC 6 Focus on Subtraction Facts
VOLUME 4	**Volume 4: Explore Addition and Subtraction** TOPIC 7 Understand Addition and Subtraction TOPIC 8 Record Addition and Subtraction TOPIC 9 Facts to 10	**Volume 4: Patterns, Shapes, Position, and Location** TOPIC 10 Patterns, Position, and Location TOPIC 11 Patterns and Shapes	**Volume 4: Add and Subtract Greater Numbers** TOPIC 7 Understand Two-Digit Addition TOPIC 8 Understand Two-Digit Subtraction TOPIC 9 Three Digit Addition and Subtraction
VOLUME 5	**Volume 5: Simple Shapes and Measurement** TOPIC 10 Simple Shapes TOPIC 11 Measurement	**Volume 5: Time and Measurement** TOPIC 12 Time TOPIC 13 Measurement	**Volume 5: Time, Money, and Measurement** TOPIC 10 Time TOPIC 11 Money TOPIC 12 Measurement
VOLUME 6	**Volume 6: Time and Money** TOPIC 12 Calendar Time TOPIC 13 Clock Time TOPIC 14 Money	**Volume 6: Explore Numbers to 100** TOPIC 14 Money TOPIC 15 Place Value	**Volume 6: Geometry, Data, and Fractions** TOPIC 13 Geometry TOPIC 14 Data, Graphing, and Fractions

Pinpoint Math

	LEVEL D	LEVEL E	LEVEL F	LEVEL G
VOLUME 1	**Volume 1: Number Sense and Place Value** **TOPIC 1** Place Value through 100 **TOPIC 2** Place Value through 10,000	**Volume 1: Number Sense and Place Value** **TOPIC 1** Place Value through 100 **TOPIC 2** Place Value through 1,000 **TOPIC 3** Place Value Beyond 1,000	**Volume 1: Number Sense and Place Value** **TOPIC 1** Place Value through 1,000 **TOPIC 2** Place Value through Millions	**Volume 1: Number Sense and Place Value** **TOPIC 1** Place Value through 1,000 **TOPIC 2** Place Value beyond 1,000 **TOPIC 3** Addition and Subtraction **TOPIC 4** Multiplication Concepts **TOPIC 5** Division Concepts
VOLUME 2	**Volume 2: Addition and Subtraction** **TOPIC 3** Addition Facts **TOPIC 4** Subtraction Facts **TOPIC 5** Add or Subtract 1-and 2-Digit Numbers **TOPIC 6** Add or Subtract Multidigit Numbers	**Volume 2: Basic Facts** **TOPIC 4** Addition and Subtraction Facts **TOPIC 5** Multiplication Facts **TOPIC 6** Division Facts	**Volume 2: Basic Facts** **TOPIC 3** Addition and Subtraction Facts **TOPIC 4** Multiplication Facts **TOPIC 5** Division Facts	**Volume 2: Whole Number Operations** **TOPIC 6** Use Addition or Subtraction to Compute **TOPIC 7** Use Multiplication to Compute **TOPIC 8** Use Division to Compute **TOPIC 9** Basic Properties of Algebra **TOPIC 10** Expressions **TOPIC 11** Equations and Inequalities
VOLUME 3	**Volume 3: Time and Money** **TOPIC 7** Understand Time **TOPIC 8** Understand Money	**Volume 3: Add and Subtract** **TOPIC 7** Add or Subtract 1- and 2-Digit Numbers **TOPIC 8** Add or Subtract Multidigit Numbers	**Volume 3: Add, Subtract, Multiply, and Divide** **TOPIC 6** Use Addition or Subtraction to Compute **TOPIC 7** Use Multiplication to Compute **TOPIC 8** Use Division to Compute **TOPIC 9** Basic Properties of Algebra **TOPIC 10** Expressions and Equations	**Volume 3: Understand Fractions** **TOPIC 12** Meaning of Fractions **TOPIC 13** Equivalence of Fractions **TOPIC 14** Addition and Subtraction of Fractions **TOPIC 15** Multiplication and Division of Fractions
VOLUME 4	**Volume 4: Multiplication and Division** **TOPIC 9** Multiplication Facts and Concepts **TOPIC 10** Division Facts and Concepts **TOPIC 11** Use Multiplication to Compute **TOPIC 12** Use Division to Compute	**Volume 4: Multiply and Divide** **TOPIC 9** Use Multiplication to Compute **TOPIC 10** Use Division to Compute **TOPIC 11** Equations and Inequalities	**Volume 4: Data, Geometry, and Measurement** **TOPIC 11** Graphing **TOPIC 12** Basic Geometric Figures **TOPIC 13** Measurement Conversion **TOPIC 14** Measure Geometric Figures	**Volume 4: Understand Decimals** **TOPIC 16** Decimals and Money **TOPIC 17** Decimal Operations **TOPIC 18** Decimal and Fraction Comparisons
VOLUME 5	**Volume 5: Geometry and Measurement** **TOPIC 13** Basic Geometric Figures **TOPIC 14** Standard Measurement **TOPIC 15** Metric Measurement **TOPIC 16** Measure Geometric Figures	**Volume 5: Data, Geometry, and Measurement** **TOPIC 12** Graphing **TOPIC 13** Basic Geometric Figures **TOPIC 14** Measurement Conversion **TOPIC 15** Measure Geometric Figures	**Volume 5: Explore Fractions** **TOPIC 15** Meaning of Fractions **TOPIC 16** Equivalence of Fractions **TOPIC 17** Addition and Subtraction of Fractions **TOPIC 18** Multiplication and Division of Fractions	**Volume 5: Data, Geometry, and Measurement** **TOPIC 19** Graphing **TOPIC 20** Basic Geometric Figures **TOPIC 21** Measurement Conversion **TOPIC 22** Measure Geometric Figures
VOLUME 6	**Volume 6: Data and Graphs** **TOPIC 17** Types of Graphs **TOPIC 18** Use Data	**Volume 6: Fractions and Decimals** **TOPIC 16** Meaning of Fractions **TOPIC 17** Equivalence of Fractions **TOPIC 18** Addition and Subtraction of Fractions **TOPIC 19** Decimals and Money **TOPIC 20** Decimal Operations and Comparisons	**Volume 6: Explore Decimals** **TOPIC 19** Decimals and Money **TOPIC 20** Decimal Operations **TOPIC 21** Decimal and Fraction Comparisons	**Volume 6: Ratios, Rates, Proportions, and Percents** **TOPIC 23** Ratios and Proportions **TOPIC 24** Use Rates **TOPIC 25** Percents

Pinpoint Math

Instructional Model

Pretest
Pretest for the first volume is given to all students, either online or by paper and pencil.

Student Action Plan
Teacher prepares individual or group objective-based assignments based on test results.

Teaching a Topic
Students complete the introduction page; teacher demonstrates, questions, and/or presents an overview of the topic based on student introduction responses.

Pinpoint Math

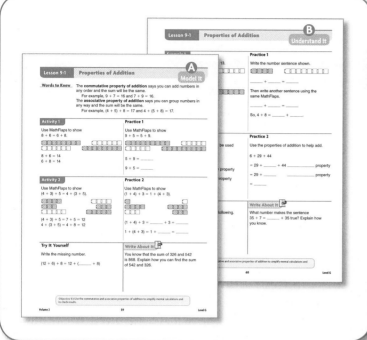

Assignments

Student booklet pages and online computer tutorials are assigned based on individual student needs.

Summary and Review

Students complete Topic Summary, Mixed Review and Progress Monitoring informal assessment pages and receive teacher feedback.

Posttest

Student takes the volume posttest and repeats the process for the next volume.

Topic Introduction

Topic Introductions These help the teacher determine how much teacher support may be needed on this particular topic for each student.

The teacher's guide suggests flexible grouping in pairs or small groups to encourage discussion and language practice while deepening student understanding.

Volume 2 — Whole Number Operations
Topic 9 — Basic Properties of Algebra
Topic Introduction

Objectives: 9.1 Use the commutative and associative properties of addition to simplify mental calculations and to check results. 9.2 Recognize and use the commutative and associative properties of multiplication. 9.3 Know and use the distributive property.

Materials ☐ MathFlaps

Distribute MathFlaps and have students create two-color rows. Have students trade with a partner and write two different addition sentences based on the MathFlaps. Repeat, having students use the MathFlaps to make equal groups and write two different multiplication sentences. Discuss whether the sum or product changes with the order of the numbers.

Informal Assessment

1. **How many MathFlaps are white?** 5 **How many are blue?** 8 **How many MathFlaps are there in all?** 13 Have students complete Part a. **What sum is illustrated by the counters?** 5 + 8 **What addition problem can you write if you start with the number of blue MathFlaps?** 8 + 5 **What is the sum?** 13 **Did the order of the addends change the sum?** No.

2. **What is the product of the first two pairs of numbers?** 21 **How could you use counters to find the missing numbers?** Sample: Arrange 21 counters in three groups of seven and skip count. **Can you use number facts to solve?** Yes, I know $3 \times 7 = 21$, so the missing number in Part a is 3.

3. **Define the distributive property.** It is a property that relates two operations on numbers, usually multiplication and addition or multiplication and subtraction. It distributes the factor outside the parentheses over the terms within the parentheses.

4. **Look at the two ways of multiplying $4 \times 2 \times 3$. How are they different?** The first two numbers that are multiplied are 4 and 2. In Part b, the first two numbers that are multiplied are 2 and 3. **Does it make any difference how you group the numbers?** No.

Student Booklet Page 58

Topic 9 — Basic Properties of Algebra — Topic Introduction

Complete with teacher help if needed.

1. Write the expression.
   a. 5 white + 8 blue =
   b. 8 blue + 5 white

2. Find the missing numbers.
   a. 3 × 7 = 21
   b. 7 × 3 = 21
   c. 8 × 2 = 16
   d. 2 × 8 = 16

Objective 9.1: Use the commutative and associative properties of addition to simplify mental calculations and to check results.
Objective 9.2: Recognize and use the commutative and associative properties of multiplication.

3. Use the distributive property to simplify the expression.
$$8 \times (6 + 5)$$
$$= (8 \times 6) + (8 \times 5)$$
$$= 48 + 40$$
$$= 88$$

4. Find $4 \times 2 \times 3$.
   a. One way:
   $(4 \times 2) \times 3 =$
   $8 \times 3 = 24$
   b. Another way:
   $4 \times (2 \times 3) =$
   $4 \times 6 = 24$
   c. This shows the associative property.

Objective 9.3: Know and use the distributive property.

Volume 2 — 58 — Level G

Another Way Suggest that students use MathFlaps to illustrate the products for Exercises 2 and 4.

Pinpoint Math

Model It

Lesson 9-1 | **Properties of Addition**

Objective 9.1: Use the commutative and associative properties of addition to simplify mental calculations and to check results.

Teach the Lesson

Materials ☐ MathFlaps

Activate Prior Knowledge
Write several addition problems on the board, such as $4 + 8$, $8 + 4$, $10 + 6$, $6 + 10$, and so on. Have students find the sums. **What do you notice about the problems you just did?** Sample: Problems such as $4 + 8$ and $8 + 4$ have the same answer. **What is the same about $4 + 8$ and $8 + 4$? What is different?** Sample: They have the same addends, operation sign, and sum but the addends are in a different order. Repeat the activity with sentences such as $4 + (5 + 8)$ and $(4 + 5) + 8$.

Develop Academic Language
Write $5 + (8 + 2)$ on the board. Point to the parentheses. **These symbols are parentheses. In math, parentheses mean you do whatever is inside of them first. What should we do first here?** $8 + 2 = 10$ **Then what?** $5 + 10 = 15$

Model the Activities
Activity 1 Provide students with MathFlaps. **Show me $8 + 6$. Then show me $6 + 8$. What is true about the sums?** They are the same. Write $8 + 6 = 6 + 8$. **What property did we illustrate?** Commutative

Activity 2 Provide students with MathFlaps and have them illustrate the sums $(4 + 3) + 5$ and $4 + (3 + 5)$. **What property are we illustrating?** Associative

Write About It ✏
ENGLISH LEARNERS If students have difficulty with this question, provide them with specific examples and ask them to explain orally what they see.

Progress Monitoring
Describe the commutative and associative properties in your own words. Sample: You can add numbers in any order and you can group the numbers in any way you want, and you will still get the same sum.

Error Analysis
If students confuse *commutative* and *associative*, help students relate *commute*, as in go back and forth, or change places, to *commutative*; and *associate*, or get together with someone, with *associative*. Relate these terms to the properties.

Volume 2 — 140 — Level G

A Model It Small groups or pairs of students use manipulatives or hands-on materials; teachers help as needed.

Activate Prior Knowledge Every lesson includes suggestions so a teacher can determine what individual students already know about a particular mathematical subject

Teacher notes recommend demonstrations using models and manipulatives to clarify the processes and concepts in a lesson.

Understand It

B
Understand It
Pairs or individual students complete pages with little teacher help.

Develop Academic Language
Again, the Teacher's Guide provides suggestions for students that will help improve their vocabulary

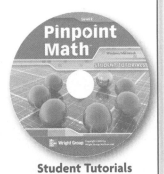
Student Tutorials

Lesson 9-1 — Properties of Addition — **B** Understand It

Objective 9.1: Use the commutative and associative properties of addition to simplify mental calculations and to check results.

Facilitate Student Understanding

Develop Academic Language
Provide a number of sentences such as $9 + 8 = 8 + 9$ and $(3 + 8) + 6 = 3 + (8 + 6)$. Have students identify which property each sentence illustrates.

ENGLISH LEARNERS To reinforce the new terminology, create a poster or chart with examples of each property labeled. As you identify the properties illustrated by the sentences, remind students that order is changed or grouping is changed.

In some languages, the property names are cognates, such as the Spanish words *commutativo* and *asociativo*.

Demonstrate the Examples
Example 1 Draw two groups of MathFlaps on the board and illustrate that $6 + 7$ and $7 + 6$ are both 13. Have students write the two sentences that you have illustrated.

Example 2 Write $13 + 38 + 7$ on the board and ask students if there are any of the numbers that would be easy to add together. **13 and 7** Have a volunteer show how to use the commutative and associative properties to get the two numbers together.

Computer Tutorial
Some students may benefit from completing a computer tutorial before they attempt the Try It page. A list of the tutorials for each lesson can be found beginning on page xii in the front of this book.

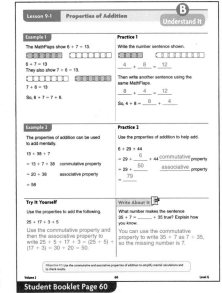
Student Booklet Page 60

⭐ Progress Monitoring
Which property of addition allows you to change the grouping without affecting the sum? **Associative property** Which property of addition allows you to change the order of the numbers being added without affecting the sum? **Commutative property**

Error Analysis
If students have difficulty remembering what the parentheses mean, give them several practice problems involving parentheses.

Try It

C Try It Students complete pages independently.

Lesson 9-1 — Properties of Addition

Objective 9.1: Use the commutative and associative properties of addition to simplify mental calculations and to check results.

Observe Student Progress

Develop Academic Language

Exercise 1 Ask students what property is shown in this example. Write the word *commutative* on the board and have students pronounce it and then restate, in their own words, what the commutative property means.

ENGLISH LEARNERS Students may find it helpful to make cards with the words *commutative* and *associative* on the fronts and an illustration of the property on the backs.

★ Error Analysis

Exercise 2 Students may have difficulty determining the correct number if they are not sure what property is used in each case. Suggest they first decide which property is being illustrated and then determine what number is missing.

Exercise 3 Some students may assume that if parentheses appear in an expression, only the associative property is being used. In each case, suggest that students first identify how the numbers change before identifying the property.

Exercise 4 Have students compare answer choices B and C. Many students make errors because they do not observe that if only the order changes, it is not an example of the associative property.

Exercise 5 Students may also suggest grouping sums of 10, so they end up adding 9 and 1, 8 and 2, 7 and 3, 6 and 4, and finally 10 and 5.

Exercise 8 If students have difficulty with this problem, ask what number can be added to 254 to make it easier to work with. Review the term *break apart* as meaning expanded notation.

Student Booklet Page 61

Volume 2 — 142 — Level G

English Learner Notes Many teaching notes point out words or phrases that may give English Learners difficulties and suggest how to make these terms more accessible to EL students.

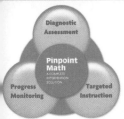

Assessment

Pinpoint Math incorporates formal and informal assessment throughout the program.

Pretests by volume and grade level
These diagnostic assessments include one or more items for all appropriate grade-level objectives. After taking the diagnostic, the teacher will be able to identify which objectives for that volume each student needs to master.

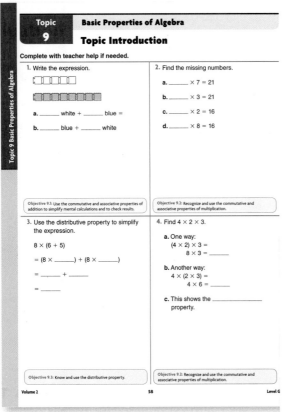

Topic Introductions
These pages are designed to provide students and teachers with an overview of students' starting knowledge on a particular topic.

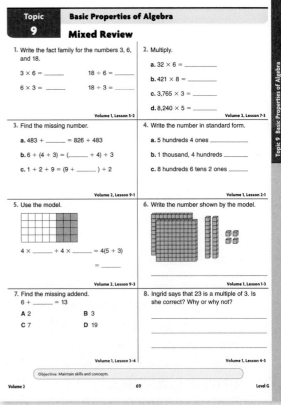

Mixed Reviews
These pages review content from earlier volumes, earlier topics, and earlier lessons within a topic.

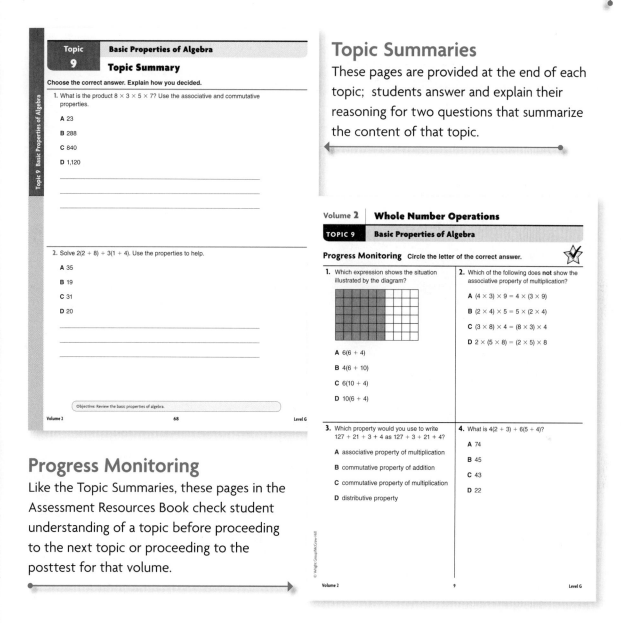

Topic Summaries
These pages are provided at the end of each topic; students answer and explain their reasoning for two questions that summarize the content of that topic.

Progress Monitoring
Like the Topic Summaries, these pages in the Assessment Resources Book check student understanding of a topic before proceeding to the next topic or proceeding to the posttest for that volume.

Teacher Notes
The Teacher's Guide pages include Ongoing Assessment, Progress Monitoring, and Error Analysis notes that the teacher can apply to those students who need additional assistance.

Posttests by Grade Level and Volume
Each test includes one or more items for all appropriate grade-level objectives and is a parallel form to the pretest. If the student answers an acceptable number of items correctly, the student is ready to move on and take the pretest for the next volume.

Pinpoint Math

Optional Online Assessment Management System

Pinpoint uses diagnostic tests to provide actionable data on student achievement thus helping educators target instruction and measure student progress.

This optional online subscription provides you with the following items:

Pretests and Posttests

If students take their formal assessments online, the system can generate reports that show individual student results. The online and print assessment are identical.

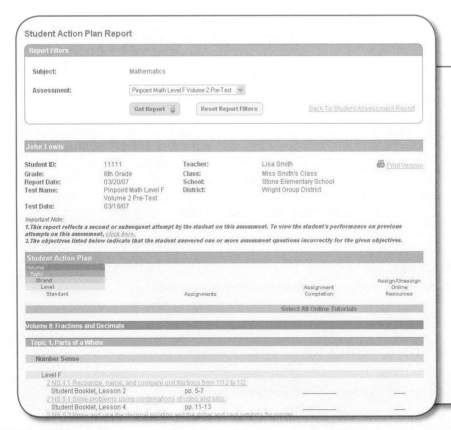

Computer-Generated Student Action Plans

After each assessment, the computer not only scores the test, but provides a listing of recommended print and computer assignments.

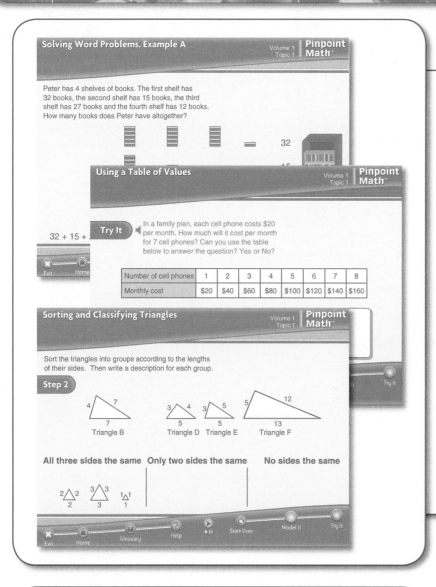

Student Tutorials

If students view these animated examples within the system, the date when the student viewed the tutorial is provided as a check for the teacher that the student completed that online assignment.

(Also available on CD-ROM)

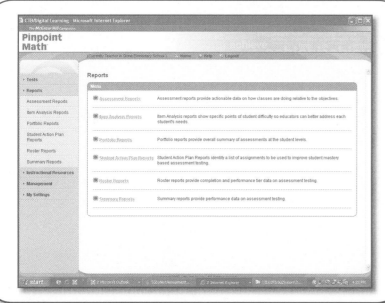

Summary Report

Teachers and administrators may want summary reports that show the assessment results of individual students, teachers, school, and school districts. The system provides several different types of reports for these purposes.

Volume 1: Number Sense and Place Value

Topic 1: Place Value through 100

Topic Introduction

Lesson 1-1 introduces students to the concept of place value with tens and ones. Lessons 1-2 and 1-3 build on this concept by having students count, read, write, compare, and order numbers to 100 using place value. Lesson 1-4 provides students with an overview of some equivalent forms that can be used to name numbers.

Lesson	Objective	Student Pages	Teacher Pages	Tutorials
Topic 1 Introduction	**1.2** Count, read, and write whole numbers to 100.	1	2	
	1.3 Compare and order whole numbers to 100 using the symbols for less than, equal to, or greater than (<, =, >).			
	1.4 Represent equivalent forms of the same number through the use of physical models, diagrams, and number expressions.			
1-1 Tens and Ones	**1.1** Count and group objects in ones and tens.	2–4	3–5	1a
1-2 Numbers to 100	**1.2** Count, read, and write whole numbers to 100.	5–7	6–8	1b
1-3 Compare and Order to 100	**1.3** Compare and order whole numbers to 100 using the symbols for less than, equal to, or greater than (<, =, >).	8–10	9–11	1c
1-4 Equivalent Forms	**1.4** Represent equivalent forms of the same number through the use of physical models, diagrams, and number expressions.	11–13	12–14	1d, 1e
Topic 1 Summary	Review place value through 100.	14	15	
Topic 1 Mixed Review	Maintain concepts and skills.	15	16	

Computer Tutorial

Some students may benefit from completing the computer tutorial before they attempt the Try It page of each lesson. If you are using the electronic components of *Pinpoint Math,* you will find a complete listing of Tutorial codes and titles when you access them either online or via CD-ROM.

Volume 1: Number Sense and Place Value

Topic 1: Place Value through 100

Topic Introduction

Objectives: 1.2 Count, read, and write whole numbers to 100. **1.3** Compare and order whole numbers to 100 by using the symbols for less than, equal to, or greater than (<, =, >). **1.4** Represent equivalent forms of the same number through the use of physical models, diagrams, and number expressions.

Provide students with base ten blocks. Ask them to model a two-digit number, such as 17. If they use 17 ones blocks to make the number, ask if there is a way to model it using fewer blocks. Elicit the idea that 1 tens block is equal to 10 ones blocks. Have students line up 10 ones blocks next to a tens block so they can see the equivalence.

Informal Assessment

1. **There are two kinds of blocks in the picture. How many little squares are there in each tall stack?** 10 **How many tens are in the picture?** 4 tens **How many ones?** 3 **How can we use two digits to write 4 tens and 3 ones?** 43

2 and 4. **What does it mean to compare two numbers?** Decide which one is greater. **These numbers have the same tens digits, so how can we compare them?** Compare the ones digits.

3. **What do the counters in the picture show?** 7 in the top row, 9 in the bottom row **Name some ways to find the sum of these numbers.** Sample answer: 7 + 7 is 14, and 2 more is 16. **What operation do we use to compare two numbers?** subtraction **How many more counters are in the bottom row?** 2

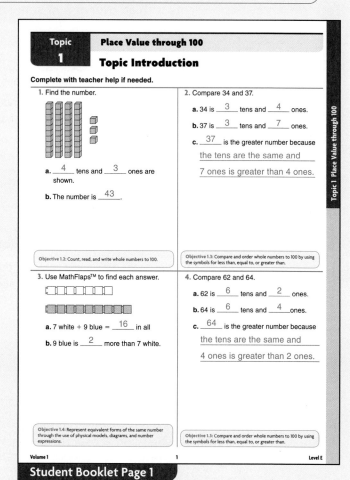

Student Booklet Page 1

Another Way Suggest that students who have difficulty with the number-line model use skip counting.

Lesson 1-1: Tens and Ones

Objective 1.1: Count and group objects in ones and tens.

Teach the Lesson

Materials ☐ Base ten blocks, tens and ones

Activate Prior Knowledge

Provide students several tens blocks and ones blocks. **What do you notice about the tens blocks?** Elicit the idea that each tens block is made of 10 ones. **How would you make the number 34?** If students suggest using 34 ones, point out that a simpler way would be to use tens blocks too.

Develop Academic Language

Write 34 on the board. Point to the 3 and the 4. **These two digits make the number thirty-four. There are 3 tens and 4 ones. The number has two places, the tens place and the ones place. Which digit is in the tens place?** 3 **Which digit is in the ones place?** 4 Write these terms on the board: two-digit number, digits, tens, ones, tens place, ones place.

Model the Activities

Activity 1 Start with the tens blocks. **Count by tens.** 10, 20, 30, 40 **Now, count on by ones from 40.** 41, 42, 43 **How many blocks are there all together?** 43 **So 4 tens and 3 ones is the same as 43.** Have students record their numbers in two ways: tens and ones and standard form.

Activity 2 Suggest that students use ones and tens blocks to model this problem. It may help students to be able to line up and count the blocks.

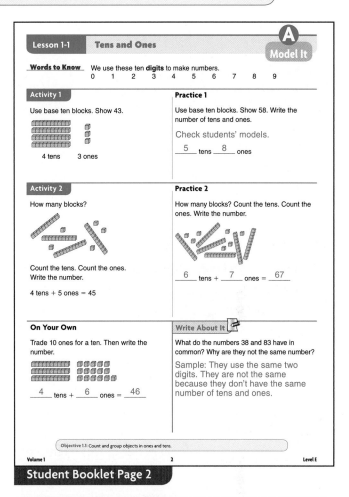

Student Booklet Page 2

Progress Monitoring

Display or draw base ten blocks for 6 tens and 3 ones. Have students explain how to find the number. **Count the tens. Then count the ones.**

Error Analysis

Students who have difficulty recording two-digit numbers for groups of tens and ones blocks may need to work with manipulatives, such as locking cubes or MathFlaps. Show students between 20 and 30 cubes or MathFlaps. Students join the ones to make tens, count the tens, and then count the ones left over.

Lesson 1-1   Tens and Ones

Objective 1.1: Count and group objects in ones and tens.

Facilitate Student Understanding

Develop Academic Language

Write 59 on the board. Explain that our number system is based on 10. Each place in a number is worth 10 times the place to the right. The 5 is worth 5 tens because it is in the tens place. Ask what the 5 would be worth if it were in the ones place. **5 ones, or 5**

Demonstrate the Examples

Example 1 Have students make four 2-digit numbers. Students record their numbers in two ways, for example:

5 tens 7 ones   7 tens 5 ones
57   75

Example 2 How do we put these numbers in order? **Compare the tens digits.** Do the ones digits help you put the numbers in order in this example? **No.**

Computer Tutorial

Some students may benefit from completing a computer tutorial before they attempt the Try It page. A list of the tutorials for each lesson can be found beginning on page x in the front of this book.

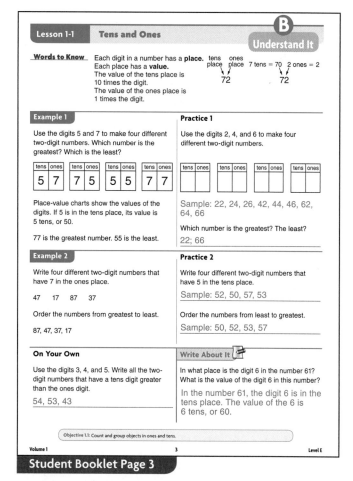

Student Booklet Page 3

Progress Monitoring

Provide pairs of students with 8 tens rods and 8 unit cubes. Have students make and record two different two-digit numbers. **Sample: 31 and 57; check students' models.**

Error Analysis

Students can use unit cubes to practice grouping by 10s before they count. Have them make as many tens as possible from a pile of cubes. Then they count by 10s; for example, 10, 20, 30, and 5 more is 35.

Lesson 1-1 Tens and Ones

Objective 1.1: Count and group objects in ones and tens.

 Error Analysis

Exercise 1 Some students may benefit from using actual base ten blocks to make sure they see that each stick is made up of 10 small cubes.

Exercise 2 Watch for students who are reversing the digits in two-digit numbers; for example, recording 64 as 4 tens and 6 ones. These students many need additional work with manipulative materials.

Exercise 3 If students forget 0, have them write the numbers from 10 to 20 and ask them to check for any digit that is not on their lists.

Exercise 6 Suggest that students make all possible numbers and put them on a number line with 50 to compare.

Exercise 7 If students choose A, have them use a place-value chart to see where the digits belong.

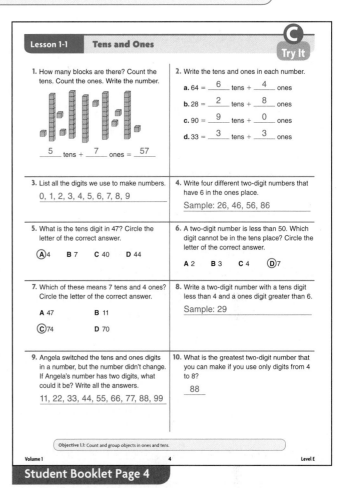

Student Booklet Page 4

Lesson 1-2: Numbers to 100

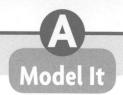
Model It

Objective 1.2: Count, read, and write whole numbers to 100.

Teach the Lesson

Materials
- ☐ Base ten blocks: tens and ones
- ☐ Teaching Aid 1 Number Lines

Activate Prior Knowledge

Write *5* and *five* on the board. **How are these alike? Both are names for the same number. How are they different?** The first is a numeral; the second is a word.

Develop Academic Language

Draw a large number line on the playground or use tape on the floor. Have students move around on the number line to represent numbers.

List the numbers 1 to 9 in a column. Have volunteers write the word names. Repeat with a second column for the numbers 10 to 19. **How are the word names for 10, 11, and 12 different from the others?** The other word names end with *-teen*.

Model the Activities

Activity 1 Provide students with tens and ones blocks. **Show me 3 ones. Now add a ten. What happened to the number?** It changed from 3 to 13. **What happened to the word name?** It changed from *three* to *thirteen*.

Activity 2 Provide Teaching Aid 1. **Make a number line to show the numbers from 0 to 20. Start at the left side.** Check students' work. **Mark 16 on the line. What are two ways to write this number?** 16 and *sixteen* Ask whether students must count from 0 to find 16. They could count on from 10.

Write About It

ENGLISH LEARNERS Define *logical* as "reasonable." Students who form numbers in this way in their first language may be able to explain Gayle's thinking to other students.

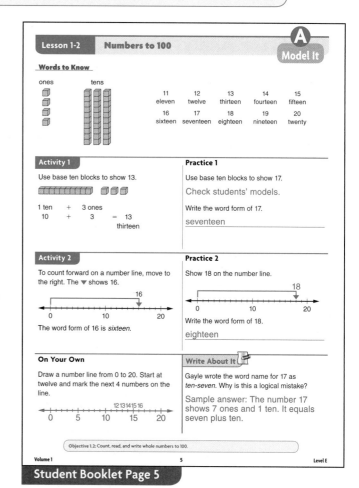

Student Booklet Page 5

⭐ Progress Monitoring

The mystery number is greater than 15. Where can the mystery number be on a number line? To the right of 15

Error Analysis

Have students make a number-line model and a block model for each number from 11 through 20.

Lesson 1-2: Numbers to 100

Objective 1.2: Count, read, and write whole numbers to 100.

Facilitate Student Understanding

Materials ☐ Base ten blocks

Develop Academic Language

List the numbers 1 to 9 in a column. Have volunteers come to the board and write the word names. Repeat with a second column for the tens numbers 10 through 90. Have volunteers try to find the complete word for the number of tens in each word. This can be done for *sixty, seventy, eighty,* and *ninety*. Discuss how the other tens words are formed.

ENGLISH LEARNERS You may want to help students make 2-language glossaries of the number words.

Demonstrate the Examples

Example 1 How many tens are there? 4 What is the value of 4 tens? Count by tens. 10, 20, 30, 40 How many ones are there? 5 What is the value of 5 ones? 5 Add 40 to 5 to find the total value of the model. 45 Help students count by tens and then by ones to verify. 40, 41, 42, 43, 44, 45 Draw a tens and ones model for 45 on the board. Have students use base ten blocks if necessary count up from the number shown. 46, 47, 48, 49, 50, 51, 52 Then have them count down. 46, 45, 44, 43, 42, 41

Example 2 How many tick marks do you see between 50 and 60? 10 Those stand for the 9 numbers between 50 and 60. Start at 50. What number comes next? 51 Have students count until they reach 80 on the number line. **How many tens are in 50?** 5 **60?** 6 **70?** 7 **How many tens are in 72?** 7 **How many ones?** 2 When you write the number with digits or with words, write the number of tens first. Make sure students do not write 70, the value of the tens. Then write the number of ones. When you write the word form of the number, use a comma to separate the tens and the ones.

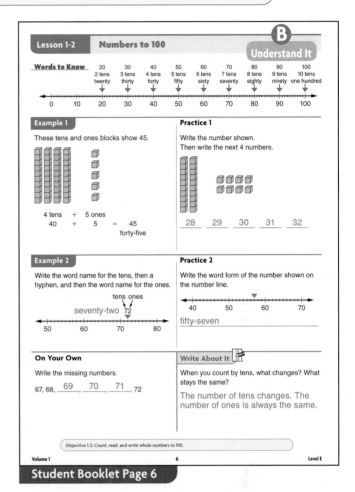

Student Booklet Page 6

Progress Monitoring

The word names for many two-digit numbers have two parts, as in *twenty-two* or *ninety-five*. **What do the parts describe?** The first part is the number of tens. The second part is the number of ones. Remind students that this is not true of the numbers 11 through 19. Write the word names for 4, 14, 40, and 44.

Error Analysis

To support students' work with base ten blocks, practice counting by tens and then counting on by ones; use dimes and pennies as models, if necessary.

Lesson 1-2: Numbers to 100

Objective 1.2: Count, read, and write whole numbers to 100.

Observe Student Progress

Computer Tutorial
Some students may benefit from completing a computer tutorial before they attempt the Try It page. A list of the tutorials for each lesson can be found beginning on page x in the front of this book.

Develop Academic Language

Exercise 1 Write 36 on the board. Ask which numeral shows the tens. **3** Which shows the ones? **6** What happens if the two parts are switched? Sample: The number changes. It becomes 63.

Write the words *tens* and *ones* on the board. Make sure students can relate these terms to the two parts of the number 36. Students who are unable to do so may benefit from extra work with place-value charts and base ten blocks.

ENGLISH LEARNERS Students may benefit from creating their own word banks. They should include words unfamiliar to them and write their own definition or example. Possible words in this lesson include *backward*, *between*, and *missing*.

Student Booklet Page 7

Lesson 1-2 Numbers to 100 — Try It

1. Show 36 in two different ways. Shade 36 blocks. Then mark the number line.
   (number line: 20, 30, 40, 50)

2. Write the missing numbers.
   a. 13, 14, __15__, __16__, __17__, 18
   b. 47, 48, __49__, __50__, __51__, 52
   c. 22, 23, __24__, __25__, __26__, 27
   d. 75, 76, __77__, __78__, __79__, 80
   e. 38, 39, __40__, __41__, __42__, 43

3. Write the word names.
   a. 93 ninety-three
   b. 15 fifteen
   c. 27 twenty-seven
   d. 49 forty-nine

4. Which number is sixty-eight?
   A 66   **B** 68
   C 86   D 608

5. Which numbers between 0 and 100 have the same number of tens and ones?
   11, 22, 33, 44, 55, 66, 77, 88, 99

6. Start at the number shown and count backward.
   30, 40, 50, 60
   47, 46, 45, 44, 43

7. The word name for 14 uses the word *four*. Which other numbers from 11 to 19 include the ones word in their names?
   16, 17, 18, 19

8. On another sheet of paper, use a number line to count by 10s from 27 to 87.
   Students' drawings should show 27, 37, 47, 57, 67, 77, 87.

★ Error Analysis

Exercise 3 Students who continue to make errors when writing word names for two-digit numbers may benefit from playing a game to provide extra practice.

Use number cubes or slips of paper labeled with the digits 0 to 9. Working in pairs or groups, students take turns making a two-digit number. The other students write the number in both numeral and word form.

Exercise 4 Have students compare answer choices B and C. Many students make errors by reversing the digits in two-digit numbers. Point out that we read numbers from left to right in the same way that we read words and sentences from left to right.

Exercises 6 and 8 Students may need to review counting backward and counting by tens.

Exercise 7 If a student answers incorrectly, have him or her write the word names from eleven through nineteen and then reread the question.

| Lesson 1-3 | Compare and Order to 100 |

Objective 1.3: Compare and order whole numbers to 100 using the symbols for less than, equal to, or greater than (<, =, >).

Teach the Lesson

Materials ☐ Base ten blocks: hundreds, tens, ones

Activate Prior Knowledge

Write the symbols >, <, and = on the board. Show how to use these symbols to compare numbers. Have volunteers come to the board and write sentences such as 8 > 2, 0 < 5, and 4 = 1 + 3. Label the comparison symbols in words. Have students read the sentences aloud and then write them in word form. **Sample: Eight is greater than two. Zero is less than five. Four equals one plus three.**

Develop Academic Language

ENGLISH LEARNERS Discuss the word *inequality*. **You know what *equal* means. When you put *in-* at the beginning of a word, it means *not*. Inequalities are comparisons of two things that are not equal.**

Model the Activity

Provide students with a container of hundreds, tens, and ones blocks. Have each student model 347. **How many hundreds, tens, and ones are in your number? 3 hundreds 4 tens 7 ones If you take away the tens blocks, what is your number? 307 If you put back the tens blocks, and then take away the ones, what is your number? 340**

ENGLISH LEARNERS Students may need some processing time to say the numbers in English. Establish a signal for wait time for everyone.

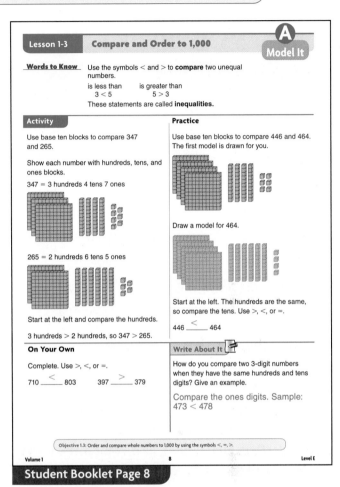

Student Booklet Page 8

 Progress Monitoring

If you are comparing a three-digit number with a two-digit number, which number is greater? Why? The three-digit number because it has at least one hundred. A two-digit number has no hundreds.

Error Analysis

Check that students are starting at the left to compare numbers by asking **When you compare 345 and 269, does it matter that 5 ones is less than 9 ones? Show or tell why.** No, it is more important that 3 hundreds are greater than 2 hundreds. Have students circle the digit that helped them decide which number is greater.

Lesson 1-3 | Compare and Order to 100

Objective 1.3: Compare and order whole numbers to 100 using the symbols for less than, equal to, or greater than (<, =, >).

Facilitate Student Understanding

Develop Academic Language

Explain that when students are counting, 8 comes before 13. Have them use a greater than or less than symbol to write this. **8 < 13** Then have students use a symbol to show that 10 comes after 7. **10 > 7** On the board, write:

comes before   is less than (<)

comes after   is greater than (>)

Explain that the arrow points to the lesser number, or the number that comes first when counting.

Demonstrate the Example

Draw a place-value chart on the board. Put the numbers in this Example into the chart. Remind students that each digit is in a place. Each place is worth ten times the place to the right. **What is the digit 4 worth in 604? 4 ones, or 4 What if the 4 was in the tens place? 4 tens, or 40**

Check that students put the numbers in increasing order by asking which one is the least. **the 2-digit number How do you know? It has no hundreds.** Then have them look for the 3-digit number with the least number of hundreds. Continue until the numbers have been put in order.

Remind students that zeros that appear in a number are important, because they help keep the other digits in the correct places.

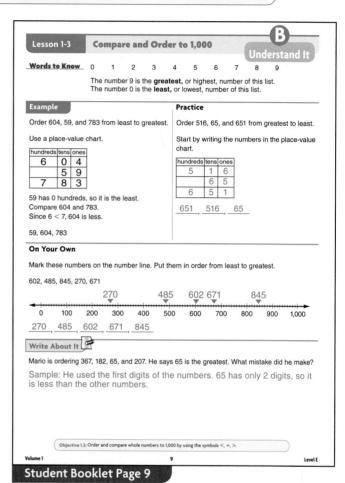

Student Booklet Page 9

Progress Monitoring

Show or explain how to compare two numbers that have the same number of digits. Sample: Start at the left. If those digits are the same, compare the digits in the next place to the right.

Error Analysis

Most students will have an easier time ordering a list of numbers if they write them in a vertical column. Lined notebook paper turned sideways is a good tool to use to make sure the digits are in the correct columns.

Lesson 1-3: Compare and Order to 100

Objective 1.3: Compare and order whole numbers to 100 using the symbols for less than, equal to, or greater than (<, =, >).

Observe Student Progress

Computer Tutorial

Some students may benefit from completing a computer tutorial before they attempt the Try It page. A list of the tutorials for each lesson can be found beginning on page x in the front of this book.

Develop Academic Language

Have students relate the meanings of *compare* and *order*. **Sample: You have to compare numbers to order them.**

ENGLISH LEARNERS Students may already know the phrase "in order", but point out that the words *compare* and *order* are used here as actions or verbs in this context. Explain that they will sometimes see these words as a direction on a test, for example, *order these numbers*.

Error Analysis

Exercise 1 Students who put the numbers in ascending order probably did not read the entire direction. Remind them that there are two ways to order a group of numbers—from least to greatest and from greatest to least. Remind English Learners that *least* means *fewest* or *smallest*.

Exercise 3 If students confuse the > and < symbols, emphasize that the smaller, pointed part of the symbol should always point to the number that is less.

Exercise 8 If students need help getting started, suggest they draw vertical lines and label them to represent the three plants.

Lesson 1-3   Compare and Order to 1,000   Try It

1. Write in order from greatest to least.
   a. 194, 409, 419
      419, 409, 194
   b. 256, 621, 97, 152
      621, 256, 152, 97

2. Write in order from least to greatest.
   a. 307, 73, 730
      73, 307, 730
   b. 250, 820, 580, 500
      250, 500, 580, 820

3. Complete the exercises below. Use >, <, or = to make the sentence true.
   a. 138 < 183   b. 921 > 721
   c. 559 < 595   d. 425 < 452
   e. 628 > 86   f. 357 < 573

4. Which inequality is true? Circle the letter of the correct answer.
   (A) 184 > 95   B 527 > 720
   C 950 < 840   D 422 < 402

5. Mark has 147 baseball cards. Sarah has 163. Jasmine has 174. Who has the greatest number of baseball cards?
   Jasmine

6. What is the greatest number that can be made using the digits 6, 2, and 8? Use each digit only once.
   862

7. Explain how to order 451, 781, and 526 by making three comparisons.
   Sample: Compare 451 and 781, 451 < 781. Compare 781 and 526, 781 > 526. Compare 451 and 526, 451 < 526. So the order is 451, 526, and 781.

8. A rose bush is 145 centimeters tall. A small palm tree is 415 centimeters. A climbing vine is 775 centimeters. Which plant is taller than 500 centimeters?
   vine

Objective 1.3: Order and compare whole numbers to 1,000 by using the symbols <, =, >.

Student Booklet Page 10

Lesson 1-4: Equivalent Forms

Objective 1.4: Represent equivalent forms of the same number through the use of physical models, diagrams, and number expressions.

Teach the Lesson

Materials ☐ MathFlaps or 2-color counters

Activate Prior Knowledge

Write on the board:

25¢ $0.25   quarter

What do you notice about these symbols and words? They are different names for the same amount of money. Tell students that numbers can also have different names.

Develop Academic Language

What are some different ways of showing the number 6? Sample: 6, *six*, 1 + 5, drawing of 6 dots Write the term *equivalent* on the board and explain to students that *equal* and *equivalent* are related. These are equivalent because they all name the same number, 6.

Model the Activities

Activity 1 Provide MathFlaps. **Show a set of 12. What are some different ways to do this?** Sample: 1 row of 12, 3 rows of 4, 2 rows of 6 **Separate your set into two groups. What sum did you show?** Sample: 4 + 8, 6 + 6, 10 + 2 Point out that these sums are also different names for 12.

Activity 2 Point out that for this exercise and for Practice 1, students will count 0 as a group. **If you use a system, you can be sure of finding all the ways of making 7. Start with 7 in one group and 0 in the other group. What sum did you show?** 7 + 0 **Now move one MathFlap from the first group to the second. What sum does this model show?** 6 + 1 Continue until you have moved all the MathFlaps to the second group. Encourage students to discuss and describe the pattern in the sums they find.

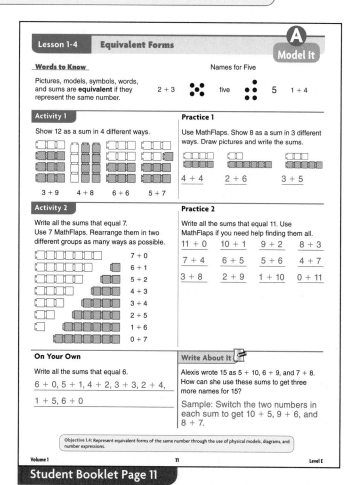

Student Booklet Page 11

✪ Progress Monitoring

Write all the sums that equal 8. 0 + 8, 1 + 7, 2 + 6, 3 + 5, 4 + 4, 5 + 3, 6 + 2, 7 + 1, 8 + 0

Error Analysis

If students write sums such as 8 + 2 and 8 + 5 when asked to write sums that equal 8, they may have misunderstood the direction as "write sums that use 8." Write the direction on the board and underline the word *equal*: Write all the sums that <u>equal</u> 8.

Lesson 1-4 Equivalent Forms

B Understand It

Objective 1.4: Represent equivalent forms of the same number through the use of physical models, diagrams, and number expressions.

Facilitate Student Understanding

Develop Academic Language

ENGLISH LEARNERS Explain that *tally* can be used as an adjective, a noun, and a verb. Add the word to their word banks. Have them work together as a group to write three sentences showing *tally* as these parts of speech. **Sample: Ed made tally marks to count his stamps. He used one tally for each stamp. He tallied twice, and the two counts agreed.**

Demonstrate the Examples

Example 1 Draw a large 0–10 number line on the board. Tell students you will show a way to use two "jumps" to get to 10. Draw arcs from 0 to 4 and 4 to 9.

Ask what sum the model shows. **4 + 5** Erase the two jumps and have volunteers come to the board and show 6 + 3 and several other sums.

Example 2 Write the term *tally marks* on the board. Say that tally marks are another way to represent numbers. Demonstrate how to make the marks by counting aloud as you draw them: one, two, three, four, *five*; six, seven, eight, nine, *ten*. **What happens after you make 4 marks? You cross them with the next mark.**

Ask for a volunteer to show 8 with tally marks. Point out that tally marks are used for counting objects in groups of 5. The marks in the example represent 5 + 5 + 4.

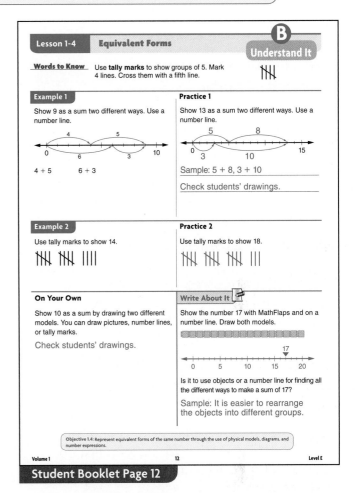

Student Booklet Page 12

Progress Monitoring

Show the number 8 with a picture model, with a word, with a numeral, and with a sum.
Sample: number line with a jump from 0 to 3 and from 3 to 8, eight, 8, 2 + 6

Error Analysis

If students have difficulty, give them 8 MathFlaps and have them find sums for 8.

Lesson 1-4 | Equivalent Forms

Objective 1.4: Represent equivalent forms of the same number through the use of physical models, diagrams, and number expressions.

Observe Student Progress

Computer Tutorial
Some students may benefit from completing a computer tutorial before they attempt the Try It page. A list of the tutorials for each lesson can be found beginning on page x in the front of this book.

Develop Academic Language
Exercise 8 You may need to remind students that *dozen* is another name for 12.

ENGLISH LEARNERS You might want to bring in an egg carton to model Exercise 8.

Error Analysis

Exercise 1 Students are to shade or draw lines on the models to separate each set into two parts. You may wish to have students first do the activity with 9 MathFlaps.

Exercise 4 To help students who don't understand the direction, ask them first to add the numbers in each answer choice. Then ask them which choice is equal to 7. That choice is *another name* for 7.

Exercise 7 Talk about numbers that are easy to work with mentally. Have students solve 6 + 4 + 8 as an example. Point out that adding 6 and 4 first makes it easy to add 8. Rewrite a problem such as 8 + 12 to show the "break apart and make ten" strategy. Since 12 = 2 + 10, the problem is equal to 8 + 2 + 10, or 10 + 10, or 20.

Student Booklet Page 13

Lesson 1-4 Equivalent Forms — Try It

1. Shade to show 4 different ways to make a sum of 9.

   Check students' drawings.

2. Show 2 different ways to make a sum of 12.

   Check students' drawings.

3. Write 3 different sums that equal each number.

   Samples are given.
   a. 14 4 + 10, 7 + 7, 9 + 5
   b. 13 6 + 7, 5 + 8, 4 + 9
   c. 10 5 + 5, 4 + 6, 7 + 3

4. What is another name for 7?
   (A) 0 + 7   B 5 + 6
   C 7 + 2   D 10 + 7

5. Look at the tally marks. Using numerals, write 4 different names for the sum.

   Sample: 13, 10 + 3,
   5 + 8, 8 + 5

6. Use subtraction. Show four different names for 3.

   Sample: 9 – 6, 8 – 5,
   7 – 4, 3 – 0

7. Margie is adding 8 to 17. Why is it helpful to write 17 as 2 + 15?

   It is easier to do the sum mentally because 8 + 2 + 15 equals 10 + 15.

8. An egg carton holds 1 dozen eggs. What sums can you model using an entire egg carton? Draw one model with your answer.

   Sums of 12; for example, 5 + 7, 3 + 9, 10 + 2. Check students' drawings.

Volume 1: Number Sense and Place Value
Topic 1: Place Value through 100

Topic Summary

Objective: Review place value through 100.

You may want to have each student analyze one answer choice, or have students work in groups of four with each student analyzing a different choice.

Ask students to share their ideas about each answer choice. Be sure they confirm the correct answer for each problem at the end of the discussion.

Answer Evaluation

1. **A** Students miscounted the ones.
   B Students miscounted the ones.
   C This choice is correct.
   D Students switched the tens digit with the ones digit.

2. **A** Students may have chosen this answer because 14 is greater than 4. Point out that they must find a single number equal to each number phrase.
   B This choice is correct.
   C Students made an error subtracting, multiplying, or both.

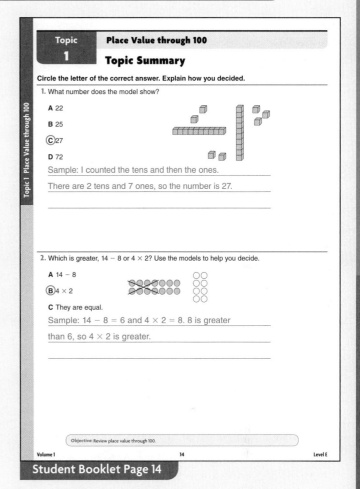

Student Booklet Page 14

Error Analysis

Exercise 1 Provide students with base ten blocks. Have them duplicate the arrangement of blocks on the page. Then have them sort the blocks into tens and ones.

Exercise 2 Review the relationships between the models on the page and the subtraction and multiplication problems. Have students make similar models for two problems of each type.

Progress Monitoring

When all assignments for this topic have been completed, assign the corresponding Progress Monitoring page for this topic (Assessment Resources Book, page 1). Be sure students complete the Progress Monitoring page before you administer the final assessment for this volume.

Volume 1: Number Sense and Place Value

Topic 1: Place Value through 100

Mixed Review

Objective: Maintain concepts and skills.

Have students complete the Mixed Review page. Work with each student individually to review results. Identify strengths and weaknesses and correct any misunderstandings.

★ Error Analysis

Exercise 1 Remind students that when the number in the tens place is the same, they have to look at the ones place and determine which digit is greater.

Exercise 2 If students are having difficulty writing out the words, suggest they make notecards with numbers and the written form of the number.

Exercise 3 Suggest that students look at the symbol first. If the symbol opens up towards the greater number, then it is a true statement. If the symbol opens up to the smaller number, then it is a false statement.

Exercise 4 Suggest that students write four sets of two blanks to represent the two digits of each number and write a 4 in the tens place. Then have them fill in the remaining digits in each number with 1, 7, and 9 to show four different numbers.

Student Booklet Page 15

Topic 1 — Place Value through 100 — Mixed Review

1. Compare 88 and 89.
   a. 88 is __8__ tens and __8__ ones.
   b. 89 is __8__ tens and __9__ ones.
   c. __89__ is greater because the tens are the same and 9 ones is greater than 8 ones.

   Volume 1, Lesson 1-3

2. Write the word form for each number.
   a. 27 twenty-seven
   b. 8 eight
   c. 66 sixty-six
   d. 50 fifty

   Volume 1, Lesson 1-2

3. Which comparison is NOT correct? Circle the letter of the correct answer.
   A 97 > 7
   B 98 > 89
   C 2 < 21
   D 25 < 24

   Volume 1, Lesson 1-3

4. Write four different numbers using 1, 4, 7, and 9 that have 4 in the tens place.

   41, 44, 47, 49

   Volume 1, Lesson 1-2

Volume 1 — Number Sense and Place Value

Topic 2: Place Value through 1,000

Topic Introduction

Lesson 2-1 provides practice in counting, reading, and writing whole numbers to 1,000 and practice in identifying the place value for each digit. This leads students into being able to use words, models, and exponential forms to represent numbers to 1,000. In Lesson 2-3, students will be practicing to order and compare whole numbers to 1,000 by using the symbols <, =, or >.

Lesson	Objective	Student Pages	Teacher Pages	Tutorials
Topic 2 Introduction	**2.1** Count, read, and write whole numbers to 1,000 and identify the place value of each digit. **2.2** Use words, models, and expanded forms to represent numbers to 1,000.	16	18	
2-1 Numbers to 1,000	**2.1** Count, read, and write whole numbers to 1,000 and identify the place value of each digit.	17–19	19–21	2a, 2b, 2c
2-2 Write Numbers to 1,000	**2.2** Use words, models, and expanded forms to represent numbers to 1,000.	20–22	22–24	2d
2-3 Compare and Order to 1,000	**2.3** Order and compare whole numbers to 1,000 by using the symbols <, =, or >.	23–25	25–27	2e, 2b
Topic 2 Summary	Review place value through 1,000.	26	28	
Topic 2 Mixed Review	Maintain concepts and skills.	27	29	

Computer Tutorial

Some students may benefit from completing the computer tutorial before they attempt the Try It page of each lesson. If you are using the electronic components of *Pinpoint Math*, you will find a complete listing of Tutorial codes and titles when you access them either online or via CD-ROM.

Volume 1: Number Sense and Place Value

Topic 2: Place Value through 1,000

Topic Introduction

Objectives: 2.1 Count, read, and write whole numbers to 1,000 and identify the place value for each digit. **2.2** Use words, models, and expanded forms to represent numbers to 1,000.

Provide students with hundreds, tens, and ones blocks. Ask them to model several 2- and 3-digit numbers. Make sure they understand that there are different ways to model each number. For example, they could model 147 with 1 hundreds block, 4 tens blocks, and 7 ones blocks, or with 14 tens blocks and 7 ones blocks, or with 147 ones blocks. Have them lay 10 tens blocks on top of 1 hundreds block so that they can see that the two are equivalent.

Informal Assessment

1. How many tens are shown in the model? **2** What number does that equal? **20** How many ones are shown? **5** What number does that equal? **5** What is the entire number shown by the model? **25** What is the number in words? **Twenty-five**

2. From left to right, what are the places of this number? **Hundreds, tens, ones**

3. How do you write the first digit of the number in words? **Eight hundred** How do you write the second two digits? **Ninety-four** Be sure students are using a hyphen in representing this number.

4. How can you tell in which place the 7 is? **It is in the far right place, which is the ones.** Have students fill in the space for the name of each digit's place value.

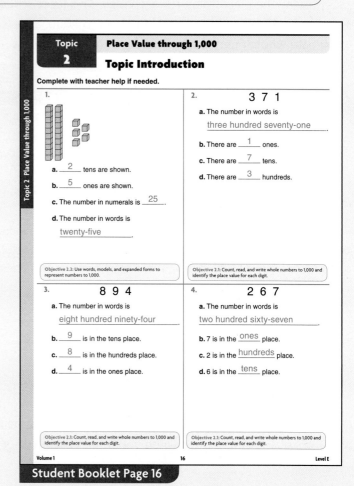

Student Booklet Page 16

Another Way For students having difficulty identifying the digit in a given place or the value of a given digit, have them set up models of each number, and then write out each digit and its value.

Lesson 2-1 Numbers to 1,000

Objective 2.1: Count, read, and write whole numbers to 1,000 and identify the place value for each digit.

Teach the Lesson

Materials ☐ Base ten blocks
☐ Teaching Aid 6 Place-Value Charts

Activate Prior Knowledge

Review the use of base ten blocks. Display 10 ones. Group them together to show that they equal one ten. Then display 10 tens and group them to show they equal one hundred. Show various groups of ones, tens, and hundreds, naming each. Then practice counting by tens and hundreds, using the blocks as models.

Develop Academic Language

On the board write the number 726. Have volunteers come to the board and change the ones digit to a different digit. Repeat with the tens place and the hundreds place.

ENGLISH LEARNERS Students may have difficulty with words that have multiple meanings. Tell them that the word *place* can be a noun and a verb.

Model the Activities

Activity 1 Look at the 3 hundreds blocks. This represents 300. **Show me 5 tens.** Students should point to 5 tens to the right of the 3 hundreds blocks. **Show me 7 ones.** Students should point to 7 ones next to the 5 tens blocks.

To write the number in the chart, write 3 in the hundreds column in the chart. **What number is in the tens column?** 5 **What number is in the ones column?** 7

Activity 2 First count by hundreds. 100, 200, 300 Now count by 10s. **What is the first number we'll say?** 310 Make sure students do not repeat 300. **Count on.** 320, 330, 340, 350 **Now what?** Count by ones: 351, 352, 353, 354, 355, 356, 357.

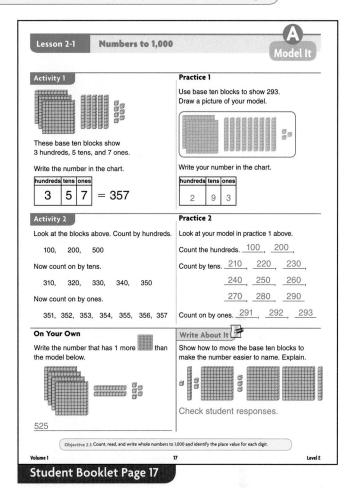

Student Booklet Page 17

Progress Monitoring

Write 512 on the board. Show how you would model this number. With 5 hundreds, 1 ten, and 2 ones

Error Analysis

Check students' models to see that they are grouping the blocks from left to right with the hundreds first, the tens next, and then the ones. Point out that when we say a number, such as three hundred forty-seven, we name the hundreds first.

Lesson 2-1: Numbers to 1,000

Objective 2.1: Count, read, and write whole numbers to 1,000 and identify the place value for each digit.

Facilitate Student Understanding

Develop Academic Language

Remind students that when writing the word form of a number that is greater than 20 and not a multiple of 10 they need to use a hyphen between the words for the tens place and the ones place. Use an example such as forty-two. Ask for other examples.

Have students write the word form for the numbers 20, 30, 40, 50, 60, 70, 80, and 90. Discuss which ones include the exact spelling of the first digit in the number's word form. For example, sixty includes the word six but forty does not include the word four. Underline those that do include the exact word.

Demonstrate the Examples

Example 1 Create two sets of large cards, one with the numerals 0-9, the other with the words *ones*, *tens*, and *hundreds*, one word per card. Write 345 on the board. Have 6 students choose the appropriate cards and arrange them in line. **Read this number.** 3 hundreds 4 tens 5 ones

Ask students how they decided where to place the cards. Have them rearrange the cards to form a new number and write it on the board. Write *ones place*, *tens place*, and *hundreds place* under each digit. After each student identifies a place, ask the class to verify the answer.

Example 2 Have students use colored markers to write numbers in expanded form and show the word equivalents below using the same color. Use a different color for each value.

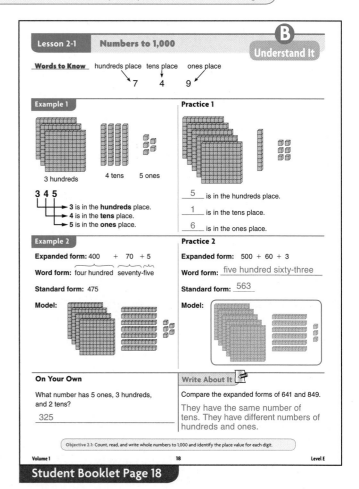

Student Booklet Page 18

Progress Monitoring

Which place value is greater, tens or hundreds? Hundreds **How do you know?** Sample: The hundreds place is to the left of the tens place; each hundred is worth 10 tens.

Error Analysis

If students have trouble correctly identifying place value, have them first write numbers in labeled hundreds, tens, and ones boxes.

If students write expanded form incorrectly, such as 7 + 2 + 9 instead of 700 + 20 + 9 for 729, have them add 7 + 2 + 9 to see that this equals 18, not 729.

Lesson 2-1 Numbers to 1,000

Objective 2.1: Count, read, and write whole numbers to 1,000 and identify the place value for each digit.

Observe Student Progress

Materials ☐ Teaching Aid 6 Place-Value Charts

Computer Tutorial

Some students may benefit from completing a computer tutorial before they attempt the Try It page. A list of the tutorials for each lesson can be found beginning on page x in the front of this book.

Develop Academic Language

Point out that when we read a number, we drop the *s* in the place value name. For example, we say, "Two hundred thirty-seven" instead of "Two hundreds thirty-seven."

ENGLISH LEARNERS Students might need to refer to a list of word forms that may be difficult to spell. Pair students and have them create posters for this purpose.

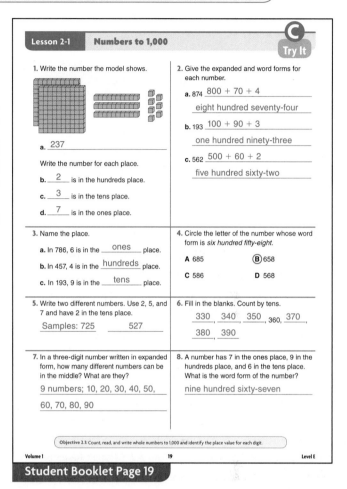

Student Booklet Page 19

⭐ Error Analysis

Exercise 1 Have the students use the place-value charts from Teaching Aid 6 to record the digits. Then write the word name for the number.

Exercise 2 Write a few 3-digit numbers on the board. Have students practice writing numbers in words.

Exercise 4 If students have difficulty with this exercise, have them first determine which answers show 8 in the ones place. **B and C** Then ask which of these two shows 6 in the hundreds place. **B**

Exercise 6 You might want to pair students to discuss these exercises. If students still have difficulty, suggest they begin with the given number and skip count up first. Then skip count back from the given number. For further assistance, have them initially model the given number, then remove (to skip count back) or add (to skip count forward) additional base ten blocks.

Exercise 7 If students answer incorrectly, have them write out the expanded form of any three-digit number. **Identify the place of the middle number. Tens** The middle number is a two-digit number in multiple of 10. List the numbers.

Lesson 2-2: Write Numbers to 1,000

Objective 2.2: Use words, models, and expanded forms to represent numbers to 1,000.

Teach the Lesson

Materials
☐ Base ten blocks
☐ Teaching Aid 6 Place-Value Charts

Activate Prior Knowledge

Write several three-digit numbers on the board. Point to various digits and ask students to identify their places and values.

Develop Academic Language

On the board, write the lists below. Have students draw lines to match the pairs.

Standard form	300 + 20 + 7
Expanded form	three hundred twenty-seven
Word form	327

Model the Activity

Activity 1 Display 3 hundreds on the board. **What number does this show?** 300 Write "300" below. Display 4 tens to the right of the 3 hundreds. **What number does this show?** 40 Write "+ 40" next to "300". Display 7 ones to the right of the tens. **What number does this show?** 7 Write "+ 7" next to "40". 300 + 40 + 7 is called the expanded form of this number. The standard form is the way you would see it if it were a page number in your book. **What is the standard form of this number?** 347 Read the standard number out loud. Three hundred forty-seven As students say the number, write the word form below the standard form.

Write About It

ENGLISH LEARNERS Students may not know other uses of the word *expanded*. Beforehand, have students act out or model the word in several ways.

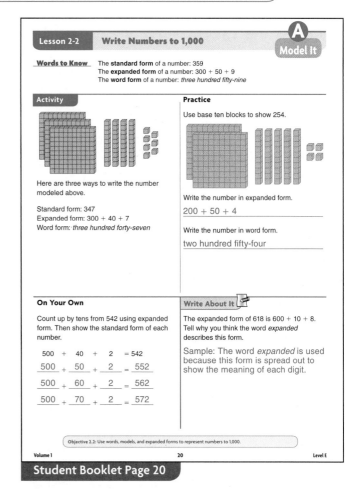

Student Booklet Page 20

⭐ Progress Monitoring

Write 632 on the board. **Show how you would write this number in expanded notation.** 600 + 30 + 2

Error Analysis

If students have difficulty understanding expanded form, model by writing 200 on a wide slip of paper, 20 on a narrower slip of paper, and 2 on a still narrower slip of paper. Stack the papers so the number 222 shows. Then spread out the papers and write a plus sign between each pair of numbers.

Lesson 2-2: Write Numbers to 1,000

Objective 2.2: Use words, models, and expanded forms to represent numbers to 1,000.

Facilitate Student Understanding

Develop Academic Language

Remind students that when writing the word form of a number that is greater than 20 and not a multiple of 10 they need to use a hyphen between the words for the tens place and the ones place. Use an example such as *forty-two*, and then ask for other examples.

Have students write the word form for the numbers 20, 30, 40, 50, 60, 70, 80, and 90. Discuss which ones include the exact spelling of the first digit in the number's word form. For example, *sixty* includes the word *six* but *forty* does not include the word *four*.

Demonstrate the Examples

Example 1 Use colored markers to write numbers in expanded form and show their word equivalents below using the same colors. Use a different color for each place value.

Example 2 Have students take turns coming to the board. Say which number form to use, and have the student write 346 in that form. Have other students follow, showing other numbers in different forms.

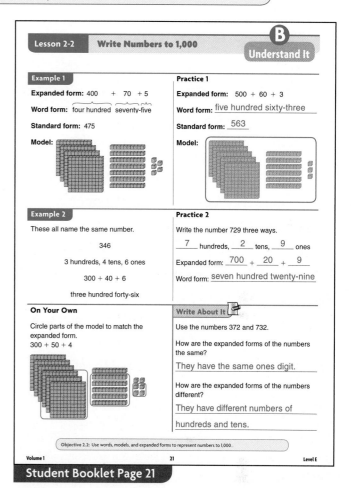

Student Booklet Page 21

Progress Monitoring

Write various forms of numbers on the board. Ask students to tell or write the name of each form. Then have them write each number in a different form.

Error Analysis

If students write numbers in expanded form incorrectly, such as 7 + 4 + 9 instead of 700 + 40 + 9 for 749, have them add 7 + 4 + 9 to see that this equals 20, not 749. Help them use place-value charts or base ten blocks to support their understanding.

Lesson 2-2: Write Numbers to 1,000

Objective 2.2: Use words, models, and expanded forms to represent numbers to 1,000.

Observe Student Progress

Materials ☐ Teaching Aid 6 Place-Value Charts

> **Computer Tutorial**
> Some students may benefit from completing a computer tutorial before they attempt the Try It page. A list of the tutorials for each lesson can be found beginning on page x in the front of this book.

Develop Academic Language

Exercises 1, 2, and 6 Have students practice writing numbers such as *thirty-seven, eighty-two,* and so on to emphasize the use of the hyphen.

ENGLISH LEARNERS Students may need to refer to a list of various word forms that may be difficult to spell. Pair students and have them create posters for this purpose.

Student Booklet Page 22

Lesson 2-2 — Write Numbers to 1,000 — Try It

1. Give the expanded and word forms for each number.
   a. 874 _800 + 70 + 4_
   eight hundred seventy-four
   b. 193 _100 + 90 + 3_
   one hundred ninety-three
   c. 562 _500 + 60 + 2_
   five hundred sixty-two

2. Write the number modeled below in three different ways.
   Standard form _293_
   Expanded form _200 + 90 + 3_
   Word form _two hundred ninety-three_

3. Write in standard form a number that has the given addend when written in expanded form. Samples are shown.
   a. a tens place value of 40 _741_
   b. a hundreds place value of 300 _342_
   c. a ones place value of 3 _863_
   d. a tens place value of 0 _405_

4. What is the standard form of *six hundred fifty-eight*?
   A 685
   B 658
   C 586
   D 568

5. In a three-digit number written in expanded form, which numbers can be in the middle?
   _10, 20, 30, 40, 50, 60, 70, 80, 90_

6. A number has 4 in the ones place, 9 in the hundreds place, and 5 in the tens place. What is the word form of the number?
   nine hundred fifty-four

 Error Analysis

Exercise 3 Give an example by writing 100 + 50 + 3 = 153 for 3a. **There are 3 addends in the expanded form of a three-digit number. Which addend represents the tens place of 153? 50**

Exercise 4 If students have difficulty with this problem, have them first determine which answers show six in the hundreds place. **A and B** Then ask which of these two shows fifty, or 5 in the tens place. **B**

Exercise 6 If students answer incorrectly, have them use place-value charts from Teaching Aid 6 to record the digits before writing the word name for the number.

hundreds	tens	ones

Lesson 2-3: Compare and Order to 1,000

Model It — A

Objective 2.3: Compare and order whole numbers to 1,000 by using the symbols <, =, >.

Teach the Lesson

Materials
- ☐ Base ten blocks
- ☐ Teaching Aid 1 Number Lines

Activate Prior Knowledge

Draw a number line from 1 to 20, numbered by 5s, on the board. **Where does 8 go?** Have a volunteer mark it. **How did you know?** *It's between 5 and 10, but closer to 10.* Repeat with other numbers.

Develop Academic Language

You may hear students using terms such as *larger*, *bigger*, *smaller*, and *fewer*. These terms are better used to describe objects or quantities than numbers. Encourage students to use *greater* and *less* when comparing numbers.

Model the Activities

Activity 1 Provide pairs of students with a container of hundreds, tens, and ones blocks. Have one student make 236. Have the other make 118. **How can you find which one of you has more blocks?** *Sample: Compare the hundreds. 2 hundreds is greater than 1 hundred, so 236 is greater than 118.*

Use the greater than and less than symbols to write two math sentences about your blocks. 118 < 236, 236 > 118

Activity 2 **When you're comparing 3-digit numbers, where do you start?** *With the hundreds* **If the hundreds are the same, what do you do?** *Compare the tens.* Point out that it doesn't make sense to start with the ones, because the tens may be different. Use Activity 1 as an example of this.

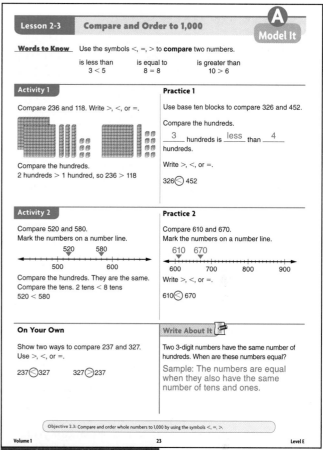

Student Booklet Page 23

✓ Progress Monitoring

What symbols can you use to compare two numbers that are not equal? >, < **What is the difference between the symbols?** > means *is greater than*, < means *is less than*.

Error Analysis

If students confuse the > and < symbols, emphasize that the smaller, pointed part of the symbol always points to the number that is less.

| Lesson 2-3 | Compare and Order to 1,000 |  Understand It |

Objective 2.3: Compare and order whole numbers to 1,000 by using the symbols <, =, >.

Facilitate Student Understanding

Develop Academic Language

Write two groups of numbers on the board.

5 8   12 4 7

Point to the left-hand group and ask which number is less. **5** Circle the 5 and write *less* below it. Point to the other group and ask which number is the least. **4** Circle the 4 and write *least* below it. Repeat for *greater* and *greatest*.

ENGLISH LEARNERS Help students with the comparative and superlative forms *great, greater,* and *greatest* as well as *few, less,* and *least.* Make two 3-column charts to collect examples.

Demonstrate the Examples

Example 1 Put students in groups of 3 and provide each group with a container of base ten blocks. Have each student make 288, 382, or 223. **How can you decide which one of you has the greatest number of blocks? The least?** Sample: Sort the blocks into hundreds, tens, and ones. Compare the hundreds. Then compare the tens. **Would it make sense to compare the ones first?** No, because then you would have 288 as the least number.

Example 2 Provide students with Teaching Aid 1 Number Lines. Have students number from 600 to 900. Suggest using counting on or counting back as strategies for labeling the numbers on the line.

Computer Tutorial

Some students may benefit from completing a computer tutorial before they attempt the Try It page. A list of the tutorials for each lesson can be found beginning on page x in the front of this book.

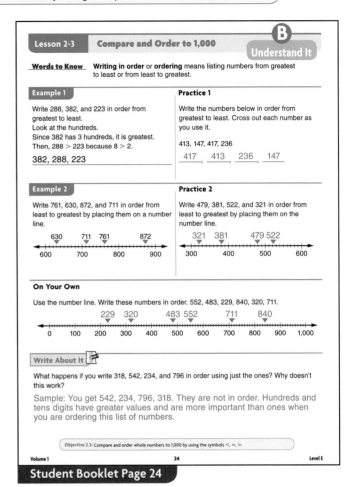

Student Booklet Page 24

⭐ Progress Monitoring

Write on the board *405, 148, 256, 793.* Ask how to find the least number. Sample: Look at the hundreds. The 148 has a 1 in the hundreds place, so it is least.

Error Analysis

Suggest that students copy the list of numbers and underline the hundreds digits.

Lesson 2-3: Compare and Order to 1,000

Objective 2.3: Compare and order whole numbers to 1,000 by using the symbols <, =, >.

Observe Student Progress

Error Analysis

Exercises 1 and 2 If students write the list in reverse order, remind them to read the problem carefully. Suggest that they underline the part of the directions that tells them the order.

Exercise 2 Students who put the numbers in ascending order probably did not read the entire direction. Remind them that there are two ways to order a group of numbers—from least to greatest and from greatest to least.

Exercise 3 Watch for students who are comparing the ones before the hundreds and tens. Point out that if the hundreds or tens are different, it doesn't matter what the ones are.

Exercise 4 Remind students that they can put the numbers on a number line to see which are false and which is true.

Exercise 5 Help students see that each comparison can be reversed. 115 < 324 leads to 324 > 115.

Student Booklet Page 25

Lesson 2-3 Compare and Order to 1,000

1. Write in order from least to greatest.
   a. 466, 243, 408, 72
   72, 243, 408, 466
   b. 85, 519, 130, 935, 246
   85, 130, 246, 519, 935
   c. 320, 852, 612, 274, 336
   274, 320, 336, 612, 852

2. Write in order from greatest to least.
   a. 776, 773, 821, 654
   821, 776, 773, 654
   b. 652, 502, 765, 48, 517
   765, 652, 517, 502, 48
   c. 601, 234, 452, 164, 578
   601, 578, 452, 234, 164

3. Complete. Use >, <, or =.
   a. 409 **>** 109   b. 733 **>** 443
   c. 619 **<** 916   d. 520 **<** 679
   e. 305 **<** 852   f. 934 **>** 892

4. Which statement is true? Circle the letter of the correct answer.
   A 340 = 430   **B** 327 < 723
   C 518 > 658   D 901 < 109

5. Use 115, 324, 257, >, and <. Write as many different statements as you can.
   115 < 324, 115 < 257, 34 < 115, 324 > 257, 257 > 115, 257 < 324

6. Name three numbers that are less than 823 and greater than 805.
   Sample: 807, 810, 822

7. Paolo owned two paintings, one worth $830 and one worth $790. How will you find the painting that is worth more?
   Sample: Compare the hundreds digits. 8 is greater than 7, so the painting worth $830 is worth more.

8. 939 is greater than 845, and 845 is greater than 608. Write two statements comparing 939 and 608. Use words.
   939 is greater than 608. 608 is less than 939.

Volume 1: Number Sense and Place Value

Topic 2: Place Value through 1,000

Topic Summary

Objective: Review skills related to place value through 1,000.

You may want to have each student analyze one answer choice, or have students work in groups of four with each student analyzing a different choice.

Ask students to share their ideas about each answer choice. Be sure they confirm the correct answer for each problem at the end of the discussion.

Answer Evaluation

1. **A** Students put the numbers in the order in which they appear in the problem.
   B This choice is correct.
   C Students do not understand place value.
   D Students switched the tens digit with the ones digit.

2. **A** Students did not split the ones and tens.
   B Students added the digits and not the values of the digits.
   C This choice is correct.
   D Students forgot to add the tens.

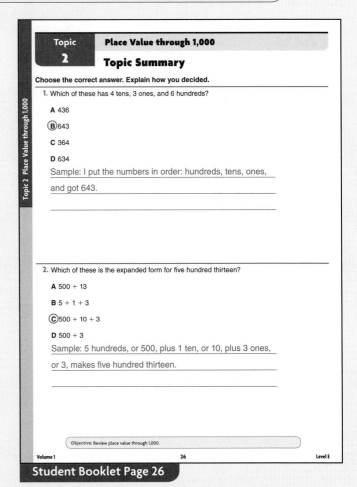

Student Booklet Page 26

 ### Error Analysis

Exercise 1 Provide students with base ten blocks for them to model each of the values given in the problem.

Exercise 2 Have students find the sum of each of the answer choices to eliminate B and D.

Progress Monitoring

When all assignments for this topic have been completed, assign the corresponding Progress Monitoring page for this topic (Assessment Resources Book, page 2). Be sure students complete the Progress Monitoring page before you administer the final assessment for this volume.

Volume 1: Number Sense and Place Value

Topic 2: Place Value through 1,000

Mixed Review

Objective: Maintain concepts and skills.

Have students complete the Mixed Review page. Work with each student individually to review results. Identify strengths and weaknesses and correct any misunderstandings.

 Error Analysis

Exercise 1 If students write the digits instead of the values, have them find the sums to check their answers.

Exercise 2 Remind students that the less than and greater than symbols always point to the lesser number.

Exercise 3 Remind students that we do not write anything for 0 when it occurs in the middle of a number like 609.

Exercise 5 Make sure that students understand that the problem is asking for the digit, and not the value of the digit.

Student Booklet Page 27

Volume 1: Number Sense and Place Value

Topic 3: Place Value beyond 1,000

Topic Introduction

Lesson 3-1 provides practice with identifying place value for each digit in numbers to 10,000. Lesson 3-2 has students use this background knowledge to write numbers in expanded form. Lessons 3-3 and 3-4 extend students' knowledge of place value and rounding whole numbers through the millions.

Lesson	Objective	Student Pages	Teacher Pages	Tutorials
Topic 3 Introduction	**3.1** Identify the place value for each digit in numbers to 10,000. **3.2** Use expanded notation to represent numbers. **3.4** Round whole numbers through the millions to the nearest ten, hundred, thousand, ten thousand, or hundred thousand.	28	31	
3-1 Place Value to 10,000	**3.1** Identify the place value for each digit in numbers to 10,000.	29–31	32–34	3a, 3b
3-2 Expanded Notation	**3.2** Use expanded notation to represent numbers.	32–34	35–37	3b
3-3 Numbers in the Millions	**3.3** Read and write whole numbers in the millions.	35–37	38–40	3c
3-4 Round Numbers through Millions	**3.4** Round whole numbers through the millions to the nearest ten, hundred, thousand, ten thousand, or hundred thousand.	38–40	41–43	3d
Topic 3 Summary	Review skills related to place value beyond 1,000.	41	44	
Topic 3 Mixed Review	Maintain concepts and skills.	42	45	

Computer Tutorial

Some students may benefit from completing the computer tutorial before they attempt the Try It page of each lesson. If you are using the electronic components of *Pinpoint Math,* you will find a complete listing of Tutorial codes and titles when you access them either online or via CD-ROM.

Volume 1: Number Sense and Place Value

Topic 3: Place Value beyond 1,000

Topic Introduction

Objectives: 3.1 Identify the place value for each digit in numbers to 10,000. **3.2** Use expanded notation to represent numbers. **3.4** Round whole numbers through the millions to the nearest ten, hundred, thousand, ten thousand, or hundred thousand.

Write several numbers in the thousands on the board, and have volunteers write the word forms of those numbers. Then, write the word forms of several numbers in the thousands, and have volunteers write the number form.

Informal Assessment

1. **In which place is the 2?** Tens **So what is the value of the 2?** 20 **Since 6 is in the hundreds place, what is the value of 6?** 600 **What place is the 7 in?** Thousands **So what is the value of 7?** 7,000 **What is the value of the 1?** 1

2. **How do you write a number in expanded form?** You find the value of the digit in the number, and write those numbers in order with addition symbols between them. **What do you do with places that have zeros in them?** You don't write those places.

3. **Read this number aloud.** Forty-two thousand, three hundred twelve **What number is in the thousands place?** 2 **What number is to the right of that place?** 3 **Is 3 large enough to round this number up?** No. **So what do we round to?** We round down to 42,000.

4. **Read this number aloud.** Nine million, eight hundred twenty-seven thousand, six hundred fifty-four **What number is in the ten thousands place?** 2 **What number is to the right of the 2?** 7 **Is 7 large enough to round this number up?** Yes. **So what do we round to?** 9,830,000

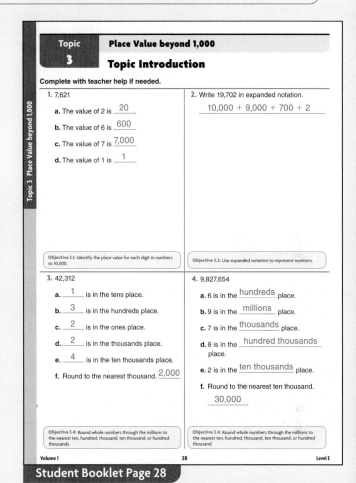

Student Booklet Page 28

Another Way You can provide Teaching Aid 7 Place-Value Charts to help students name the places. Building the number in Exercise 1 with base ten blocks may help students remember the meaning of *value*.

Lesson 3-1 — Place Value to 10,000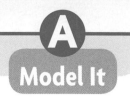

Objective 3.1: Identify the place value for each digit in numbers to 10,000.

Teach the Lesson

Materials ☐ Base ten blocks: thousands, hundreds, tens, ones

Activate Prior Knowledge

Review the use of ones, tens, and hundreds to represent three-digit numbers. Have students give the value of various digits in a given number.

Develop Academic Language

Point out that for numbers in the thousands, we use a *comma* to separate the hundreds, tens, and ones digits from the thousands digit.

Model the Activities

Activity 1 Look at the 1 thousands block. **This represents 1,000. Show me 2 hundreds.** Students should point to the 2 hundreds. **Show me 2 tens.** Students should point to the 2 tens. **Are there any ones?** No.

There is 1 thousands block, so write 1 in the thousands column in the chart. How many hundreds blocks are there? 2 **Where do you write 2?** In the hundreds column **What number belongs in the tens column?** 2 **What number is in the ones column?** 0

Activity 2 Display 2 thousands blocks on the board. **What number does this show?** 2,000 Write "2,000" below. Display the 1 hundreds block on the board. **What number does this show?** 100 Write "+ 100" next to "2,000". Display 3 tens blocks on the board. **What number does this show?** 30 Write "+ 30" next to "+ 100". Display the 5 ones blocks on the board. **What number does this show?** 5 Write "+ 5" next to "+ 30". **2,000 + 100 + 30 + 5 is the expanded form of this number. What is the standard form of this number?** 2,135 As students say the number, write the word form below the standard form.

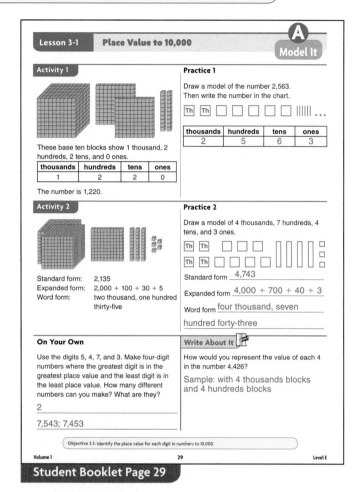

Student Booklet Page 29

Write About It

ENGLISH LEARNERS Allow students to respond orally or by drawing a picture.

Progress Monitoring

Write 7,625 on the board. Point randomly to digits and ask students to identify the value of each digit.

Error Analysis

If students have trouble identifying random place values, rework the problem beginning in the ones place and moving to the left.

Lesson 3-1: Place Value to 10,000

Objective 3.1: Identify the place value for each digit in numbers to 10,000.

Facilitate Student Understanding

Materials ☐ Base ten blocks: thousands, hundreds, tens, ones

Develop Academic Language

Remind students that the *value* of a digit is based on the value of the place it is in. Thus, in the number 6,423, the 4 is in the hundreds place and its value is 4 hundreds. Skip count to see that this means 400.

Demonstrate the Examples

Example 1 Have students write *thousands, hundreds, tens,* or *ones* in large letters on a piece of paper. Have students hold up the card naming the place of each digit you call. **What place is the 6 in?** Students with the *thousands* card should hold up the card. Continue until there is a digit for each place.

Example 2 Display 3 thousands blocks. **What is this value?** 3,000 Display 4 hundreds blocks. **What is this value?** 400 Display 2 tens blocks. **What is this value?** 20 Display 1 ones block. **What is this value?** 1 Have students combine the values and write the number 3,421 in standard form.

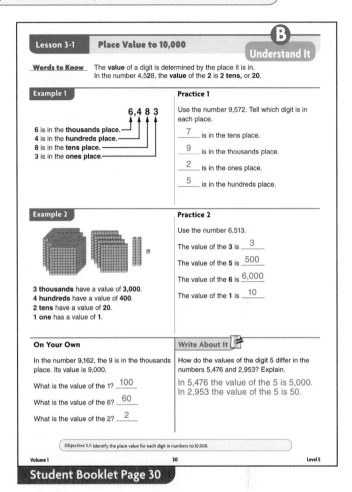

Student Booklet Page 30

✓ Progress Monitoring

Write several four-digit numbers on the board, each with a 7. Have students identify the place and value of the 7 in each number.

Error Analysis

For a number such as 7,352, if students give the value of the 3 as 100, remind them that there are 3 hundreds, which equal 300, not 100.

Lesson 3-1 — Place Value to 10,000

Objective 3.1: Identify the place value for each digit in numbers to 10,000.

Observe Student Progress

Computer Tutorial

Some students may benefit from completing a computer tutorial before they attempt the Try It page. A list of the tutorials for each lesson can be found beginning on page x in the front of this book.

 Error Analysis

Exercise 1 If students answer incorrectly, have them start with the digit farthest to the right (5) in the standard form of the number and determine what place it is in. **Ones** Then have them write 5 in the correct blank. Continue having them identify the digit in each place in order, right to left.

Exercises 2, 3, and 8 Remind students to read the problems carefully, so they know that while Exercises 2 and 3 are asking about value, Exercise 8 is asking about place.

ENGLISH LEARNERS You may need to explain the word *bold* so students understand the term *bold digit* in Exercise 2.

Exercise 4 Have students identify the value of 4 in each of the answer choices. Clarify that there can be more than one correct answer, and that students should find all correct answers.

Exercise 5 Students can use models to help them. Remind them that in expanded form, the values of the digits are added in order from greatest to least. Help students identify the value of each digit.

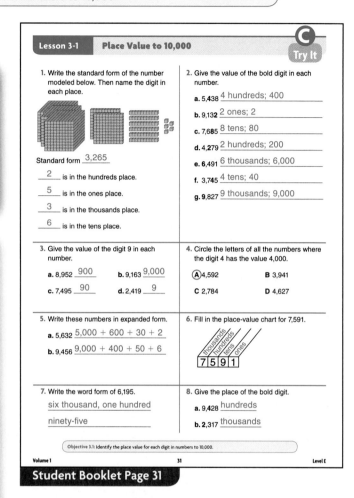

Lesson 3-2 | Expanded Notation with Zeros

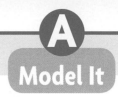

Objective 3.2: Use expanded notation to represent four-digit numbers.

Teach the Lesson

Materials
☐ Teaching Aid 6 Place-Value Charts
☐ Base ten blocks

Activate Prior Knowledge

Review the use of base ten blocks to represent four-digit numbers. Have students give the place and the value of various digits in a given number.

Develop Academic Language

Clarify what a place-value chart is. Make sure students understand that each section represents the place value shown at the top and that only one digit can be written in each section.

Model the Activities

Activity 1 Write the number in a place-value chart. **Where do you write the zero? In the tens place** What does a model with 0 tens look like? **There are no tens blocks.** Have students model the number. **How many thousands are there? 3 How many hundreds? 4 How many tens? 0 How many ones? 2** Write the number in standard form. **3,402** Now use another place-value chart. Write a number with 6 thousands, 5 hundreds, 4 tens, and 0 ones. **6,540** Write 654 on the board as you read the numbers of thousands, tens, and ones again. **Why is this incorrect? Sample: 6 is in the hundreds place.**

Activity 2 Model the number 4,997. **What number does this show? 4,997** Write the number on the board. **What is the expanded form of this number? 4,000 + 900 + 90 + 7** Have a student write the expanded form next to the number. Remove one ten at a time from the model. Have students count down, and have a volunteer update the expanded form. When there are no tens, ask, **How has this changed the number? There are no tens in the new number. What number does this show? 4,907** Erase "+ 0" from the expanded form. **Adding 0 doesn't change the sum, so we don't need to write "+ 0."**

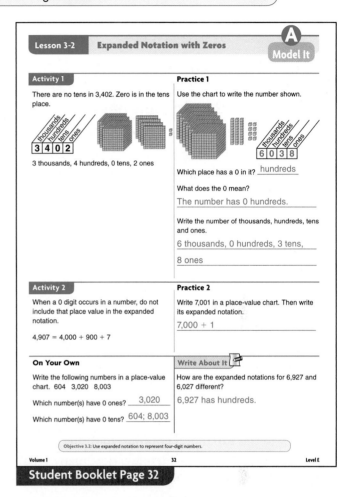

Student Booklet Page 32

Write About It

ENGLISH LEARNERS Allow students to write the expanded notations and circle the differences.

Progress Monitoring

Write various four-digit numbers that include a 0 digit in a place-value chart and have students write the expanded form of each number.

Error Analysis

Remind students that a four-digit number with a 0 digit will only have three numbers in its expanded form. Continue to relate the base ten block models to expanded form, and have students add to check expanded form.

Lesson 3-2: Expanded Notation with Zeros

Objective 3.2: Use expanded notation to represent four-digit numbers.

Facilitate Student Understanding

Develop Academic Language

You may want to have students practice increasing and decreasing numbers by various amounts of ones, tens, and so on, to ensure they understand the meaning of the terms *increasing* and *decreasing*.

ENGLISH LEARNERS Remind students that when writing numbers in standard form, a comma follows the number of thousands. Students who have attended school outside of the United States may use a period instead of a comma as is the convention in some countries.

Demonstrate the Examples

Example 1 How many thousands are there? **9** How do you show this in expanded notation? **Write 9,000.** How many hundreds are there? **0** How do you show in the expanded notation that there are no hundreds? **Don't write anything.** How many tens? **5** How do you show this? **+ 50** How many ones? **6** How do you show this? **+ 6** You may want to have students add to check. Now read the word form of the number. If students struggle to write numbers with zeros in word form, remind them to read the digits on either side of the comma as if that part of the number stood alone.

Example 2 Use the words *thousands, hundreds, tens* and *ones* to read the expanded notation. It begins "5 thousands plus..." **5 thousands plus 3 tens plus 7 ones** How many thousands are there? **5** Write 5 in the thousands place. Write the thousands comma after it. How many hundreds? **0** What should you write in the next place, the hundreds place? **0** How many tens? **3** What should you write in the tens place? **3** The ones? **7** Have students read the number.

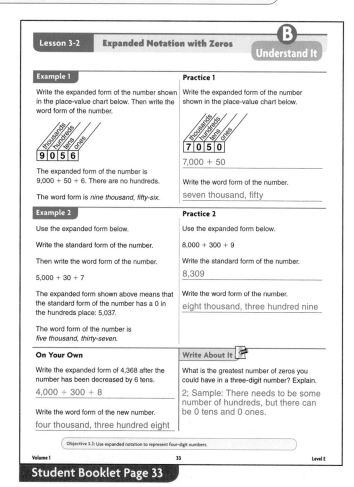

Student Booklet Page 33

✓ Progress Monitoring

What is the expanded form for 9,035?
9,000 + 30 + 5

Error Analysis

If students have difficulty correctly writing expanded form for numbers with zeros, have them model the number with base ten blocks to demonstrate that there are no blocks for the 0 digit's place value.

Lesson 3-2   Expanded Notation with Zeros

Objective 3.2: Use expanded notation to represent four-digit numbers.

Observe Student Progress

Computer Tutorial

Some students may benefit from completing a computer tutorial before they attempt the Try It page. A list of the tutorials for each lesson can be found beginning on page x in the front of this book.

Error Analysis

Exercises 1, 3, and 4 Some students may need to be reminded that a 0 in the standard form of a number means that that place is not represented in its expanded form.

Exercise 2 It may be helpful to begin by reading each addend as a number of thousands, hundreds, tens, or ones; drawing or making block models may also help. Remind students that the first number in an expanded form tells the value of the left-most digit. If the first number is 300, then the number is a three-digit number; if the first number is 3,000, then the number is a four-digit number; and so on. Therefore if the expanded form is 6,000 + 30 + 7, the standard form would not be 6,37 because the number must have four digits. It should be 6,037.

Exercises 5 and 6 ENGLISH LEARNERS Students may need options for responding to the word *explain*. An oral or demonstrated response or a diagram may be more appropriate than a written explanation to evaluate student comprehension.

Review the use of the word *reverse*. Practice reversing the digits of several numbers, changing 347 to 743; 5,106 to 6,015; and so on. Write the word *reverse* with an example for students to include in their word banks.

Student Booklet Page 34

Lesson 3-2   Expanded Notation with Zeros   **Try It**

1. Write the expanded form of each number.
   a. 5,209   5,000 + 200 + 9
   b. 730   700 + 30
   c. 8,007   8,000 + 7
   d. 1,036   1,000 + 30 + 6

2. Write the standard form of each number.
   a. 300 + 50   350
   b. 7,000 + 800 + 40 + 3   7,843
   c. 2,000 + 5   2,005
   d. 4,000 + 70   4,070

3. Write the expanded form of the number shown in the place-value chart below.

thousands	hundreds	tens	ones
3	6	0	2

   3,000 + 600 + 2

4. Janine drove 3,072 miles on Saturday. Circle the letter of the expanded form of 3,072.
   A 3,000 + 700 + 2
   B 3,000 + 200 + 7
   C 300 + 70 + 2
   D 3,000 + 70 + 2

5. Reverse the digits of the number 4,020. Is the new number still a four-digit number? Explain.
   No. A four-digit number has a digit other than zero in the thousands place.

6. Reverse the digits of the number 7,805. Is the new number still a four-digit number? Explain.
   Yes. It is the four-digit number 5,087.

7. The expanded form of a four-digit number has only two addends. What is the greatest number it could be? What is the least number?
   9,900; 1,001

8. Write all the numbers that include only 8,000 and a ten-place value in their expanded forms.
   8,010; 8,020; 8,030;
   8,040; 8,050; 8,060; 8,070;
   8,080; 8,090

Objective 3.2: Use expanded notation to represent four-digit numbers.

Exercise 7 Write _,000 + _00 + _0 + _ to help students get started. Tell them they can fill in only two blanks. **Which blanks should you fill in to get the greatest answer?** The first two **What digit should you use in each blank to get the greatest answer?** 9 For the least number, point out that the first blank must be filled in to ensure that the number has 4 digits.

Exercise 8 Show students that since the expanded form of the number contains only thousands and tens, the hundreds and ones places are shown with a 0.

Lesson 3-3 — Numbers in the Millions

Objective 3.3: Read and write whole numbers in the millions.

Teach the Lesson

Materials ☐ Teaching Aid 7 Place-Value Charts

Activate Prior Knowledge
Review the ones, tens, hundreds, and thousands places. Have students read and write several four-digit numbers, identifying each digit's value.

Develop Academic Language
Standard form is the way we write a number with digits and commas. Give an example. *Standard* means "in the usual way."

ENGLISH LEARNERS Explain the difference between a period at the end of a sentence and the mathematical meaning.

Model the Activities

Activity 1 Draw a place-value chart that shows six places. **What are the names of the places starting from the right?** Ones, tens, hundreds, thousands, ten thousands, hundred thousands **What comes after hundred thousands?** Millions Add three places to the left. Help students identify and use the "one, ten, hundred" pattern to label these places. **Starting at the right, each group of 3 places is a period.** Underline the ones period. **This is the ones period.** Underline the thousands period. **What do you see in each of these places' name?** Thousand **This is the thousands period.** Repeat for the millions. **When you read the number, read each period separately. At the end of the period, say the name of the last place in the period.** Model.

Activity 2 Separate the ones period and the thousands period with a large comma. **This comma comes after the thousands. We call it the thousands comma. Where should the next comma be?** Between the hundred thousands and the millions **What can we call it?** The millions comma Draw the second comma. Help students read the number.

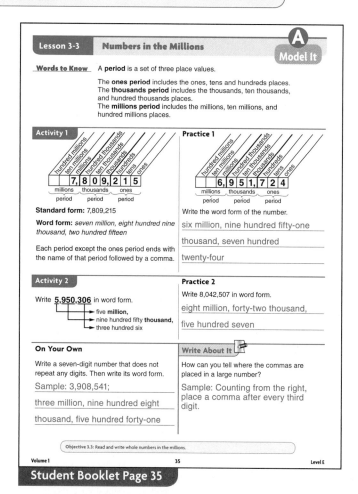

Student Booklet Page 35

Progress Monitoring

Write a seven-digit number on the board. Point to various digits and have students name each digit's place. Then have students read the number.

Error Analysis
Watch for students who have trouble with numbers that contain the digit 0. Suggest they start at the left, circle periods, and read each separately.

Lesson 3-3 — Numbers in the Millions

Objective 3.3: Read and write whole numbers in the millions.

Facilitate Student Understanding

Develop Academic Language

Have students list the first seven place-value names beginning with ones.

Remind students that when writing the word form of a number, although the words *million* and *thousand* are used for the larger two periods, the ones period does not include the word *ones*.

Demonstrate the Examples

Example 1 Write 4,750,692 on the board. Using three different colors, draw a ring around each of the three periods. Emphasize the commas that separate the periods. Ask students where the name of the period comes in the word form of the number. **Before the comma** Allowing students to label the commas may be helpful as they read the number.

Practice 1 You can provide a frame like this one to help students: __ __ __, __ __ __, __ __ __. Labeling the commas may also be helpful.

Example 2 Cover the millions and hundred thousands in each number. What is the first ten thousands digit you see? **6** Then start with 6 ten thousands, or a value of 60,000. Add 10,000. **70,000** Point out that adding 10,000 means to increase the digit in the ten thousands place by 1. When students reach a value of 90,000, say **60,000, 70,000, 80,000, 90,000... What's next? 100,000** Then add 1 hundred thousand, and reset the ten thousands at 0. Have students uncover the hundred thousands place and see the 46, 47, 48, 49, 50 ten thousands pattern.

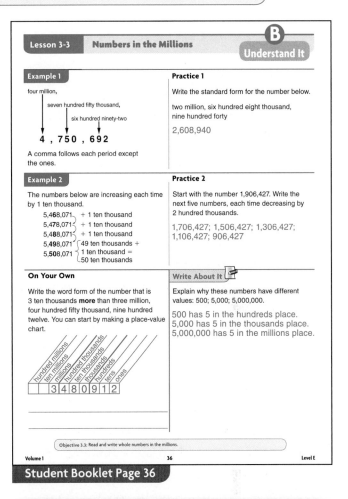

Student Booklet Page 36

 Progress Monitoring

Write any seven-digit number on the board and have students increase it by thousands until they reach an added 10,000.

Error Analysis

If students have difficulty with Example 2, start with 10,000. Then have students work with 410,000, and then with 5,410,000 before completing the Example.

Lesson 3-3: Numbers in the Millions

Objective 3.3: Read and write whole numbers in the millions.

Observe Student Progress

Computer Tutorial

Some students may benefit from completing a computer tutorial before they attempt the Try It page. A list of the tutorials for each lesson can be found beginning on page x in the front of this book.

Error Analysis

Exercise 3 Be sure that students are looking at the correct place for each problem. Suggest that they underline the place that is changing. You may want to point out that the original number can be read as having 46 hundred thousands.

Exercise 4 Have students write the standard form for each answer option before selecting the correct one. Helping students write place-value charts or blanks with labeled commas may also be helpful.

Exercise 5 If students put the commas in the wrong places, tell them that the commas in the standard form go in the same places as in the word form.

Exercise 6 Suggest that students write the problem in vertical form, with blanks for the digits of the unknown number. Help students identify 9,999,999 as the solution, since it is the greatest seven-digit number. Have students test various numbers by adding each to 8,999,999. If one does not work, encourage them to erase it and try again.

Student Booklet Page 37

Lesson 3-3 — Numbers in the Millions — Try It

1. Write the word form for each number.
   a. 4,900,567
      four million, nine hundred thousand, five hundred sixty-seven
   b. 732,040
      seven hundred thirty-two thousand, forty

2. Write the standard form for each number.
   a. seven million, six hundred six thousand, forty-one
      _7,606,041_
   b. one million, four hundred thirty-five thousand, six hundred ninety-eight
      _1,435,698_
   c. two million, twenty thousand, twenty
      _2,020,020_

3. Change the number 4,607,511 by each amount. Write the new number.
   a. increase by 2 ten thousands
      _4,627,511_
   b. decrease by 3 hundreds
      _4,607,211_
   c. decrease by 5 hundred thousands
      _5,107,511_

4. Which is the word form of 8,005,032?
   A eight thousand, five hundred, thirty-two
   B eight million, five hundred, thirty-two
   Ⓒ eight million, five thousand, thirty-two
   D eight million, fifty thousand, thirty-two

5. In 2006, the population of Dallas, Texas, was one million, two hundred eighty thousand, five hundred people. Write that in standard form.
   _1,280,500_

6. What is the only seven-digit number that you could add to 8,999,999 and still have a seven-digit number? Explain.
   _1,000,000; 1,000,000 + 8,999,999 = 9,999,999_

Objective 3.3: Read and write whole numbers in the millions.

Lesson 3-4 Round Numbers through Millions

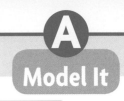

Objective 3.4: Round whole numbers through the millions to the nearest ten, hundred, thousand, ten thousand, or hundred thousand.

Teach the Lesson

Activate Prior Knowledge
Review rounding two-digit numbers to the nearest ten. Ask students to share their strategies for identifying the 2 closest tens they must choose between. Continue by rounding three-digit numbers to the nearest hundred and the nearest ten.

Develop Academic Language
Ask for words that indicate rounding: *about, almost, nearly, approximately, close to,* and so on.

ENGLISH LEARNERS Have students add the terms collected above, *round* and *rounded,* to the word bank in the back of their booklets. Explain that this meaning of *round* is different from "curved."

Model the Activities

Activity 1 To help students identify the 2 closest tens, you can have them count by 1,000s until they hear part of 5,346. They can write 5,000 below 5,346, continue counting, and write 6,000 above. **Which numbers shown on the number line are thousands?** 5,000 and 6,000 Help students count the tick marks by 50s to verify the placement of 5,346. **What thousand is 5,346 closest to?** 5,000

Activity 2 You can to start by helping students identify the two possible answers. **Underline the place you're rounding to. Which place is that?** Hundred thousands **Circle the next place. Which place is that?** Ten thousands **This place tells us whether to round up or down. If it is 5 or greater, round the underlined digit up. Is the circled digit 5 or greater?** No. **Round up?** No. Round down. **The underlined digit stays the same, and everything after it becomes 0. In most numbers, the digits *before* the underlined digit stay the same.**

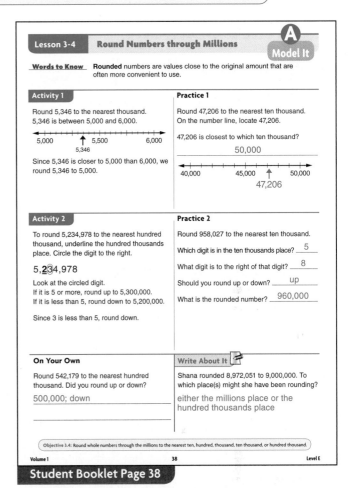

Student Booklet Page 38

Progress Monitoring
Round 2,581,359 to the following places: tens, hundreds, thousands, ten thousands, hundred thousands, and millions. 2,581,360; 2,581,400; 2,581,000; 2,580,000; 2,600,000; 3,000,000

Error Analysis
If students need help identifying the possible answers, another way to think about rounding is either to leave the underlined digit the same, or to write "+ 1" below and increase the digit by 1. This may help students understand what happens when a 9 is rounded up, since they can think of "carrying" to the next place. Remind students to change the following digits to 0.

Volume 1   41   Level E

Lesson 3-4: Round Numbers through Millions

Understand It

Objective 3.4: Round whole numbers through the millions to the nearest ten, hundred, thousand, ten thousand, or hundred thousand.

Facilitate Student Understanding

Develop Academic Language

Write the following equivalences on the board.
1 hundred = 10 tens
2 hundreds = 2 × 10 tens = 20 tens
3 hundreds = 3 × 10 tens = 30 tens

Help students see that when they are rounding 496 up to 500, they are really rounding 49 tens to 50 tens.

Demonstrate the Examples

Example 1 What tens are shown on this number line? **370, 380, 390, 400, 410** Between which two tens does 397 belong? **390 and 400** Help students count tick marks from 390 to confirm the location of 397. **Is 397 closer to 390 or 400? 400**

To reinforce the "5 or greater" rule, ask students to locate 385. Help them see that it is equidistant from 380 and 390, and remind them that there is a rule for the situation.

Example 2 Circling the digit to the right of the place to be rounded to can help students focus on what number will determine the direction of their rounding. You may want to encourage them to write the answer choices for each exercise above and below the number.

Student Booklet Page 39

⭐ Progress Monitoring

Round 358,713 to the nearest ten and to the nearest ten thousand. 358,710; 360,000

Error Analysis

Remind students that when rounding the digit 9 up they need to make it a zero and increase the next digit to the left by 1. Have them practice rounding the following number to several places: 4,979,695. **To millions, 5,000,000; to hundred thousands, 5,000,000; to tens, 4,929,700**

Use number lines to help students see that rounding down only affects the digits after the underlined digit. It does not decrease the value of the underlined digit.

Lesson 3-4 Round Numbers through Millions

Objective 3.4: Round whole numbers through the millions to the nearest ten, hundred, thousand, ten thousand, or hundred thousand.

Observe Student Progress

Computer Tutorial

Some students may benefit from completing a computer tutorial before they attempt the Try It page. A list of the tutorials for each lesson can be found beginning on page x in the front of this book.

Error Analysis

Exercise 1 Have students use underlining and circling to decide which digit determines how to round each number.

Exercise 3 For students having trouble getting started with this problem, ask them which numbers will be at the far left and right of their number line. **3,000; 4,000** They might find that locating and labeling the number that is halfway is helpful.

Exercise 4 Have students first round each answer option to the nearest thousand and then select the correct answer.

Exercise 6 If students suggest using exact numbers, ask them how Tamika would know exactly how many hamburgers people would buy. Help them to realize exact numbers aren't possible in this situation.

Student Booklet Page 40

Lesson 3-4 Round Numbers through Millions

1. Round each number.
   a. 4,765 to the tens ___4,770___
   b. 306,973 to the thousands ___307,000___
   c. 2,050,030 to the ten thousands ___2,050,000___
   d. 8,999,999 to the millions ___9,000,000___
   e. 68,317 to the hundreds ___68,300___

2. When you round to the nearest thousand, does the given number round to 3,540,000?
   a. 3,540,468 ___yes___
   b. 3,540,975 ___no___
   c. 3,539,421 ___no___
   d. 3,539,539 ___yes___

3. Show how you would use a number line to round 3,376 to the nearest thousand. Then write the rounded number.

   3,376 (shown on number line between 3,000 and 4,000)

   ___3,000___

4. Which number rounds to 6,780,000 when rounded to the nearest thousand?
   A 6,781,000   B 6,780,500
   C 6,780,972   (D) 6,779,500

5. Give three numbers that will round to 27,000 when rounded to the nearest thousand.
   Sample: 26,841; 27,068; 27,499
   ___hundred thousands place___

6. Sports Star baseball stadium holds 23,000 people. Tamika is in charge of ordering hamburgers and hot dogs for Saturday's game. When she places her order, would she be likely to use exact numbers or rounded numbers? Why?
   rounded numbers; She won't know exact numbers before the game.

Volume 1 — Number Sense and Place Value

Topic 3: Place Value beyond 1,000

Topic Summary

Objective: Review skills related to place value beyond 1,000.

Have students complete the student summary page. You may want to have students work in groups of four with each student analyzing a different choice.

Ask students to share their ideas about each answer choice. Be sure they confirm the correct answer for each problem at the end of the discussion.

Answer Evaluation

1. **A** Students omitted the zeros for the hundred thousands and the hundreds places.

   B This choice is correct.

   C Students have the zero in the incorrect position in the thousands period.

   D Students have the zero in the incorrect position in the thousands and the ones periods.

2. **A** This choice is correct.

   B Students rounded up instead of down.

   C Students rounded down by decreasing the value of the thousands digit.

   D Students rounded to the nearest hundred thousand.

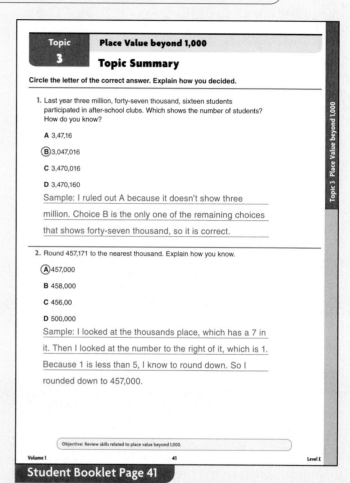

Student Booklet Page 41

 Error Analysis

If students made errors in Exercise 1, use Teaching Aid 7 Place-Value Charts to review.

For Exercise 2, help them use number lines, underline and circle, or list the two possible answers above and below the number.

Progress Monitoring

When all assignments for this topic have been completed, assign the corresponding Progress Monitoring page for this topic (Assessment Resources Book, page 3). Be sure students complete the Progress Monitoring page before you administer the final assessment for this volume.

Volume 1 — Number Sense and Place Value
Topic 3 — Place Value beyond 1,000
Mixed Review

Objective: Maintain concepts and skills.

Have students complete the Mixed Review page. Work with each student individually to review results. Identify strengths and weaknesses and correct any misunderstandings.

Error Analysis

Exercise 1 For students having difficulty, suggest they use Teaching Aids 6–8 Place-Value Charts.

Exercise 3 Remind students that in expanded form they do not include anything for place values where the digit is 0.

Exercise 5 Remind students that the smaller side of the greater than or less than symbol points to the lesser number.

Exercise 6 Have students rewrite the clues in order from greatest place value to least place value. Afterward, they can write the number in expanded form, and then standard form.

Exercise 7 Remind students that for the number to round down to 3,684,000, the digit in the hundreds place must be less than 5. Give examples of numbers that do not round down to 3,684,000.

Exercise 8 If students round to an incorrect place, suggest that they circle the place that they're rounding to in the original number.

Student Booklet Page 42

Topic 3 — Place Value beyond 1,000 — Mixed Review

1. Give the place of each underlined digit.
   a. 3<u>7</u>9 _tens_
   b. <u>5</u>,930 _thousands_
   c. <u>9</u>22 _hundreds_
   Volume 1, Lesson 2-1

2. Write the standard form for six thousand, seventy.
   6,070
   Volume 1, Lesson 2-2

3. Give the expanded form for each number.
   a. 3,427 _3,000 + 400 + 20 + 7_
   b. 20,908 _20,000 + 900 + 8_
   d. 4,902,050 _4,000,000 + 900,000 + 2,000 + 50_
   Volume 1, Lesson 3-2

4. Choose the word form for 8,181
   A eighty-one thousand, eighty-one
   B eight-one, eighty-one
   Ⓒ eight thousand, one hundred eighty-one
   D eighty-one thousand, eight hundred one
   Volume 1, Lesson 2-2

5. Compare with <, =, >.
   a. 194 _>_ 149
   b. 723 _>_ 372
   Volume 1, Lesson 1-3

6. Combine these values. Write the number in standard form.
   3 hundreds   8 millions
   0 ten thousands 1 ten
   6 hundred thousand 2 ones
   7 thousands
   _8,607,312_
   Volume 1, Lesson 3-3

7. What digits could be in the blank place if the number rounded to the nearest thousand is 3,684,000?
   3,684,_29
   _0, 1, 2, 3, 4_
   Volume 1, Lesson 3-4

8. Round 5,841,087 to the nearest hundred thousand.
   A 6,800,000   B 5,841,100
   C 6,000,000   Ⓓ 5,800,00
   Volume 1, Lesson 3-4

Objective: Maintain concepts and skills.

Volume 2 — Basic Facts
Topic 4: Addition and Subtraction Facts

Topic Introduction

Lesson 4-1 introduces problems in which students will have to choose addition or subtraction. Lesson 4-2 teaches students some algebraic properties and gets them thinking about number patterns. Lessons 4-3 and 4-6 teach different strategies students can use to solve addition and subtraction problems. Lesson 4-4 has students find the sum of three 1-digit numbers. Finally, in Lessons 4-5 and 4-7, students relate addition and subtraction and use fact families to solve problems.

Lesson	Objective	Student Pages	Teacher Pages	Tutorials
Topic 4 Introduction	4.3 Use different thinking strategies to add numbers.	1	47	
	4.5 Illustrate the relationship between addition and subtraction.			
	4.6 Use different thinking strategies to subtract numbers.			
4-1 Add or Subtract to Solve Problems	4.1 Solve problems by choosing the operation and show the meanings of addition and subtraction.	2–4	48–50	4a, 4b, 4c, 4d, 4e
4-2 Properties of Addition	4.2 Show number patterns and apply various properties including the commutative, associative, and identity properties.	5–7	51–53	4f, 4g
4-3 Addition Strategies	4.3 Use different thinking strategies to add numbers.	8–10	54–56	4h, 4i, 4j, 4k
4-4 Add Three 1-Digit Numbers	4.4 Find the sum of three 1-digit numbers.	11–13	57–59	4i
4-5 Relate Addition and Subtraction	4.5 Illustrate the relationship between addition and subtraction.	14–16	60–62	4f, 4c
4-6 Subtraction Strategies	4.6 Use different thinking strategies to subtract numbers.	17–19	63–65	4f, 4c
4-7 Fact Families	4.7 Write addition and subtraction fact families.	20–22	66–68	4f
Topic 4 Summary	Review addition and subtraction skills.	23	69	
Topic 4 Mixed Review	Maintain concepts and skills.	24	70	

Computer Tutorial

Some students may benefit from completing the computer tutorial before they attempt the Try It page of each lesson. If you are using the electronic components of *Pinpoint Math,* you will find a complete listing of Tutorial codes and titles when you access them online or via CD-ROM.

Volume 2 — Basic Facts
Topic 4: Addition and Subtraction Facts
Topic Introduction

Objectives: 4.3 Use different thinking strategies to add numbers. **4.5** Illustrate the relationship between addition and subtraction. **4.6** Use different thinking strategies to subtract numbers.

Materials ☐ Base ten blocks

Have students practice making groups of numbers using the base ten blocks. They should start with single-digit numbers using the ones blocks, and then begin grouping blocks in the tens.

Informal Assessment

1. **How many counters are in the model?** 6 **How many counters would you need to add to make it a group of 10 counters?** 4

   Suppose you separated the 6 counters into two smaller groups. If one group has 1 counter, how many would the other group have? 5 Repeat for 2 and 3 counters.

2. Draw the number line on the board. **Where does the arrow start?** 5 **Which direction is it pointing?** Right **What operation does moving to the right indicate?** Addition **What number does it end on?** 10 **How many places did we move to get from 5 to 10?** 5

3. **How many circles are there in all?** 4 **How many circles are crossed out or subtracted?** 3 **How many circles are left?** 1 **So, 3 subtracted from 4 equals what?** 1 Write the equation on the board. **What's another way to say this?** 4 minus 3 equals 1.

4. **How many blue circles are there?** 3 **How many white circles are there?** 3 **How many circles are there when we add the blue circles to the white circles?** 6 **How can we write that?** 3 + 3 = 6 If we wanted to write a subtraction problem based on this model, **what number would we start with?** 6 **What number would we subtract?** 3 **What are we left with?** 3

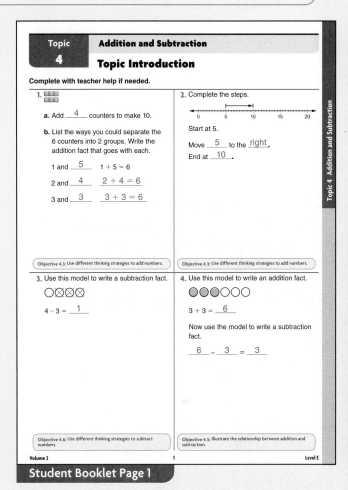

Student Booklet Page 1

Another Way You may want to have students model the problems with base ten blocks.

Lesson 4-1: Add or Subtract to Solve Problems

Objective 4.1: Solve problems by choosing the operation and show the meanings of addition and subtraction.

Teach the Lesson

Materials ☐ 2-color counters

Activate Prior Knowledge

Ask students to tell you different kinds of addition stories. They should mention joining groups, getting more, and so on. Then ask students to tell you different kinds of subtraction stories. Make sure they mention taking away, comparing quantities, and separating categories. If they do not mention a certain situation, give them an example, like "There are 12 cats on the lawn. Five of them are gray. How many are not gray?"

Develop Academic Language

Write on the board: *adding, addition, adds, added*. **How are these words alike?** They all begin with the word *add*. Have volunteers use the words in sentences. Repeat for the words *subtracting, subtraction, subtracts, subtracted*.

ENGLISH LEARNERS Students will benefit from the strategy used here to locate the root word.

Model the Activities

Activity 1 Provide two-color counters. **Put 5 red counters together, all in one group. Then put 8 yellow counters together, all in another group. Combine both groups. How many counters are there all together?** 13 **What operation did you model?** Addition

Activity 2 Have students use two-color counters. **Put 7 red counters together. Now take 4 counters away. How many counters are left?** 3 **What operation did you model?** Subtraction **Now put 4 yellow counters together in a row. Then put 7 red counters together in another row. Compare the yellow counters to the red counters. How many more red counters are there than yellow counters?** 3 **What operation did you model?** Subtraction

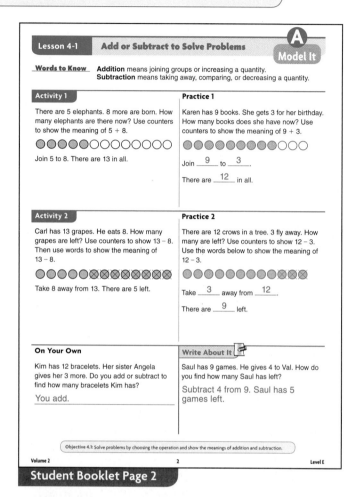

Student Booklet Page 2

Progress Monitoring

Write *add* and *subtract* on the board. These are operations you can perform on a pair of numbers. **Which one means you are joining two groups?** Add **Taking part of a group away?** Subtract

Error Analysis

If students cannot describe the meanings of addition and subtraction in general terms, have them use counters to show you specific examples.

Lesson 4-1: Add or Subtract to Solve Problems

Understand It

Objective 4.1: Solve problems by choosing the operation and show the meanings of addition and subtraction.

Facilitate Student Understanding

Materials
- ☐ Teaching Aid 1 Number Lines
- ☐ 2-color counters

Develop Academic Language

Define *difference* as the solution to a subtraction problem. Help students relate this to the nonmathematical meaning of difference.

Demonstrate the Examples

Example 1 Provide students with 2-color counters. Have them line up 11 red counters in a row, and 5 yellow counters underneath that. **What do we need to do in this problem?** Compare the costs. **How can you compare the two rows of counters to find the difference?** Match up the two rows and find the difference between the two. Tell students to put the yellow counters on top of the red counters. **How many red counters don't match up?** 6 6 is the difference between 11 and 5.

Example 2 Provide copies of Teaching Aid 1. **Is the number of dogs increasing or decreasing in this problem?** Decreasing **Where do we start?** 13 **How do you know?** We started with 13 dogs. **How many units are we going to move?** 6 **How do you know?** 6 dogs left. **What direction should we move in?** Left **How do you know?** The numbers decrease as you move left, and since some dogs went away, we will end with a lesser number than we started with.

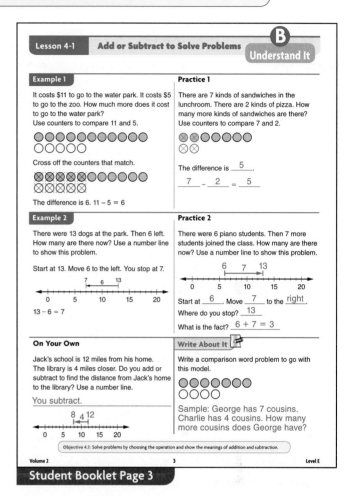

Student Booklet Page 3

ENGLISH LEARNERS Have students who cannot explain addition and subtraction in words make drawings that illustrate the operations.

 Progress Monitoring

When you're subtracting, what direction do you move on the number line? Left **Why?** Sample: Numbers decrease when you move left, and subtracting involves decreasing a number.

Error Analysis

If students are confused by what direction to move on the number line, ask them to think about whether the answer will be greater or less than the number they started with.

Lesson 4-1 — Add or Subtract to Solve Problems

Objective 4.1: Solve problems by choosing the operation and show the meanings of addition and subtraction.

Observe Student Progress

Materials ☐ MathFlaps or 2-color counters

Computer Tutorial

Some students may benefit from completing a computer tutorial before they attempt the Try It page. A list of the tutorials for each lesson can be found beginning on page x in the front of this book.

⭐ Error Analysis

Exercise 1 To help students write the facts, have them start by counting the MathFlaps or 2-color counters. **How many white MathFlaps are shown?** 9 **How many blue?** 5 **How can you compare these numbers?** Subtract. **How can you find the number in all?** Add. Have students cross out one white and one blue and repeat the questions.

Exercises 2 and 4 Students who need more practice making number-line models may benefit from making a whole page of models that show different sums to 10. Then have them make a whole page of models showing different numbers subtracted from 15.

Exercise 3 Use the question to point out that subtraction does not always mean *How many are left?* In this problem, it means *How much more do you need?* Point out to students that this is a comparison problem. They are comparing how much rope Dan needs with how much he has.

Exercise 8 Pair students to ask each other to explain the addition and subtraction phrases (decreased by, increased by, take away, more than, and so on) using different numbers.

Student Booklet Page 4

Lesson 4-1 Add or Subtract to Solve Problems Try It

1. Use this model to write an addition and a subtraction fact.

   Sample: 9 − 5 = 4 and
   9 + 5 = 14

2. There are 12 monkeys at the zoo. Eight are adults. How many are **not** adults? Draw a number line model.

3. Dan has 12 yards of rope, but he needs 20 yards. How can he find out how much more rope he needs to buy? Circle the letter of the correct answer.

   A Add 12 to 20.   **B** Subtract 12 from 20.
   C Add 20 to 12.   D Subtract 20 from 12.

4. Lani has 7 yards of fabric. She buys 9 more yards. How many yards does she have now? Draw a number line model.

5. Sue is 15 years old. Bob is 6 years younger than Sue. Show the number fact you use to figure out how old Bob is.

   15 − 6 = 9

6. Darrell has 8 fish. Kari has 5 fish. What problem can you use to find how many fish they have all together? Circle the letter of the correct answer.

   A 8 − 5   **B** 8 + 5
   C 5 − 8   D 85

7. Jake baked 12 muffins. Then Janine gave him 4 more. What operation do you use to find how many muffins Jake has all together? What fact can you write for this problem?

   Addition; 12 + 4 = 16

8. Explain the difference between these phrases.

   8 decreased by 5   8 increased by 5

   Sample: 5 is subtracted from 8;
   5 and 8 are added.

Lesson 4-2 | Properties of Addition

Objective 4.2: Apply various addition properties including the associative, commutative, and identity properties.

Teach the Lesson

Materials ☐ MathFlaps

Activate Prior Knowledge

Write several addition problems on the board, such as 4 + 8, 8 + 4, 10 + 6, 6 + 10, and so on. Have students find the sums. **What do you notice about the problems you just did? Sample: Problems such as 4 + 8 and 8 + 4 have the same answer. What is the same about 4 + 8 and 8 + 4? What is different?** Sample: They have the same addends, operation sign, and sum but the addends are in a different order. Repeat the activity with sentences such as 4 + (5 + 8) and (4 + 5) + 8.

Develop Academic Language

Write 5 + (8 + 2) on the board. Point to the parentheses. **These symbols are parentheses. In math, parentheses mean you do whatever is inside of them first. What should we do first here?** 8 + 2 = 10 **Then what?** 5 + 10 = 15

Model the Activities

Activity 1 Provide students with MathFlaps. **Show me 8 + 6. Then show me 6 + 8. What is true about the sums?** They are the same. Write 8 + 6 = 6 + 8. **What property did we illustrate?** Commutative

Activity 2 Provide students with MathFlaps and have them illustrate the sums (4 + 3) + 5 and 4 + (3 + 5). **What property are we illustrating?** Associative

Write About It ✏️

ENGLISH LEARNERS If students have difficulty with this question, provide them with specific examples and ask them to explain orally what they see.

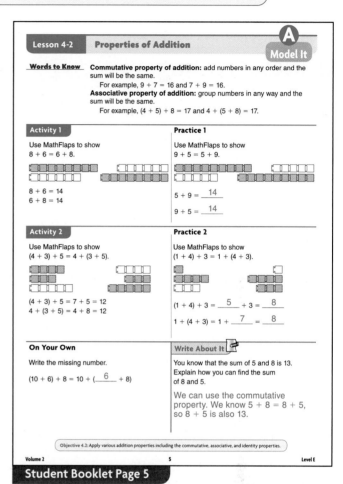

Student Booklet Page 5

Progress Monitoring

Describe the commutative and associative properties in your own words. **Sample:** You can add numbers in any order and you can group the numbers in any way you want, and you will still get the same sum.

Error Analysis

If students confuse *commutative* and *associative*, help students relate *commute*, as in go back and forth, or change places, to *commutative*; and *associate*, or get together with someone, with *associative*. Relate these terms to the properties.

Lesson 4-2 Properties of Addition

Objectives 4.2: Apply various addition properties including the associative, commutative, and identity properties.

Facilitate Student Understanding

Develop Academic Language

Provide a number of sentences such as 9 + 8 = 8 + 9 and (3 + 8) + 6 = 3 + (8 + 6). Have students identify which property each sentence illustrates.

ENGLISH LEARNERS To reinforce the new terminology, be sure to include a poster or chart with examples of each property labeled. As you identify the properties illustrated by the sentences, remind students that order is changed or grouping is changed.

In some languages, the property names are cognates, such as the Spanish words *commutativo* and *asociativo*.

Demonstrate the Examples

Example 1 Have students work with MathFlaps to see that the identity property holds true for any number added to 0. Make sure they understand that 0 can be the first or second addend.

Example 2 Write 3 + 0 + 7 on the board and ask students if there are any of the numbers that would be easy to add together. **3 and 7** Have a volunteer show how to use the commutative and associative properties to get the two numbers together. Then have them use the identity property to find the sum.

Computer Tutorial

Some students may benefit from completing a computer tutorial before they attempt the Try It page. A list of the tutorials for each lesson can be found beginning on page x in the front of this book.

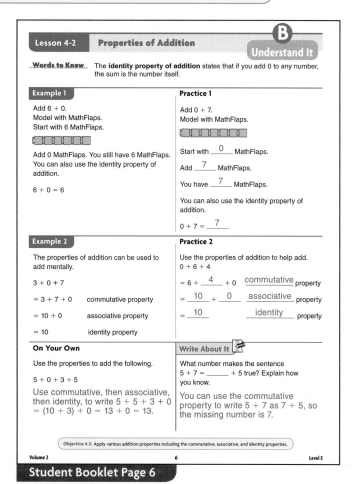

Student Booklet Page 6

⭐ Progress Monitoring

Which property of addition allows you to change the grouping without affecting the sum? **Associative property** Which property of addition says you can add any number and 0 and you get the number itself? **Identity property**

Error Analysis

If students have difficulty remembering what the parentheses mean, give them several practice problems involving parentheses.

Lesson 4-2 Properties of Addition

Objectives 4.2: Apply various addition properties including the associative, commutative, and identity properties.

Observe Student Progress

Develop Academic Language

Exercise 1 Ask students what property is shown in this example. Write the word *commutative* on the board and have students pronounce it and then restate, in their own words, what the commutative property means.

ENGLISH LEARNERS Students may find it helpful to make cards with the words *commutative*, *associative*, and *identity* on the fronts and an illustration of the property on the backs of the cards.

 Error Analysis

Exercise 2 Students may have difficulty determining the correct number if they are not sure what property is used in each case. Suggest they first decide which property is being illustrated and then determine what number is missing.

Exercise 4 Have students compare answer choices B and C. Many students make errors because they do not observe that if only the order changes, it is not an example of the associative property.

Exercise 6 If students are overwhelmed by the large number, remind them that the identity property works for any number.

Exercise 7 Allow students to use base ten blocks or counters to help them do these additions. It is more important for this problem that they understand how the properties are being used than that they know all their basic facts.

Exercise 8 Some students may assume that if parentheses appear in an expression, only the associative property is being used.

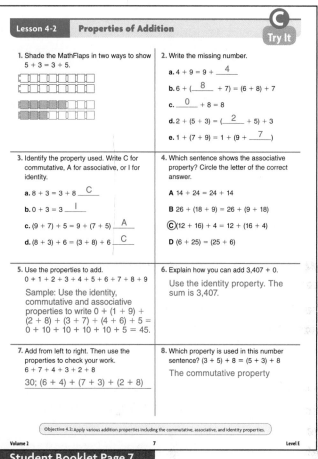

Student Booklet Page 7

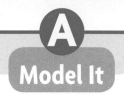

Lesson 4-3 Addition Strategies

Objective 4.3: Use different thinking strategies to add numbers.

Teach the Lesson

Materials ☐ MathFlaps

Activate Prior Knowledge

Practice doubles facts and counting on 1, 2, 3, and 4 from given numbers to sums of 10 or less.

Develop Academic Language

Review parts and the whole by having students use MathFlaps to make a group of 2 and a group of 3. **Each of these groups is a part. If I put both parts together I get a whole.** Have students put their MathFlaps into one group. **How many are in the whole? 5**

Model the Activities

Activity 1 Write 7 + _____ = 10 on the board. Tell students to make a row of 7 MathFlaps next to a row of 10 MathFlaps. **Look at the row of 7. How many more are needed to make 10? 3** Have students connect 3 more flaps to make 10. Students can use the row of 10 for comparison in Practice 1 and Activity 2.

Activity 2 Have students make a row of 6 red MathFlaps and a row of 5 yellow MathFlaps. **Look at the row of 6. How many more are needed to make 10? 4 Where can we get those 4? From the row of 5** Have students take 4 MathFlaps from the row of 5, and add them to the row of 6. Students should now have a row of 10 and a row of 1. **How many are in the row that had 6? 10 How many are left in the row that had 5? 1 What is 10 + 1? 11 So what is 6 + 5? 11** Tell students that this is also called the "plus 10" method to showing 10.

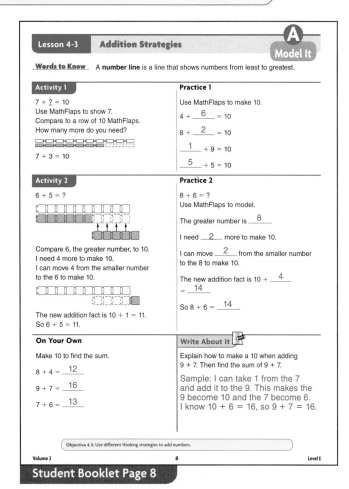

Student Booklet Page 8

⭐ Progress Monitoring

Give an example of an addition fact that makes 10. **Samples: 2 + 8 = 10, 4 + 6 = 10, 9 + 1 = 10** Have students continue to write as many examples as they can think of.

Error Analysis

Help students use the MathFlaps to model making 10. It may be helpful to line up the two addends in a long row, break apart the lesser addend into single MathFlaps, and add as many of those as needed to the greater addend to make 10.

Lesson 4-3: Addition Strategies

Objective 4.3: Use different thinking strategies to add numbers.

Facilitate Student Understanding

Develop Academic Language

Write 2 + 2 on the board. **What is the sum?** 4 When two of the same number are added together, we call this a *doubles fact*. Generate a list of doubles facts.

ENGLISH LEARNERS Model doubles with objects by showing two pencils and then doubling to make four. Have students double a number of items.

Demonstrate the Examples

Example 1 Start with the greater number in 3 + 12. **Which is the greater number?** 12 Tell students they are going to count on from 12. **How many numbers do we have to add to 12?** 3 Let's start with the number that comes right after 12. **What are the next three numbers?** 13, 14, 15 **What is 3 + 12?** 15

Example 2 Have students model 5 + 5 with 2 rows of 5 MathFlaps. **What is the sum?** 10 Now add one more MathFlap to the second row. This model shows 5 + 5 + 1 more, or 5 + 6. Since we added 1 to the doubles fact, how will the sum change? It will be 1 more. **What is 5 + 6?** 11 If you increase one addend, the sum increases by the same amount. If students base their choice of doubles fact on the lesser addend, they'll always add 1 to the sum.

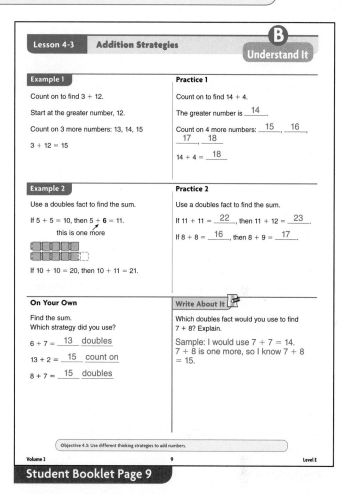

Student Booklet Page 9

★ Progress Monitoring

Would you count up or use a doubles fact to find 6 + 7? Explain. Sample: I would use the doubles fact 6 + 6 = 12 and add 1 to the answer. If I counted on, I would have to count up 6.

Error Analysis

Remind students not to repeat the number from which they are counting on. For example, when solving 14 + 2, students should count *15, 16*, not *14, 15*. Also make sure students start with the greater number.

To provide practice with doubles-plus-one facts, you may want to have students model all doubles facts from 2 + 2 to 10 + 10, add 1 to each model, and write the resulting facts.

Lesson 4-3 Addition Strategies

Objective 4.3: Use different thinking strategies to add numbers.

Observe Student Progress

Computer Tutorial

Some students may benefit from completing a computer tutorial before they attempt the Try It page. A list of the tutorials for each lesson can be found beginning on page x in the front of this book.

Error Analysis

Exercise 1 Remind students that they can make a row of 10 MathFlaps to compare to a row representing the known addend.

Exercise 3 Students may start counting up using the first number in the addition fact. Remind students that if they start with the larger number, they will be able to count up by a smaller amount. This will be quicker and might result in fewer errors.

Exercise 4 For Part b, students might use the doubles fact 5 + 5 because 5 is the first number in the addition fact. Tell students that 5 + 5 will yield an answer greater than 5 + 4 because one of the numbers is larger. So if students use 5 + 5 as the doubles fact, they must subtract 1 from the answer.

Exercise 7 Students might try to guess which problem has a sum of 14. Encourage students to solve all of the problems to make sure they have the correct answer.

Exercise 8 Students might simply add the two numbers that they see in the problem. Remind students to use the words as clues to determine what to do with the numbers.

Student Booklet Page 10

Lesson 4-3 Addition Strategies — Try It

1. Find the number needed to make 10.
   a. 4 + __6__ = 10
   b. 2 + __8__ = 10
   c. 3 + __7__ = 10

2. Make 10 to solve. Write the "plus 10" problem.
   a. 8 + 7 = __15__
      10 + 5 = 15
   b. 9 + 4 = __13__
      10 + 3 = 13
   c. 6 + 5 = __11__
      10 + 1 = 11

3. Count on to solve. Show how you counted.
   a. 2 + 7 = __9__
      Count on: 8, 9
   b. 11 + 4 = __15__
      Count on: 12, 13, 14, 15
   c. 3 + 4 = __7__
      Count on: 5, 6, 7

4. Use a doubles fact to solve. Write the doubles fact you used.
   a. 8 + 9 = __17__
      8 + 8 = 16
   b. 5 + 4 = __9__
      4 + 4 = 8
   c. 7 + 8 = __15__
      7 + 7 = 14

5. Jen downloaded 9 songs on Monday. She downloaded 7 songs on Tuesday. How many songs did Jen download in all? Write a fact to solve.
   __9__ + __7__ = __16__
   Which strategy did you use? make 10

6. Explain how to find 8 + 5.
   Sample: I know 8 + 2 = 10. If I take 2 away from 5, I am left with 3. So 8 + 5 can be written as 10 + 3. I know 10 + 3 = 13, so 8 + 5 = 13.

7. Which of the following has a sum of 14?
   A 8 + 8   B 1 + 4
   C 4 + 4   **D** 9 + 5

8. Cory has 7 pencils. How many more pencils does Cory need to make 10?
   A 3   B 4
   C 17   D 13

ENGLISH LEARNERS Remind students that the word *sum* means the answer to an addition problem. Differentiate from the homophone *some*. For Spanish speakers, relate to the cognate *suma*.

Lesson 4-4   Add Three 1-Digit Numbers

Objective 4.4: Find the sum of three one-digit numbers.

Teach the Lesson

Materials ☐ MathFlaps

Activate Prior Knowledge

To review basic addition facts, have each student write a number from 1 to 10. Call on students two at a time to share their numbers, then call on a third student to give the sum.

Develop Academic Language

What are you told to do in Activity 1? **Find the sum.** What does *sum* tell you to do? **Add.**

Model the Activities

Activity 1 Have students model the problem with MathFlaps. **When you add, you can put the numbers together in any order. Try putting 5 and 1 together. Now what addition problem does your model show?** 6 + 6 **What is the sum?** 12 **Now model the original problem again. What is a different way to put two of these addends together?** Add 6 and 1, or add 5 and 6. Have students try the other ways to confirm that the sum is the same for each. **Which way do you think was easiest?** Discuss. Point out that students can focus on starting with numbers that they can add mentally, save the smallest numbers for last, or use both of these strategies.

Activity 2 Have students model the problem. **What are some different ways to add these numbers?** First do 7 + 3, 7 + 5, or 5 + 3. **Look at the example shown on the page. Which numbers were added first?** 7 and 3 **Why is 7 + 3 a good choice to start with?** 7 and 3 make 10, and 10 is easy to add to. **Looking for a way to make 10 is a good strategy to solve problems quickly. Add 7 and 3. What else do you have to add?** 5 **What is the sum?** 15

ENGLISH LEARNERS Explain that *mentally* means "without pencil and paper (or other tools)," or "in your head."

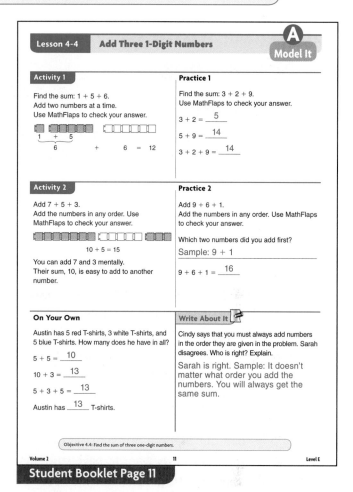

Student Booklet Page 11

 Progress Monitoring

Have students show 2 different ways to find the sum 8 + 3 + 2. 8 + 2 = 10, 10 + 3 = 13 and 8 + 3 = 11, 11 + 2 = 13

Error Analysis

Some students might add only two of the addends and record their answers without including the third. Suggest that students cross out the addends as they add to help them remember that there are three addends to be added. Provide drill in sums of 10 if necessary.

Volume 2   57   Level E

Lesson 4-4 Add Three 1-Digit Numbers

Objective 4.4: Find the sum of three one-digit numbers.

Facilitate Student Understanding

Develop Academic Language
Remind students that the numbers that are added together in an addition expression or sentence are called *addends*.

Demonstrate the Examples

Example 1 What are some different ways to add these numbers? First do 6 + 3, 6 + 4, or 4 + 3. Does any of those addition problems have a sum of 10? Yes, 6 + 4. Start with 6 + 4. Cross out those numbers in the problem. What is the other number to be added to their sum? 3 Rewrite the problem. 10 + 3 What is the sum? 13 What does this tell us about the sum of the original problem? It is 13. Suggest that students add in a different order to check.

Example 2 Point out that the problem has one unnecessary number. You may want to have students draw a picture to model the problem. What does the question ask us to find? The total number of flowers Which flowers are named in the problem? Roses, carnations, and daisies How many of each type of flower are there? 5 roses, 8 carnations, and 9 daisies Which number in the problem is not part of the total number of flowers? 3, because it tells the number of vases Add 5, 8, and 9 to find the total number of flowers. Do you see a way to make 10? No. Then start with any two numbers that are part of a basic fact you know. Then you can add the third number in your head, or use pencil and paper. 5 + 8 + 9 = 22 flowers

ENGLISH LEARNERS Tell students that roses, carnations, and daisies are types of flowers.

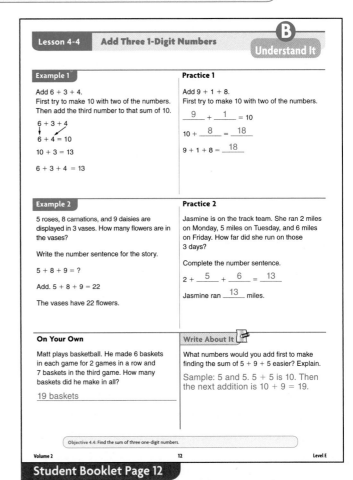

Student Booklet Page 12

Progress Monitoring
Show how to make 10 when finding the sum 3 + 9 + 1. Add 9 and 1.

Error Analysis
When adding the numbers out of the given order, students may add one of the numbers twice. For example, when adding 5 + 4 + 6, the student may add 4 + 6 + 10, 10 + 6 = 16. Here they added the 6 twice and did not add the 5 at all. Suggest that students cross out the addends as they use them.

Lesson 4-4 | Add Three 1-Digit Numbers

Objective 4.4: Find the sum of three one-digit numbers.

Observe Student Progress

Computer Tutorial

Some students may benefit from completing a computer tutorial before they attempt the Try It page. A list of the tutorials for each lesson can be found beginning on page x in the front of this book.

Develop Academic Language

Exercise 4 Point out that this exercise uses the term *not*. Students need to identify the addends with which making 10 is not possible. Caution students to read carefully when a question involves the word *not*.

Exercise 8 **ENGLISH LEARNERS** You may wish to demonstrate tossing number cubes. Point out the phrase "rolled two 6s." Remind students that the number cube has the numbers 1 to 6, and that to have rolled two 6s means that the number 6 shows on the tops of two of the number cubes.

Error Analysis

Exercises 2 and 3 Encourage students to add the addends for each exercise in two different orders to check the sum.

Exercise 5 Point out that this problem has one unnecessary number. Help students make a diagram of the projects in the rooms to help them determine the addends.

Exercises 6 and 8 To help students get started, suggest that they act out the problem with cubes.

Student Booklet Page 13

Lesson 4-4 Add Three 1-Digit Numbers — Try It

1. Find the sum. Write the addition sentence for the picture.
   $4 + 3 + 5 = 12$

2. Add.
   a. $5 + 2 + 4 = \underline{11}$
   b. $10 + 3 + 3 = \underline{16}$
   c. $4 + 2 + 6 = \underline{12}$
   d. $5 + 8 + 1 = \underline{14}$

3. Find these sums.
   a. $2 + 5 + 8 = \underline{15}$
   b. $7 + 7 + 7 = \underline{21}$
   c. $6 + 9 + 2 = \underline{17}$
   d. $8 + 6 + 6 = \underline{20}$

4. In which problem could you **not** make 10 when adding?
   A $3 + 7 + 1$
   B $9 + 1 + 6$
   Ⓒ $4 + 9 + 2$
   D $4 + 8 + 2$

5. At an art show, Room 1 has 6 projects. Room 2 has 3 projects. Room 3 has as many projects as Room 1. How many projects are in the 3 rooms?
   15 projects

6. Ryan is playing a game. He rolls three number cubes and looks at the numbers—1, 2, 3, 4, 5, or 6—on top. Then he finds the sum of the three numbers. What is the greatest total he can get?
   18

7. Tiffany adds $4 + 7 + 2$. She says $4 + 7 = 10$ and $10 + 2$ is 12, so the sum is 12. Is she right?
   No, the sum should be 13, because $4 + 7$ is not 10. It is 11, and $11 + 2 = 13$.

8. Celeste tossed 3 number cubes with 1, 2, 3, 4, 5, and 6 on each. The sum of the numbers she rolled is less than 15. Could she have rolled two 6s? Explain.
   Yes, $6 + 6$ is 12. The third cube could have been 1 or 2, since $12 + 1 = 13$ and $12 + 2 = 14$. Both 13 and 14 are less than 15.

Objective 4.4: Find the sum of three one-digit numbers.

Lesson 4-5 Relate Addition and Subtraction

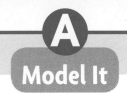

Objective 4.5: Illustrate the relationship between addition and subtraction.

Teach the Lesson

Materials ☐ Two-color counters

Activate Prior Knowledge

Form a group of 13 students, and ask 4 students to step away from the main group. **What addition or subtraction fact does this show?** $13 - 4 = 9$ Bring the groups together. **What addition or subtraction fact does this show?** $9 + 4 = 13$

Develop Academic Language

Refer to *parts* and *totals* to help students see the role each number plays.

Model the Activities

Activity 1 To find how many were taken, draw the missing circles. **How will you know when there are enough?** When there are 12 **How many did you draw?** 5 You added the missing part to the remaining part in order to find the total. **What addition fact shows what you did?** $7 + 5 = 12$ Look back at the problem. **If some circles were taken away, were they subtracted from or added to the total?** Subtracted As you say 12 minus some number equals 7, write $12 - \_\_ = 7$. Use your model. **What number is missing?** 5 **How many circles were taken away?** 5

Activity 2 Start with the total, 13. **How many should you put to the side?** 8 Now your model shows two parts of 13. We want to find the other part of 13 without having to count. You can add two parts to find a total. Think: 8 plus another number equals 13. **What is the missing number?** 5 $8 + 5 = 13$, so 5 is the other part of 13. **$13 - 8$ leaves how many?** 5

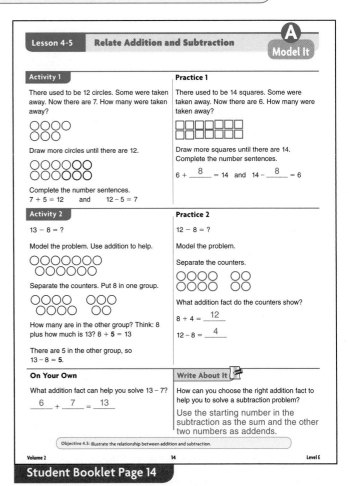

Student Booklet Page 14

Progress Monitoring

What addition fact will help you solve $20 - 5$?
$15 + 5 = 20$

Error Analysis

If students count the remaining counters rather than use addition in Practice 2, model the separation quickly and cover the remaining part.

ENGLISH LEARNERS Students who learned *minuend* and *subtrahend* in their first language may want to know that those are the English terms for the total in subtraction and the number being subtracted.

Lesson 4-5: Relate Addition and Subtraction

Objective 4.5: Illustrate the relationship between addition and subtraction.

Facilitate Student Understanding

Develop Academic Language

Subtract often means to take away. When else do we use subtraction? **When we compare or find the difference between things; when we separate groups**

Demonstrate the Examples

Example 1 You can model the problem by starting with 15 counters in your hand and removing 7 without showing the rest of the counters. **You know a total and one part of the total. One part is missing. Write __ + __ = __ on the board. Where does the total go?** In the single blank after the equal sign **Where do we put the part we know?** In either of the other blanks **Think of basic addition facts. What number is the other part of 15?** 8 Then 15 − 7 = 8. Allow students to confirm the solution with counters or drawings.

Example 2 Ask yourself: should I combine, separate, or compare? **Compare**. What part of the question was a clue? **"How many more"** Do you add or subtract to compare? **Subtract**. Point out that comparing can mean finding the *difference* between numbers. **Write the subtraction problem.** 18 − 15 **Now use addition to solve it. Start with 15. How many more pages does Thomas have to read to equal Miguel?** 3 **What addition fact describes Thomas starting with 15 pages and reading 3 more?** 15 + 3 = 18 **How does that help you solve the subtraction problem?** 3 is the missing part.

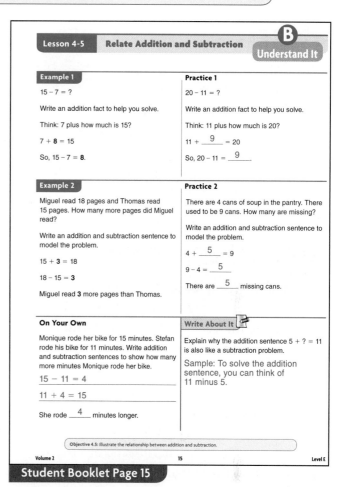

Student Booklet Page 15

Progress Monitoring

Samara ate 14 grapes and Kate ate 7. How many more grapes did Samara eat? **7** What addition problem can you use to solve (or check)? **7 + 7 = 14**

Error Analysis

If students write the numbers in the wrong places in subtraction problems, have them support their work with counters or drawings. Help them see that they can "turn around and add" to subtract, or they can remember to start with the total when subtracting and to end with the total when adding.

Lesson 4-5 — Relate Addition and Subtraction

Objective 4.5: Illustrate the relationship between addition and subtraction.

Observe Student Progress

Computer Tutorial

Some students may benefit from completing a computer tutorial before they attempt the Try It page. A list of the tutorials for each lesson can be found beginning on page x in the front of this book.

Error Analysis

Exercise 1 Watch for students who draw 11 more triangles. Clarify that the total must be 11. Rephrase the question: **We started with 11 triangles. Some were taken away. Now 4 are left.**

Exercise 2 Students might choose D since the number 7 was mentioned before 9 in the problem. Show that 7 − 9 will not equal 2. Provide practice with comparison problems so students learn to subtract the lesser from the greater number.

Exercise 3 This problem is the opposite of what students did on previous pages, but students should be able to solve it by recognizing that one part and the total are given. If not, have students solve the addition problem and then turn it into a subtraction problem.

Exercise 7 The first number given is not the total Melanie started with. Call attention to the key words *now* and *used to*. Encourage students to use models or draw a diagram.

Exercise 8 Have students model the addition and then consider how the model could show the subtraction problem, or have them identify the parts and total. Remind students that part + part = total, and total − part = part.

Student Booklet Page 16

Lesson 4-5 — Relate Addition and Subtraction — Try It

1. There used to be 11 triangles. Now there are 4.
   △△△△
   △△△△
   Draw more triangles until there are 11.
   Complete the number sentences.
   4 + __7__ = 11
   11 − 4 = __7__

2. Margie scored 7 points and Henry scored 9 points. To find how many more points Henry scored than Margie, which number sentence could be used?
   A 9 + 7 = 16   B 16 − 9 = 7
   C 9 − 7 = 2   D 7 − 9 = 2

3. Write a subtraction problem that can help you solve 2 + ___ = 11.
   __11__ − __2__ = __9__

4. Write an addition fact to help you solve 12 − 3. Then solve.
   3 + __9__ = 12
   12 − 3 = __9__

5. Fill in the blank to make the number sentence true.
   17 − __7__ = 10

6. Write an addition fact that will help you solve 20 − 5.
   5 + __15__ = __20__
   20 − 5 = __15__

7. Melanie has 12 pencils now. She used to have 18. How many are missing? Write an addition and subtraction sentence to model the problem. Then solve.
   12 + __6__ = __18__
   __18__ − 12 = __6__
   __6__ pencils are missing.

8. Rosa knows that 4 + 5 = 9. How will this help her find the difference of 9 − 5?
   Sample: If she knows that 4 + 5 = 9, then 9 − 5 = 4 since 4 is the other part of 9.

ENGLISH LEARNERS Make sure students distinguish *parts* and *total*, general names for the roles numbers play in problems, from the specific names *addend*, *sum*, and *difference*.

| Lesson 4-6 | Subtraction Strategies |

Objective 4.6: Use different thinking strategies to subtract numbers.

Teach the Lesson

Materials ☐ Counters

Activate Prior Knowledge

Have students practice counting backward from a given number. Continue with numbers up to 20.

Develop Academic Language

Write 10 − 4 = 6 on the board. **When you subtract, the answer is called the *difference*. What is the difference of 10 and 4?** 6

Model the Activities

Activity 1 Have students show 7 counters. Then have them remove three of them. **What is the subtraction fact represented by the counters?** 7 − 3 **How many counters are left?** 4 **What is 7 minus 3?** 4 **How can you add to check your work?** 4 + 3 = 7

Activity 2 Tell students that counting backward can be used when subtracting small numbers. **For the subtraction fact 9 − 2, what number do you start at?** 9 **How do you know to count backward?** Subtraction decreases the total; we need a lesser number as the answer. **How many numbers will you say when you count backward?** 2 **What numbers do you say?** 8, 7 **What is 9 − 2?** 7

ENGLISH LEARNERS Students can write the numbers 1–20 to help them practice counting backward. Have students place their fingers on the total in the subtraction fact, and then count back.

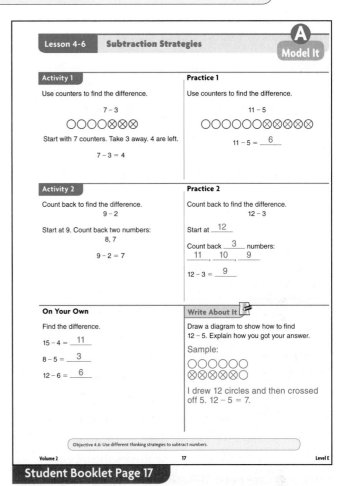

Student Booklet Page 17

Progress Monitoring

How can counting backward help you find the difference 12 − 4? Sample: I start with 12 and then count back 4 numbers, 11, 10, 9, 8. 12 − 4 = 8.

Error Analysis

When counting backward, some students include the number from which they are subtracting. Be sure students understand that when counting back, they should start with one less than the number they're counting from.

Lesson 4-6 | Subtraction Strategies

Objective 4.6: Use different thinking strategies to subtract numbers.

Facilitate Student Understanding

Develop Academic Language

Use the terms *part* and *total* to talk about the function of each number in addition and subtraction facts. Emphasize that we begin with the total when we subtract, and we find the total when we add. Remind students of the meanings of *addend*, *sum*, and *difference*.

Demonstrate the Examples

Example 1 Tell students that subtraction problems can be thought of as addition problems missing one of the parts. **In the first problem, some number plus 6 equals 8. What is the missing addend?** 2 **We can start with the total and subtract. What is the difference of 8 − 6?** 2

Example 2 Students may be familiar with this concept as the method of "adding up" to check subtraction. **A subtraction problem shows the total and one part. The other part is missing. Using the total, 14, and the given part, 8, what addition question can you ask?** 8 plus what number equals 14? **What is the missing addend?** 6 **What is 14 − 8?** 6

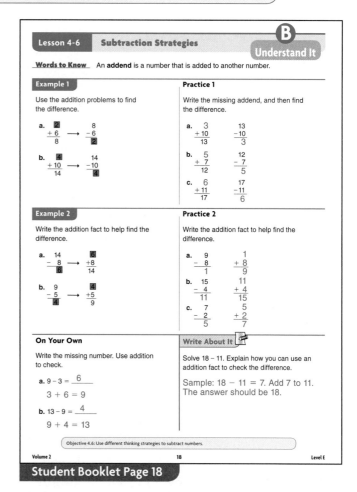

Student Booklet Page 18

Progress Monitoring

How will addition help you find the missing number in the subtraction fact 15 − 8? Sample: I can complete the related addition fact, ? + 8 = 15, with the missing part, 7.

Error Analysis

When finding the missing addend in ? + 4 = 10, students might add the 4 and 10. Have students repeat the problem in words to help them understand. "What number plus four equals 10?"

Lesson 4-6: Subtraction Strategies

Objective 4.6: Use different thinking strategies to subtract numbers.

Observe Student Progress

Computer Tutorial

Some students may benefit from completing a computer tutorial before they attempt the Try It page. A list of the tutorials for each lesson can be found beginning on page x in the front of the book.

Error Analysis

Exercise 1 Although the first and last problem can be solved by counting back, the middle problem involves subtracting a large number. Remind students they can think of a related addition fact (10 + 8 = 18) rather than counting back.

Exercise 3 Remind students that when asked to explain, they must use words to show their thought process.

Exercise 4 Point out that the correct answer is the problem that can be used to check the sum of the addition fact. The answer choice must contain all the numbers in the original addition fact.

Exercise 5 Point out that when zero is subtracted from a number, the answer stays the same, since nothing was taken away.

Student Booklet Page 19

Lesson 4-6 — Subtraction Strategies — Try It

1. Find the difference.
   a. 8 − 2 = __6__
   b. 18 − 10 = __8__
   c. 13 − 4 = __9__

2. What subtraction fact is shown?
   ○○○○○⊗⊗⊗
   8 − 3 = 5

3. Joe wants to find the difference of 17 and 3 by counting back. Explain how Joe can do this.
   Joe can count back 3 numbers. He will say 16, 15, 14. The difference is 14.

4. Which subtraction fact is related to the addition fact 7 + 2 = 9?
   (A) 9 − 7 = 2
   B 11 − 2 = 9
   C 16 − 9 = 7
   D 7 − 2 = 5

5. Write the missing number.
   a. 18 − 9 = __9__
   b. 8 − 0 = __8__
   c. 12 − 9 = __3__

6. Write the addition fact to help find the difference.
   7 − 3 = __4__   __4__ + __3__ = __7__
   16 − 10 = __6__   __6__ + __10__ = __16__

7. Explain how you can use addition to solve 7 − 3.
   Sample: Instead of thinking of 7 minus 3, think of some number plus 3 is equal to 7. Because 3 + 4 = 7, I know 7 − 3 = 4.

8. Find the difference.
   a. 20 − 4 = __16__
   b. 16 − 8 = __8__
   c. 17 − 3 = __14__

Objective 4.6: Use different thinking strategies to subtract numbers.

Lesson 4-7 Fact Families

Objective 4.7: Write addition and subtraction fact families.

Teach the Lesson

Materials ☐ MathFlaps

Activate Prior Knowledge
Draw two groups of circles on the board, one group with 5 circles in it, the other with 4 circles. **How many circles are in the first group?** 5 circles **How many are in the second group?** 4 circles Have students count the total number of circles as you point to each circle and record the addition fact.

Develop Academic Language

Model the Activities

Activity 1 Ask students to put 8 white MathFlaps together. Then put 3 blue MathFlaps together to the white ones. **What fact family can we write based on these MathFlaps?** If students miss any of the number facts in the fact family, guide them to look back at the MathFlaps and see if there are other ways they can put the numbers together to make a fact.

Activity 2 Look at the fact. **How can we start to sketch the MathFlaps?** Start with 6 white MathFlaps. **What do we need next?** 4 blue MathFlaps Check your work by counting the total number of MathFlaps. There are 10. The drawing is correct. Then guide the students to find the rest of the fact family.

ENGLISH LEARNERS Explain the term *fact family* as being like a family in that all the numbers are related to each other.

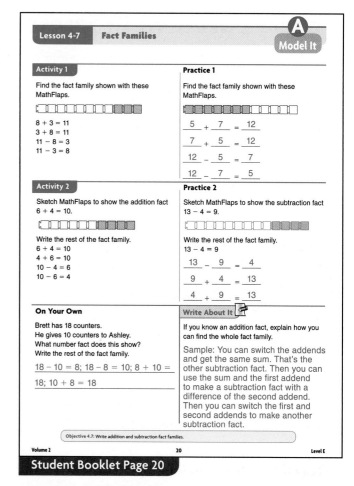

Student Booklet Page 20

Progress Monitoring
What is the fact family for 6 and 5? $6 + 5 = 11$, $5 + 6 = 11$, $11 - 6 = 5$, $11 - 5 = 6$

Error Analysis
If students write only addition or only subtraction facts for their fact families, remind them that another operation could make a new fact with the same numbers.

Lesson 4-7 Fact Families

Objective 4.7: Write addition and subtraction fact families.

Facilitate Student Understanding

Develop Academic Language

Use the term *total* to describe the sum, the minuend, and the number at the top of each fact triangle.

Demonstrate the Examples

Example 1 **What are the addends in this fact family?** 6 and 7 **What is the sum?** 13 Remind students that numbers can be added in any order, so the sum of 6 and 7 is represented by two addition facts. **Which of those numbers is the total in this fact family?** 13 Circle 13. **Now subtract. What number will you start with?** 13 Take away one part. Repeat for the other subtraction fact. Explain that on the fact triangle shown, the total belongs in the circle at the top. The numbers below can be added to equal the total, or one number can be subtracted from the total to yield the other number.

Example 2 12 minus 5 equals the missing number. **What is another way to subtract using the 2 known numbers shown in the triangle and the one missing number?** 12 − the missing number = 5 Have students refer to the triangle to write 2 addition problems with missing numbers. **Which of the blanks is it easy to fill in?** Answers will vary. If you know that 7 + 5 = 12, then you know that those three numbers belong together in a family. If 5 and the total 12 are in any addition or subtraction problem, the other number must be 7.

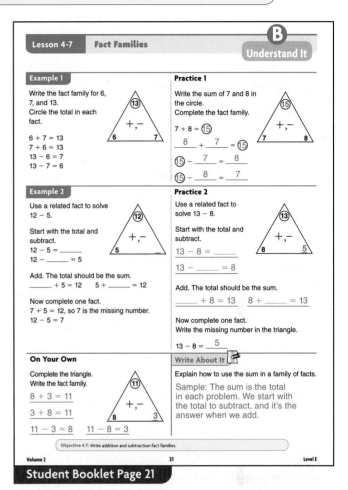

Student Booklet Page 21

★ Progress Monitoring

Make a fact triangle to help you solve 16 − 7. What are the facts in this family? 16 − 7 = 9, 16 − 9 = 7, 9 + 7 = 16, 7 + 9 = 16

Error Analysis

Make sure students understand that 9 + 7 = 16 and 7 + 9 = 16 are considered different facts even though they have the same sum. Provide further practice in using given fact triangles and in using a given fact to fill in a fact triangle and complete the fact family.

Lesson 4-7 Fact Families

Objective 4.7: Write addition and subtraction fact families.

Observe Student Progress

Computer Tutorial

Some students may benefit from completing a computer tutorial before they attempt the Try It page. A list of the tutorials for each lesson can be found beginning on page x in the front of this book.

Error Analysis

Exercise 1 If students have difficulty answering this question based on the visual information, they may have more success with a concrete model. Suggest that they use MathFlaps to make a model of the drawing.

Exercise 2 Encourage students to use a MathFlap model or a fact triangle to help them get started.

Exercises 3 and 4 Make sure students realize that the triangles will be different. Help students identify the total for each.

ENGLISH LEARNERS Highlight *alike* in one color and *different* in another color. Have students color-code the fact triangles with the same highlighters to show the differences and similarities.

Exercise 7 Only two sentences are needed because the addends are the same.

Student Booklet Page 22

Volume 2 — Basic Facts
Topic 4: Addition and Subtraction Facts

Topic Summary

Objective: Review skills related to addition and subtraction facts.

Have students complete the students summary page. You may want to have students work in groups of four with each student analyzing a different choice.

Ask students to share their ideas about each answer choice. Be sure they confirm the correct answer for each problem at the end of the discussion.

Answer Evaluation

1. **A** The student gave the number more.
   B The student gave the number Tiffany has now.
   C The student added instead of subtracting.
   D This choice is correct.

2. **A** The student chose an equation that is part of the same fact family.
   B The student chose an equation that is part of the same fact family.
   C The student chose an equation that is part of the same fact family.
   D This choice is correct.

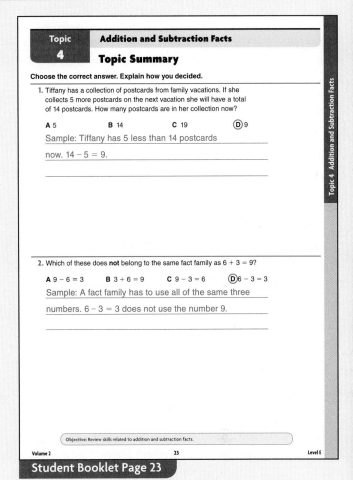

Student Booklet Page 23

 Progress Monitoring

When all assignments for this topic have been completed, assign the corresponding Progress Monitoring page for this topic (Assessment Resources Book, page 4). Be sure students complete the Progress Monitoring page before you administer the final assessment for this volume.

Volume 2: Basic Facts

Topic 4: Addition and Subtraction Facts
Mixed Review

Objective: Maintain concepts and skills.

Have students complete the Mixed Review page. Work with each student individually to review results. Identify strengths and weaknesses and correct any misunderstandings.

Error Analysis

Exercise 1 Provide copies of Teaching Aid 6 Place-Value Charts for students to use to write the expanded notation.

Exercise 3 Review rounding rules and the names of the places with students.

Exercise 4 Check to see that students express and support their solutions in a clear and logical manner. Be sure they use the appropriate terms, notations, and symbols and provide evidence to support their solutions.

Exercise 5 Students may first think of 1,199 as the greatest number that rounds to 1,200. It is the greatest number that rounds up to 1,200, but remind students that there are numbers greater than 1,200 that round down to 1,200.

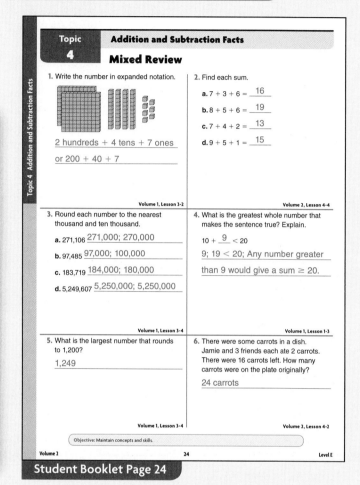

Student Booklet Page 24

Volume 2 — Basic Facts
Topic 5 — Multiplication Facts

Topic Introduction

Topic 5 introduces students to multiplication. Lesson 5-1 begins with the basic meaning of multiplication. Lesson 5-2 starts students on learning the multiplication tables with multiplication by 2, 5, and 10. Lesson 5-3 introduces the basic properties of multiplication. Lesson 5-4 teaches students different strategies to use when multiplying. Finally, Lesson 5-5 builds on the work begun in Lesson 5-2 to teach students all the multiplication facts for numbers 1 through 10.

Lesson	Objective	Student Pages	Teacher Pages	Tutorials
Topic 5 Introduction	**5.1** Use repeated addition, arrays, and counting by multiples to do multiplication. **5.3** Recognize and use the properties of zero and one and the commutative and associative properties of multiplication. **5.5** Memorize to automaticity the multiplication table for numbers 1 through 10.	25	72	
5-1 Meaning of Multiplication	**5.1** Use repeated addition, arrays, and counting by multiples to do multiplication.	26–28	73–75	5a, 5b
5-2 Multiply by 2, 5, and 10	**5.2** Know the multiplication tables of 2s, 5s, and 10 (to "times 10").	29–31	76–78	
5-3 Properties of Multiplication	**5.3** Recognize and use the basic properties of multiplication.	32–34	79–81	5c, 5a, 5b, 5d
5-4 Multiplication Strategies	**5.4** Apply multiplication strategies such as skip counting and doubles.	35–37	82–84	
5-5 Basic Multiplication Facts	**5.5** Memorize to automaticity the multiplication table for numbers 1 through 10.	38–40	85–87	5e
Topic 5 Summary	Review multiplication concepts and skills.	41	88	
Topic 5 Mixed Review	Maintain concepts and skills.	42	89	

Computer Tutorial

Some students may benefit from completing the computer tutorial before they attempt the Try It page of each lesson. If you are using the electronic components of *Pinpoint Math,* you will find a complete listing of Tutorial codes and titles when you access them either online or via CD-ROM.

Volume 2 — Basic Facts
Topic 5 — Multiplication Facts
Topic Introduction

Objectives: 5.1 Use repeated addition, arrays, and counting by multiples to do multiplication. **5.3** Recognize and use the properties of zero and one and the commutative and associative properties of multiplication. **5.5** Memorize to automaticity the multiplication table for numbers 1 through 10.

Draw a 3 by 5 array on the board. Ask students how they could find the total. **Samples: count them, add 3 five times or 5 three times, multiply 3 and 5.**

Informal Assessment

1. **How many MathFlaps are in the first row? 7 How many rows are there? 2** Have students write a sum that fits the picture in the space provided for Example 1.

   What product can you write to model the sum? 2 × 7 = 14 How did you choose the product? There are 2 rows with 7 in each row (2 × 7) and 14 MathFlaps in all.

2. **How can you use the array to find a sum? There are 4 stars in a row and 4 rows, so we add 4 + 4 + 4 + 4. How can you use the array to find the product? There are 4 stars in a row and 4 rows, so multiply 4 × 4.**

3. What do you know about any number multiplied by 0? **The product is 0.** What do you know about any number multiplied by 1? **The product is the number itself.**

4. If students do not yet know their basic multiplication facts, allow them to use counters or draw arrays to model these problems.

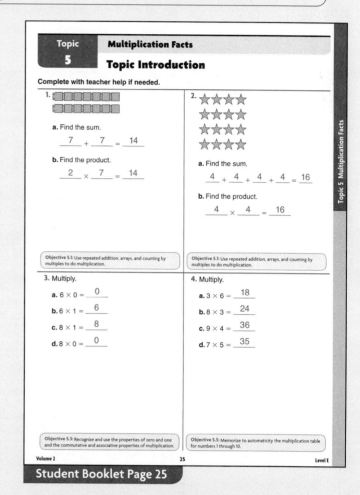

Student Booklet Page 25

Another Way It may be helpful to have students model the arrays in these problems using counters or other manipulatives. This will be especially useful for Exercise 3. Have students model zero rows of 6 to prove to them the multiplication property of zero.

Lesson 5-1: Meaning of Multiplication

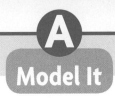

Objective 5.1: Use repeated addition, arrays, and counting by multiples to do multiplication.

Teach the Lesson

Materials
- ☐ MathFlaps or counters
- ☐ Teaching Aid 3 $\frac{1}{4}$-in. grid paper

Activate Prior Knowledge

Draw 3 groups of 4 hearts on the board. **How can you find the total number for all three groups?** Count, add, or multiply. Have volunteers write the addition sentence and the multiplication sentence for the drawing. $4 + 4 + 4 = 12$, $3 \times 4 = 12$

Addition means we are finding the total of two or more groups. The groups can have different numbers of things. Write $5 + 7 = 12$. **Multiplication is used when the groups are equal.** Write *4 groups of 2* and beneath it $4 \times 2 = 8$, and draw a model.

Model the Activities

Activity 1 Provide each group of students with MathFlaps. **Make 3 equal groups with 5 MathFlaps in each group to show a multiplication fact. How many times did you make a row of 5?** 3 times **You model shows 3×5. How can you add to find the total?** $5 + 5 + 5 = 15$ **How many fives did you add?** 3 You added 5 three times, so you found 3×5. Have each group model another sentence, and have students take turns saying the fact for other students' models.

Activity 2 Provide Teaching Aid 3. **Shade one row of 6 squares. Shade another row. Shade a third row. How many times did you shade a row of 6 squares?** 3 times **Your model shows 3×6. How can you find the total without counting every square?** Add: $6 + 6 + 6 = 18$. **How do you know that is the same as the total of 3×6?** Sample: We added 6 three times. Have each student shade in a rectangle. **How many rows does your rectangle have? How many squares in each row? How many squares in all? What multiplication sentence does your rectangle show?** Sample: 7 rows, 5 in each row, 35 in all, $7 \times 5 = 35$

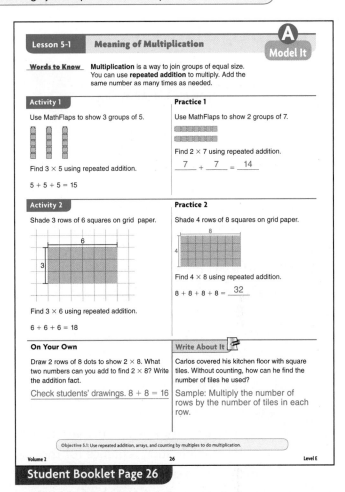

Student Booklet Page 26

Progress Monitoring

Write *addition* and *multiplication* on the board. **Which one means you are joining two or more groups, whether the groups are equal or unequal?** Addition **Which one always means you are joining two or more equal groups?** Multiplication

Error Analysis

To help students explain the contrast in the meanings of addition and multiplication in general terms, have them use counters to show you specific examples of these two operations.

Lesson 5-1: Meaning of Multiplication

Objective 5.1: Use repeated addition, arrays, and counting by multiples to do multiplication.

Facilitate Student Understanding

Materials ☐ Teaching Aid 1 Number Lines

Develop Academic Language

When we skip count by 2s, we name the *multiples* of 2. When we skip count by 5s, we name the *multiples* of 5. What word do you know that is related to *multiple*? **Sample: multiply** Multiples are numbers you find when you multiply the same number by different whole numbers. List the first few multiples of 2. **2, 4, 6, 8...**

ENGLISH LEARNERS Students who speak Spanish may know the cognates *multiplicar, multiplicando, múltiplos, multiplicado,* and *multiplicación.*

Demonstrate the Examples

Example 1 Provide students with copies of Teaching Aid 1 Number Lines. Have them make jumps of 4 and write the multiplication sentence for the model. **How many jumps of 4 should you make? 6** Jump over 4 spaces each time. Count by 4s as you jump. **4, 8, 12, 16, 20, 24** You jumped over 4 spaces 6 times. **What is 4 × 6? 4 × 6 = 24** Repeat with jumps of different sizes.

Example 2 Point out that students have modeled multiplication using groups of counters, rectangles on grid paper, and number lines. Ask for ideas on how coins could be used for models. **Sample: Use coins of the same type and find the total number of cents.**

On Your Own Students can show the hours on a time line and write *5 miles* in each space to help them visualize the problem.

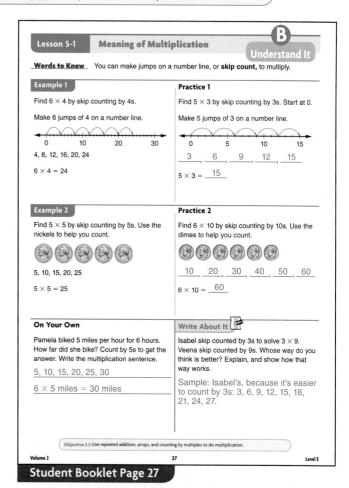

Student Booklet Page 27

Progress Monitoring

On the board draw a number line that shows jumps of 8, 4, and 10. Ask students to discuss why they cannot write a multiplication sentence for the jumps. **The jumps, or groups, are not equal. What number sentence can you write? 8 + 4 + 10 = 22** Have students change the model to show multiplication.

Error Analysis

If students cannot explain that multiplication means joining equal groups, provide more activities with counters. Students make 3 equal groups, write the multiplication sentence, remove one counter, and write the addition sentence.

Lesson 5-1 Meaning of Multiplication

Objective 5.1: Use repeated addition, arrays, and counting by multiples to do multiplication.

Observe Student Progress

Materials ☐ MathFlaps

> **Computer Tutorial**
>
> Some students may benefit from completing a computer tutorial before they attempt the Try It page. A list of the tutorials for each lesson can be found beginning on page x in the front of this book.

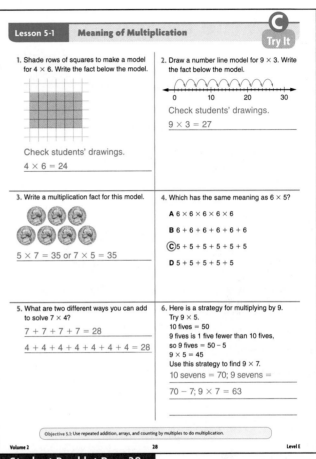

Student Booklet Page 28

★ Error Analysis

Exercise 1 Remind students that the drawing must be a rectangle with the same number of squares in each row.

Exercise 3 Help students see each nickel as an equal group of cents. You can replace each nickel with a stack of 5 pennies to show that counting nickels can be seen as a way of skip counting cents.

Exercise 4 If students are looking for only 6 + 6 + 6 + 6 + 6, point out that the problem can mean "6 groups of 5" or "5 groups of 6." Have students verify with MathFlaps that both interpretations lead to the same total.

Exercise 5 Have students shade a 7-by-4 rectangle on grid paper, and show that they can add the numbers in the columns or in the rows.

Exercise 6 A number-line model can help convey this strategy to students. Show 10 jumps of 5 and say **10 fives are 50.** Cover up the last jump and say, **I've subtracted one jump of five. How many jumps of five are left? 9 fives**

Lesson 5-2: Multiply by 2, 5, and 10

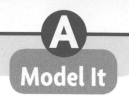

Objective 5.2: Know the multiplication tables of 2s, 5s, and 10s (to "times 10").

Teach the Lesson

Materials
- ☐ Teaching Aid 5 Multiplication Facts Table
- ☐ Index cards

Activate Prior Knowledge

Write × on the board. **What does this sign mean when you put it between two numbers?** Write the expression 2 × 7. Sample: joining equal groups, adding the same number, making equal jumps on a number line

Develop Academic Language

Skip count by 2s. 2, 4, 6, 8, 10, … 20 These are *multiples* of 2. They are, in order, the products you find when you multiply 2 × 1, 2 × 2, 2 × 3, and so on. If you want to find 2 × 6, you can count the first 6 multiples of 2. The sixth number you say will be the product. Have students verify that this method works.

Model the Activities

Activity 1 Direct students' attention to the facts table. **This table shows the multiplication facts for 2, 5, and 10. The numbers in blue are numbers to multiply, and numbers in white are the products. Find the second row of the table.** Read the numbers aloud starting with zero. **What are these numbers?** Numbers you get when counting by 2s Relate skip counting on the table to skip counting on a number line: 2 represents the end of the first jump, so 2 is the product of 2 × 1. Repeat for the rows for 5 and 10. Have students skip count to multiply and then use the table to check.

Activity 2 Use the table to count by tens. 10, 20, 30, … 100 Now state the multiplication facts the table shows. 10 × 1 = 10, 10 × 2 = 20, … 10 × 10 = 100 **What pattern do you see in the products? The product is the other factor with a zero at the end.** Students can interpret 10 × 8 as 8 tens, meaning that there should be an 8 in the tens place and a 0 in the ones place.

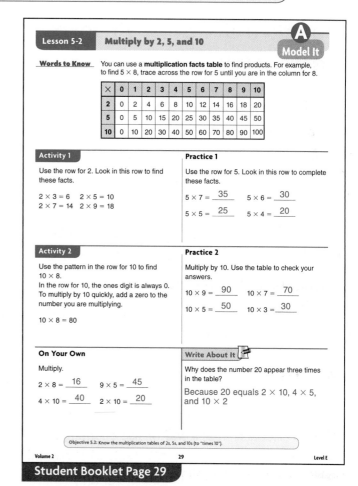

Student Booklet Page 29

⭐ Progress Monitoring

Give each small group of students a 10-part spinner or slips of paper labeled with the digits 0–9. Have each student spin the spinner and multiply the result by 2. Repeat using 5 and 10 as the multiplier.

Error Analysis

To help students memorize the multiplication facts for 2, 5, and 10, have them work with partners to make flash cards out of index cards. On the front of each card have them put the problem and a dot model. The complete fact—not just the product—should be written on the back of the index card.

Lesson 5-2: Multiply by 2, 5, and 10

Objective 5.2: Know the multiplication tables of 2s, 5s, and 10s (to "times 10").

Facilitate Student Understanding

Develop Academic Language

Write the numbers 1–6 on the board. Have students count by 2s as you point to each number. Write *six 2s make 12* and then $6 \times 2 = 12$. Point to the 6 and the 2. Explain that these are the factors. **Factors are numbers we multiply.** Point to the 12. Tell students this is the product. **The answer to a multiplication problem is called the *product*.**

Demonstrate the Examples

Example 1 Write on the board $2 \times 8 = \underline{} \times 2$. Explain that the quantities on the two sides of the equal sign must name the same number or the equation isn't true. Ask how to make the equation true. **Use 8 for the missing factor. 2×8 means 2 groups of 8 OR 8 groups of 2. You can write the problem either way and find the same answer by skip counting by 2s. How many 2s do you need to count to solve each problem? 8**

Example 2 Help students follow the directions given in the example. Then have students write all the facts with 2, 5, or 10 as factors. Then have them reverse the order of each pair of factors.

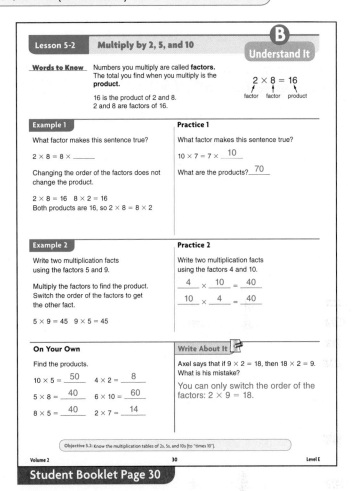

Student Booklet Page 30

Progress Monitoring

Show or describe how you could use $5 \times 7 = 35$ to write a related multiplication fact. Switch the two factors to get $7 \times 5 = 35$.

Error Analysis

Students who have interpreted the $\times$ symbol as "groups of" may need help seeing that switching the order of the factors does not change the product. Demonstrate this property of multiplication by drawing an array of dots; for example, 6 rows of 10. Turn the paper 90 degrees to show that the model now shows 10 rows of 6. Have students give the multiplication fact for each model. $6 \times 10 = 60$, $10 \times 6 = 60$

Lesson 5-2: Multiply by 2, 5, and 10

Objective 5.2: Know the multiplication tables of 2s, 5s, and 10s (to "times 10").

Observe Student Progress

Computer Tutorial

Some students may benefit from completing a computer tutorial before they attempt the Try It page. A list of the tutorials for each lesson can be found beginning on page x in the front of this book.

Error Analysis

Exercise 1 Suggest that students think of counting 6 nickels to find 6 × 5.

Exercise 5 Make sure students can identify the ones digit in a two-digit number. Help them write a few sample products.

Exercise 6 Have students multiply to check.

Exercise 8 Suggest that students use the facts table on page 29 to help them identify the numbers with a factor of 2.

Student Booklet Page 31

Lesson 5-2 Multiply by 2, 5, and 10 Try It

1. Write two multiplication facts using the factors 6 and 5.

   _6_ × _5_ = _30_
   _5_ × _6_ = _30_

2. Multiply each of these numbers by 2.

   a. 8 _16_   b. 9 _18_   c. 4 _8_
   d. 3 _6_   e. 7 _14_   f. 1 _2_

3. Find the products.

   a. 5 × 8 = _40_   b. 2 × 5 = _10_
   c. 10 × 4 = _40_ d. 6 × 5 = _30_
   e. 5 × 1 = _5_   f. 10 × 8 = _80_
   g. 7 × 10 = _70_ h. 2 × 9 = _18_

4. 9 × 5 =

   A 14
   Ⓑ 45
   C 54
   D 95

5. When you multiply a number by 5, what is the ones digit in the product?

   0 or 5

6. Fill in the missing number.

   4 × 10 = 10 × _4_

7. Explain how counting by 5s can help you find 7 × 5.

   Count 5, 10, 15, 20, 25, 30, 35. The seventh number in the list is the answer.

8. Think about the numbers from 10 to 20. Which numbers are products you can make when you use 2 as a factor? Write the multiplication facts for these products.

   10, 12, 14, 16, 18, 20
   2 × 5 = 10, 2 × 6 = 12,
   2 × 7 = 14, 2 × 8 = 16,
   2 × 9 = 18, 2 × 10 = 20

Lesson 5-3: Properties of Multiplication

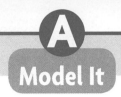

Objective 5.3: Recognize and use the basic properties of multiplication.

Teach the Lesson

Activate Prior Knowledge

Write $5 + 8 = 8 + 5$ and $(2 + 4) + 9 = 2 + (4 + 9)$ on the board. Have students verify that the equations are correct. **What properties do these problems show?** Commutative and associative properties of addition

Develop Academic Language

Continue to provide practice with identifying the commutative and associative properties of addition to be sure students know when to use the terms.

Model the Activities

Activity 1 Draw a diagram to illustrate 4×6 on the board. **What product does this illustrate?** $4 \times 6 = 24$ Then draw a diagram to illustrate 6×4. **What product does this illustrate?** $6 \times 4 = 24$ **What can we conclude about 4×6 and 6×4?** They are equal. You can also display one model and then rotate it 90° to show the other fact. **What property does that illustrate?** Commutative property of multiplication When we multiply, the order of the factors is not important.

Activity 2 Write $5 \times 9 \times 2$ on the board. **Which two numbers would be easy to multiply?** 5×2 If students suggest 5×9, point out that 45 will be difficult to work with. **What property can you use to get 5 and 2 together?** First the commutative, then the associative Change the order of the factors so 5 and 2 are next to each other. Next, use the associative property. Use parentheses to group the numbers you want to multiply first. Then solve. Students may also suggest that they could write $9 \times 5 \times 2$ using the commutative property and then use the associative property to group 5 and 2.

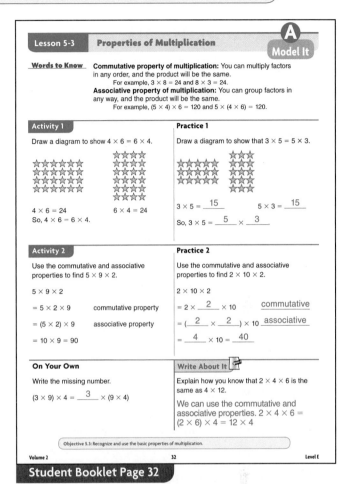

Student Booklet Page 32

★ Progress Monitoring

If you change the order of two factors, what property are you using? Commutative property of multiplication Have students explain why this can be helpful.

Error Analysis

ENGLISH LEARNERS Relate *associative* to associate and association. Relate *commutative* to commute.

Lesson 5-3 — Properties of Multiplication

Objective 5.3: Recognize and use the basic properties of multiplication.

Facilitate Student Understanding

Materials ☐ MathFlaps

Develop Academic Language
Continue to encourage students to use the word *product* to describe the answer in a multiplication fact.

Demonstrate the Examples

Example 1 Write 5×1 on the board. **How can you use repeated addition to find the product?** $1 + 1 + 1 + 1 + 1 = 5$ Draw five counters on the board as students model the problem with counters or MathFlaps. **What is the product of 5×1?** 5 **What is another product you could model with these counters?** 1×5 **What is the product of 1×5?** 5

Practice 1 Have students model and complete Practice 1. **What conclusion can you draw from these two problems?** If you multiply a number by 1, the answer is the number.

ENGLISH LEARNERS Explain that to *draw conclusions* is to use the facts to describe a rule.

Example 2 Help students interpret the model. **There are 8 zeros. How many pounds is that in all?** 0 **What other related multiplication fact has a product of 0?** $8 \times 0 = 0$

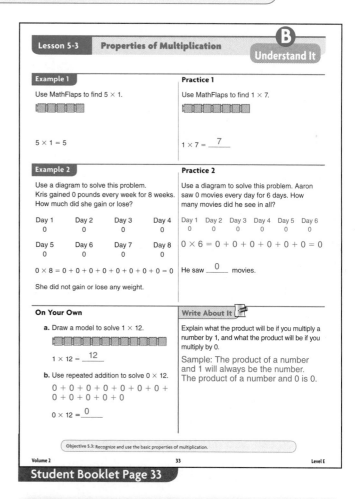

Student Booklet Page 33

⭐ Progress Monitoring
Draw a picture to show 1×9 and find the product. 9 Then draw a picture to show 5×0 and find the product. 0

Error Analysis
For students who give the product of a number and 1 as 1, continue to suggest they use models to find the products.

| Lesson 5-3 | **Properties of Multiplication** |

Objective 5.3: Recognize and use the basic properties of multiplication.

Observe Student Progress

Computer Tutorial

Some students may benefit from completing a computer tutorial before they attempt the Try It page. A list of the tutorials for each lesson can be found beginning on page x in the front of this book.

Develop Academic Language

Exercise 1 Ask students what property is shown in this example. Have students share the drawings they used to illustrate the commutative property.

ENGLISH LEARNERS Help students make diagrams or drawings to remember the meaning of *commutative* or *associative*.

Student Booklet Page 34

| Lesson 5-3 | **Properties of Multiplication** | Try It |

1. Draw a diagram to show that $6 \times 3 = 3 \times 6$.
   Sample:

2. Write the missing number.
   a. $12 \times 6 = 6 \times \underline{12}$
   b. $9 \times (\underline{4} \times 8) = (9 \times 4) \times 8$
   c. $\underline{5} \times 12 = 12 \times 5$
   d. $12 \times (9 \times 5) = (\underline{12} \times 9) \times 5$
   e. $35 \times (12 \times 6) = 35 \times (6 \times \underline{12})$

3. Identify the property used. Write C for commutative, A for associative, or B for both.
   a. $5 \times (4 \times 2) = (5 \times 4) \times 2$ __A__
   b. $15 \times 9 \times 4 = (15 \times 4) \times 9$ __B__
   c. $15 \times 74 = 74 \times 15$ __C__
   d. $(17 \times 6) \times 6 = 17 \times (6 \times 6)$ __A__

4. Which sentence shows the commutative property of multiplication?
   (A) $18 \times 9 \times 5 = 18 \times 5 \times 9$
   B $24 \times 1 = 12 \times 2 \times 1$
   C $(13 \times 9) \times 6 = 13 \times (9 \times 6)$
   D $(4 + 13) + 8 = 4 + (13 + 8)$

5. Multiply.
   $12 \times 5 \times 1 \times 2 \times 0$
   0

6. Use the commutative and associative properties to simplify the calculation.
   $2 \times 4 \times 5 \times 10$
   400

7. Multiply from left to right. Then use the properties to check your work.
   $2 \times 8 \times 5 \times 5 \times 2 \times 1$
   800

8. Write the rule for multiplying by 1. Give an example.
   The product is the other factor: $9 \times 1 = 9$.

Objective 5.3: Recognize and use the basic properties of multiplication.

 Error Analysis

Exercise 2 If students misidentify the property being used, it may be difficult for them to find the number. Be sure they first decide if order has been changed or if grouping has changed before determining the missing number.

Exercise 3 Some students may assume that if parentheses appear in an expression, only the associative property is being used. In each case, suggest that students first identify whether the numbers' order has changed before identifying the property.

Exercise 5 Help students look for numbers that are easy to multiply. If they don't identify 0 as the most significant number in this problem, allow them to multiply until they reach 0 and realize that it determines the value of this expression. Then have them write a few similar equations with 0 as a factor to solidify their understanding.

Exercise 7 If students struggle to multiply the larger numbers, allow them to use the properties first and then use a calculator to check the answer. Discuss how using the properties made the problem easier to solve.

Lesson 5-4: Multiplication Strategies

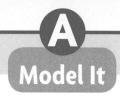

Objective 5.4: Apply multiplication strategies such as skip counting and doubles.

Teach the Lesson

Activate Prior Knowledge
Have students practice skip counting by 2s, 5s, and 10s. Clarify that they are finding *multiples* of 2, 5, or 10.

Develop Academic Language
Review the terms *factor* and *product*.

ENGLISH LEARNERS After Activate Prior Knowledge, explain: **We call this *skip counting* because we skip over certain numbers as we count.**

Model the Activities

Activity 1 The problem asks us how many are in **4 groups of 6.** You can model the problem by drawing 4 groups of 6 circles on the board; point out that counting every circle takes a lot of time. You can also point out that we could also find 6 groups of 4. **How many are in one group of 4? 4** If we want to find how many there are if we have another group of 4, how many numbers do we count on? **4** Count together: **5, 6, 7, 8** What is the fourth number when we count on from 4? **8** 8 is the next multiple of 4. This tells us that 4 × 2 = 8. Keep going. If there are 6 groups of 4, how many times should you count on? **6** What is the sixth number? **24** What is 4 × 6? **24**

Activity 2 Earlier we practiced skip counting by 2s, 5s, and 10s. Now we will skip count by 3s. **How many groups of 3 does the problem call for? 6** How many numbers do we need to say as we skip count by 3s? **6** Allow students to whisper the numbers to be skipped. **When you count by 3s, what is the sixth number? 18**

Student Booklet Page 35

Progress Monitoring
How can you use skip counting to find 9 × 8? Sample: Start with 9 and count on 9 eight times; 9 × 8 = 72.

Error Analysis
Point out that skip counting is most useful with 10, 5, and smaller numbers. In On Your Own, counting by 3s may be easier than counting by 8s. For 8s, students could use a "count on 10, go back 2" method.

Volume 2   82   Level E

Lesson 5-4 Multiplication Strategies

Objective 5.4: Apply multiplication strategies such as skip counting and doubles.

Facilitate Student Understanding

Materials ☐ Centimeter connecting cubes

Develop Academic Language

Have students practice *doubling* numbers you name. Then have each student make 5 stacks of 3 cubes to model 3 × 5 and write the fact. Then put two models in a bag. **The total has *doubled*. How can we find the new total?** Double 30: 30 + 30 = 60. **What problem have you solved?** Empty the bag to show. 3 × 10 3 × 10 is *double* 3 × 5.

ENGLISH LEARNERS Show that an *array* is an orderly arrangement of objects in rows and columns.

Demonstrate the Examples

Example 1 Have students use counters to make an array 6 units wide and 8 units long. **Separate the left and right halves of the array. What is one half of 8?** 4 **What fact does each of the new arrays show?** 6 × 4 = 24 **If each half of the total array represents 24, how can we find the total?** Double 24. Point out that halving and doubling are inverse operations. **How do we double 24?** Add it to itself: 24 + 24 = 48. **Then what is the product of 6 × 8?** 48 We can double the product of 6 × 4 to find the product of 6 × 8.

Example 2 We will cut one factor in half, multiply, and double the product. **Why is it a good idea to choose 12 as the factor to cut in half?** Breaking apart the larger factor will make smaller numbers that are easier to multiply. **What is half of 12?** 6 **So first solve 4 × 6.** Help students solve. **That is half of the total. How do you find the total?** Double 24: 24 + 24 = 48. **What is 4 × 12?** 64

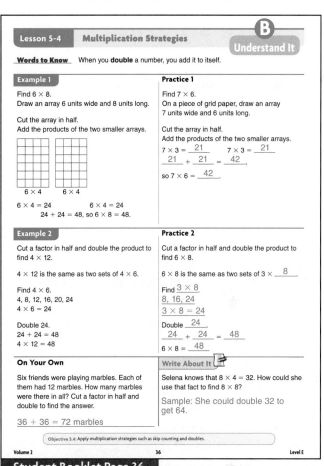

Student Booklet Page 36

Progress Monitoring

Describe all the different ways you can think of to solve 10 × 8. Sample: Skip count by 10s or by 8s, double the product of 5 × 8, double the product of 10 × 4.

Error Analysis

Be sure that students cut only one factor in half. You can show that the process in Example 2 is the same as rewriting the problem to read 4 × 6 × 2. Emphasize the purpose of the method: since doubling can be done by addition, often mentally, saving the "times 2" till last simplifies the work.

Lesson 5-4 | Multiplication Strategies

Objective 5.4: Apply multiplication strategies such as skip counting and doubles.

Observe Student Progress

Computer Tutorial

Some students may benefit from completing a computer tutorial before they attempt the Try It page. A list of the tutorials for each lesson can be found beginning on page x in the front of this book.

Error Analysis

Exercise 1 Help students see that since the list starts 2, 4, … 2 is the number to skip count by. If students need more support to skip count, encourage them to counters or other manipulatives.

Exercise 3 ENGLISH LEARNERS Clarify that students are being asked to show the first step of the "cut one factor in half and double the product" method. They are looking for a multiplication fact in which one factor is half of a factor in 9×8.

Exercise 4 Allow students to count on from a known, nearby fact.

Exercise 8 Encourage students to draw a picture to help them find the factors.

Student Booklet Page 37

Lesson 5-5 Basic Multiplication Facts

Objective 5.5: Memorize to automaticity the multiplication table for numbers between 1 and 10.

Teach the Lesson

Materials   ☐ Counters or MathFlaps
☐ Teaching Aid 5 Multiplication Facts Table

Activate Prior Knowledge

Draw 4 groups of 5 stars on the board. **How many groups are there? 4 How many are in each group? 5** Have students count by 5s to find the total as you point to each group: 5, 10, 15, 20. Write $4 \times 5 = 20$ to show the multiplication fact.

Develop Academic Language

Ask students to describe different terms they might find in a problem that indicate that they need to find a product. **Sample: product, times**

Draw an array on the board for $4 \times 6 = 24$. **Arrays are made up of rows and columns. One factor tells the number of rows. The other factor tells the number of columns. How many rows are there? 4 How many columns? 6 How many squares all together? 24 That tells the product.**

Model the Activities

Activity 1 Write $3 \times 6 = ?$ on the board. Have students use MathFlaps or counters to model the problem as you draw the corresponding model on the board. **How many rows are there? 3 How many are in the first row? 6 How many are in the second row? 6 So how many are in the first two rows? $6 + 6 = 12$ How many are in the third row? 6 How can we find the total? Add $12 + 6 = 18$.** Write $3 \times 6 = 18$ on the board. Point out that 18 is the *product* of 3 and 6.

Activity 2 Draw 6 rows of 4 MathFlaps or counters. **How can we find the number in the top three rows? $3 \times 4 = 12$ How can we find the number in the bottom three rows? $3 \times 4 = 12$ How do we find the total? Add the products: $12 + 12 = 24$, so $6 \times 4 = 24$.**

ENGLISH LEARNERS Reinforce terminology by creating a poster of an array with *rows* and *columns* labeled.

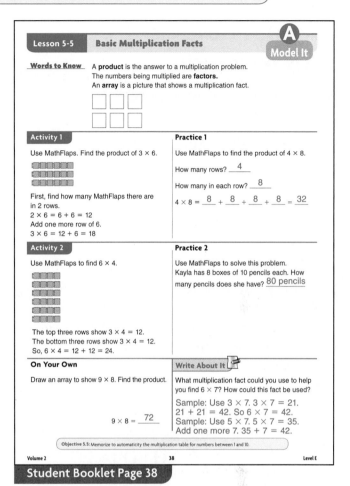

Student Booklet Page 38

Progress Monitoring

Show how to find the product of 3×9 using an array. Make an array of 3 rows of 9 counters, for a product of 27. Can you find the product another way? Accept any reasonable answer.

Error Analysis

For students who have difficulty making arrays to show multiplication, remind them that they can read the first number as the number of rows and the second number as the number of counters in each row.

Lesson 5-5 Basic Multiplication Facts

Objective 5.5: Memorize to automaticity the multiplication table for numbers between 1 and 10.

Facilitate Student Understanding

Materials ☐ Teaching Aid 5 Multiplication Facts Table

Develop Academic Language

In a number sentence such as $8 \times 9 = 72$, have students identify the product, and encourage them to use *product* rather than *answer* to describe the result of multiplication.

Demonstrate the Example

Example Provide students with copies of Teaching Aid 5 Multiplication Facts Table. Have them practice using the table to find products. If students have difficulty, encourage them to use two pieces of paper to help them line up the factors and the product.

×	0	1	2
0	0	0	0
1	0	1	2
2	0	2	4

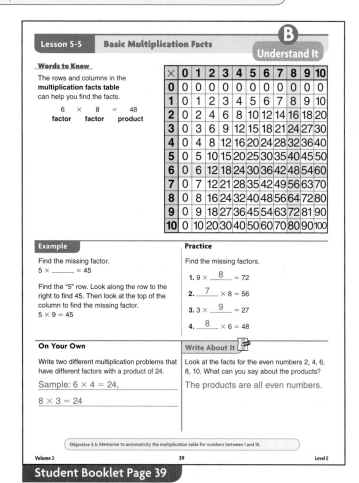

Student Booklet Page 39

⭐ Progress Monitoring

What is the missing number in the multiplication sentence $? \times 7 = 56$? 8 **How can you use the table to help answer the question?** Find the 7 column, look down until you find the product 56, and then look back to find the row you're in.

Error Analysis

If students have difficulty finding the missing factor, have them work with MathFlaps or counters. For example, for $6 \times ? = 24$, have them separate 24 MathFlaps or counters into equal groups until they find a number of rows and columns that give the answer.

Lesson 5-5 | Basic Multiplication Facts

Objective 5.5: Memorize to automaticity the multiplication table for numbers between 1 and 10.

Observe Student Progress

Computer Tutorial

Some students may benefit from completing a computer tutorial before they attempt the Try It page. A list of the tutorials for each lesson can be found beginning on page x in the front of this book.

Develop Academic Language

Exercise 7 Discuss the meaning of *twice* in this example. Be sure students understand that 8 is twice 4 because $4 \times 2 = 8$.

 Error Analysis

Exercise 2 Make flash cards to help students practice facts they find most difficult.

Exercise 3 If students don't yet have the facts memorized, allow them to refer to the Multiplication Facts Table.

ENGLISH LEARNERS Make sure students understand that finding the missing number means to fill in the missing factor.

Exercise 7 If students have trouble deciding how to approach the problem, suggest that they double 4 and find 4×8 and then double 4 both times and find 8×8.

Student Booklet Page 40

Lesson 5-5 Basic Multiplication Facts

1. Write a multiplication fact for the array.

   $6 \times 7 = 42$

2. Multiply.
   a. $9 \times 8 = \underline{72}$
   b. $7 \times 8 = \underline{56}$
   c. $4 \times 9 = \underline{36}$
   d. $6 \times 7 = \underline{42}$

3. Find the missing factor.
   a. $6 \times \underline{5} = 30$
   b. $8 \times \underline{6} = 48$
   c. $4 \times \underline{9} = 36$
   d. $8 \times \underline{8} = 64$

4. Which has a product of 60? Circle the letter of the correct answer.
   A 5×10
   Ⓑ 10×6
   C 9×6
   D 9×7

5. Write two multiplication facts using the factors 6 and 9.
   $6 \times 9 = 54, 9 \times 6 = 54$

6. Put these facts in order from greatest product to least.
   4×9 8×7
   6×5 3×9
   $8 \times 7 = 56; 4 \times 9 = 36;$
   $6 \times 5 = 30; 3 \times 9 = 27$

7. Maeve says 8×8 is 32 because $4 \times 4 = 16$, and 8 is twice 4. Do you agree? Explain.
   No; Each factor of 8 is twice 4, so 8×8 is 4 times 16, or 64.

8. Describe two different ways to find the product of 7×6.
   $7 + 7 + 7 + 7 + 7 + 7 = 42$ or draw an array of 7 rows of 6.

Objective 5.5: Memorize to automaticity the multiplication table for numbers between 1 and 10.

Volume 2 — Basic Facts
Topic 5 — Multiplication Facts

Topic Summary

Objective: Review skills related to multiplication facts.

Have students complete the student summary page.

You may want to have students work in groups of four with each student analyzing a different choice.

Ask students to share their ideas about each answer choice. Be sure they confirm the correct answer for each problem at the end of the discussion.

Answer Evaluation

1. **A** Students gave a factor instead of the product.
   B Students do not know basic multiplication facts.
   C This answer is correct.
   D Students added instead of multiplying.

2. **A** Students multiplied by 2 instead of 1.
   B This answer is correct.
   C Students incorrectly thought $15 \times 1 = 1$ instead of 15.
   D Students incorrectly thought $15 \times 1 = 0$ instead of 15.

Student Booklet Page 41

Progress Monitoring

When all assignments for this topic have been completed, assign the corresponding Progress Monitoring page for this topic (Assessment Resources Book, page 5). Be sure students complete the Progress Monitoring page before you administer the final assessment for this volume.

Volume 2 — Topic 5: Basic Facts

Multiplication Facts

Mixed Review

Objective: Maintain concepts and skills.

Have students complete the Mixed Review page. Work with each student individually to review results. Identify strengths and weaknesses and correct any misunderstandings.

Develop Academic Language

Exercise 2 Have students read each problem in two different ways. For example, *310 is greater than 299, and 299 is less than 310.*

★ Error Analysis

Exercise 3 If students add 6 and 11 to find the answer, have them write the problem as ___ + 6 = 11. When they identify subtraction as the method they used to find the missing addend, have them apply that to the problem.

Exercise 5 Review the multiplication properties to make sure students understand the difference between commutative and associative.

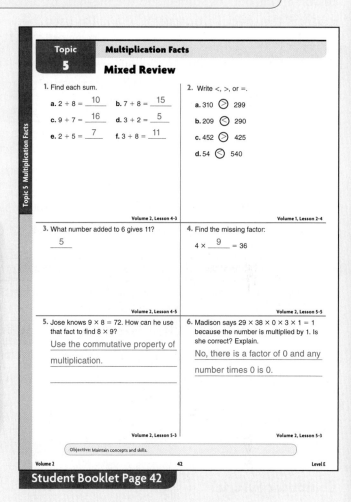

Student Booklet Page 42

Volume 2   89   Level E

Volume 2 — Basic Facts
Topic 6: Division Facts

Topic Introduction

Lesson 6-1 introduces students to the meaning of division by relating it to multiplication. Lesson 6-2 teaches students the special properties of 0 and 1 as used in multiplication and division. Lessons 6-3 and 6-4 give students practice in the basic division facts to division by 9. Finally, Lesson 6-5 applies students' knowledge of multiplication and division to building fact families.

Lesson	Objective	Student Pages	Teacher Pages	Tutorials
Topic 6 Introduction	**6.1** Define division by showing the relationship between multiplication facts and division facts. **6.2** Understand the special properties of 0 and 1 in multiplication and division. **6.3** Show the patterns of dividing by 2, 3, 4, or 5. **6.5** Show and apply the families of facts for multiplication and division.	43	91	
6-1 Meaning of Division	6.1 Use models and multiplication to understand division.	44–46	92–94	6a, 6b
6-2 Properties of Zero and One	6.2 Understand the special properties of 0 and 1 in multiplication and division.	47–49	95–97	
6-3 Divide by 2, 3, 4, or 5	6.3 Find the basic facts involving division by 2, 3, 4, or 5.	50–52	98–100	6c
6-4 Divide by 6, 7, 8, or 9	6.4 Find basic facts involving division by 6, 7, 8, or 9.	53–55	101–103	6c
6-5 Relate Multiplication and Division	6.5 Write multiplication and division fact families.	56–58	104–106	6d
Topic 6 Summary	Review division skills.	59	107	
Topic 6 Mixed Review	Maintain concepts and skills.	60	108	

Computer Tutorial

Some students may benefit from completing the computer tutorial before they attempt the Try It page of each lesson. If you are using the electronic components of *Pinpoint Math*, you will find a complete listing of Tutorial codes and titles when you access them either online or via CD-ROM.

Volume 2 — Basic Facts
Topic 6: Division Facts
Topic Introduction

Objectives: 6.1 Define division by showing the relationship between multiplication facts and division facts. **6.2** Understand the special properties of 0 and 1 in multiplication and division. **6.3** Show the patterns of dividing by 2, 3, 4, or 5. **6.5** Show and apply the families of facts for multiplication and division.

Materials ☐ MathFlaps

Write 4 × 2 on the board. Have students model the fact with MathFlaps. **What does multiplication mean?** Joining equal groups So when we model of 4 × 2, we make 4 rows with 2 in each row, for a total of 8. Repeat for other multiplication facts.

Informal Assessment

1. **How many boxes are in each circle?** 5 **How many circles are there?** 2 **How does the picture show multiplication?** There are two groups of 5. Have students write the multiplication problem in the space provided in Problem 1. **How does the picture show division?** There are 10 items divided into 2 groups. Have students write the division problem in the space provided in Problem 1.

2. **What does division mean?** Splitting something into equal groups So if we make 1 group of 6, how many are in each group? 6 If we make 4 groups of 0, how many are in each group? 0

3. If students are not yet confident in their division facts, suggest that they use MathFlaps or draw a sketch to model each problem.

4. **What operations will be used in this fact family?** Multiplication and division **Why?** Multiplication and division are inverse operations. **What numbers will be used in this fact family?** 15, 3, and 5 **How do you know?** The same numbers are used in all facts in a fact family.

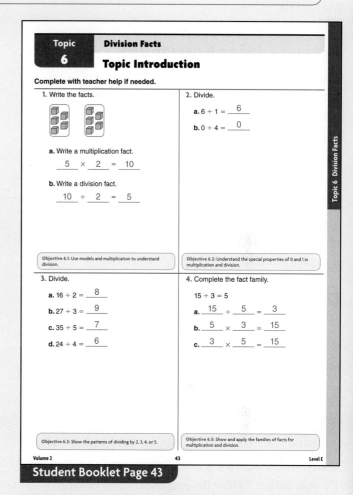

Student Booklet Page 43

Another Way For any of these exercises, suggest that students use MathFlaps or other manipulatives, or make a drawing similar to that in Exercise 1, to help them understand the problem.

Lesson 6-1: Meaning of Division

Objective 6.1: Use models and multiplication to understand division.

Teach the Lesson

Materials ☐ MathFlaps

Activate Prior Knowledge

Draw 21 stars on the board or use 21 magnets. Ask a volunteer to divide the stars into equal groups. **How many groups are there? 3 or 7 How many are in each group? 7 or 3** Have students identify the division problem and then write 21 ÷ 3 = 7 or 21 ÷ 7 = 3 to show the division fact.

Ask students to describe different situations they might find in a problem that indicate they need to divide. **Sample: separating into equal groups, finding an average**

Develop Academic Language

Encourage students to use the words *quotient*, *divisor*, and *dividend*. Have them write and label the problem in 8 ÷ 2 = 4 format at the top of the page.

Model the Activities

Activity 1 Write 8 ÷ 2 = ? on the board. Have students use MathFlaps to model the problem. **How many MathFlaps are there? 8 How many stacks do you make? 2 How do you know? The divisor is 2.** Show how to count the number of rows. Point out that each row is an equal group. **How do you determine the quotient? Count the number of rows.** Write 8 ÷ 2 = 4.

Have students work with MathFlaps to model other division problems.

Activity 2 **Which operation will you use to solve the problem? Division How do you know? There is a total to be divided in equal groups. What is the total to be divided? 32 flowers How many equal groups must there be? 8 What division problem will you solve? 32 ÷ 8** Have students use MathFlaps to solve.

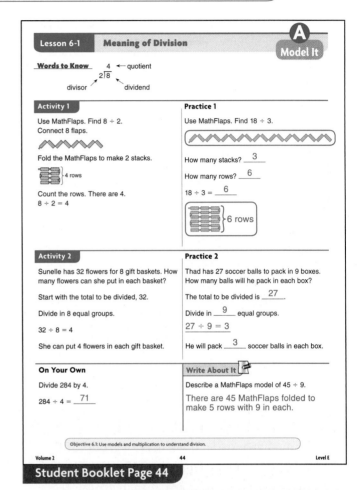

Student Booklet Page 44

⭐ Progress Monitoring

Show how to find the quotient 27 ÷ 9 using MathFlaps. Hook 27 MathFlaps together. Make 9 stacks. Count the rows. Can you find the quotient another way? Accept any reasonable answer.

Error Analysis

For students who have difficulty making the appropriate number of stacks, remind them to count off the number of MathFlaps in the divisor and then just fold the remaining MathFlaps back and forth over the first stacks.

Lesson 6-1: Meaning of Division

Objective 6.1: Use models and multiplication to understand division.

Facilitate Student Understanding

Develop Academic Language

Relate *dividend* and *product* to the meaning "total." In a number sentence such as $35 \div 5 = 7$, ask students to identify the divisor and the quotient. Encourage them to use the correct terms as they discuss the problems. Have students make a poster to hang in the classroom to remind them of the labels of each part of the problem.

Demonstrate the Examples

Example 1 What is the total number of MathFlaps in the model? **15** In how many rows is that total divided? **3** What division problem describes 15 items divided in 3 equal rows or groups? **15 ÷ 3** How many are in each row or group? **5** What is the completed division fact? **15 ÷ 3 = 5** Now write a multiplication problem. How many MathFlaps are in each row? **5** How many rows or groups of 5 are there? **3** What multiplication problem describes 5 rows of 3 MathFlaps? **5 × 3** What is the total? **15** What is the completed multiplication fact? **5 × 3 = 15** When you divide, you start with the total. When you multiply, you find the total. Multiplication and division are opposites.

Example 2 What is the total in this division problem? **10** Write a multiplication problem that has the same total and also includes 2. You'll need to use a question mark or a blank. **? × 2 = 10** Recall the "times 2" facts. What number times 2 equals 10? **5** Then the missing number in both problems is 5. Use MathFlaps to check that $10 \div 2 = 5$. Stress that $10 \div 5 = 2$ because of the related multiplication problem: $5 \times 2 = 10$. That is, division is multiplication with a missing factor.

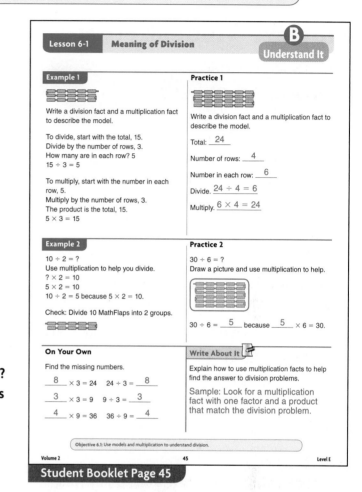

Student Booklet Page 45

Progress Monitoring

What is the quotient $81 \div 9$? **9** What is the multiplication fact you can use to help answer the question? **9 × 9 = 81**

Error Analysis

If students are familiar with the "add up" method of checking subtraction, relate this procedure to that method. You can also separate and combine unconnected MathFlaps to make the inverse relationship of multiplication and division more clear.

Lesson 6-1 | Meaning of Division

Objective 6.1: Use models and multiplication to understand division.

Observe Student Progress

Computer Tutorial

Some students may benefit from completing a computer tutorial before they attempt the Try It page. A list of the tutorials for each lesson can be found beginning on page x in the front of this book

Error Analysis

Exercise 2 Allow students to use MathFlaps, or help them write related multiplication facts.

Exercise 3 Point out that the multiplication and division problems are related.

Exercise 4 If students select C, remind them that they are looking for a multiplication problem, not a division problem. However, point out that recalling that fact is another valid way of solving the problem.

Exercise 6 If students have trouble answering this question, suggest that they first write the division problem and then find a related multiplication problem to help decide on the answer.

Exercise 7 Help students see that each hour represents an equal group of money.

Student Booklet Page 46

Lesson 6-1 — Meaning of Division — Try It

1. Find the quotient. Write a multiplication sentence to help.
   56 ÷ 7 = ?
   8 × 7 = 56
   56 ÷ 7 = __8__

2. Find the quotient.
   a. 45 ÷ 9 = __5__
   b. 32 ÷ 4 = __8__
   c. 64 ÷ 8 = __8__
   d. 30 ÷ 3 = __10__

3. Find the product or quotient.
   a. 30 ÷ 6 __5__
   b. __5__ × 6 = 30
   c. 42 ÷ 7 = __6__
   d. __6__ × 7 = 42

4. Which multiplication fact can you use to help find 48 ÷ 6?
   A 48 × 6 = 288
   B 12 × 4 = 48
   C 48 ÷ 8 = 6
   (D) 8 × 6 = 48

5. Write a division problem you can solve if you know that 8 × 5 = 40.
   40 ÷ 5 = 8 or 40 ÷ 8 = 5

6. Use multiplication to explain why 5 divided by 0 does not equal 0.
   Sample: 5 ÷ 0 could not equal 0 because 0 × 0 is not equal to 5.

7. A waiter wants to earn $36 in tips during his 6-hour shift. How much does he have to earn each hour?
   $6

8. There are 7 empty darkrooms. The photography teacher has 21 students, and she wants to assign the same number to work in each darkroom. How many students will work in each darkroom?
   3 students

Objective 6.1: Use models and multiplication to understand division.

Lesson 6-2 Properties of Zero and One

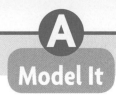

Objective 6.2: Understand the special properties of 0 and 1 in multiplication and division.

Teach the Lesson

Activate Prior Knowledge
Write 3 × 0 on the board. **How can you use repeated addition to find the product?** 0 + 0 + 0 = 0

Develop Academic Language
Continue to encourage students to use the words *product* and *quotient*.

Model the Activities

Activity 1 It may be helpful to present the problem in story form: **If you have 0 dollars to share among 4 friends, how much does each friend get?** 0 dollars Look at Example 1. The space below the first sentence represents zero. **How is that space divided?** Into 4 sections **How much is in each section?** 0 0 ÷ 4 = 0

Practice 1 Have students complete Practice 1. **What conclusion can you draw from these two problems?** If you divide 0 by any number, the answer is 0.

ENGLISH LEARNERS Explain that to *draw conclusions* is to use the facts to describe a rule.

Activity 2 One of these problem can't be solved. Start with the multiplication problems. **What does the first one mean?** Sample: 5 groups of 0 **The second?** Sample: 0 groups of 5 Show 1 group of 0. Point out that students can draw a dashed circle as they did in Activity 1. **Now show 5 groups of 0 in all. How could you add to solve this problem?** 0 + 0 + 0 + 0 + 0 = 0 **What is the total of 5 zeros?** 0 You can also ask students to imagine a group of 5, and then to draw 0 groups of 5; students will not draw anything. Since multiplication is commutative, the two multiplication facts must have the same answer, 0. **What is a rule for multiplying by 0?** The answer will be 0.

Try 0 ÷ 5. Find a blank space to represent 0. Divide it in 5 groups. **How much is in each group?** 0 Try 5 ÷ 0. Draw a total of 5 objects. Now try to separate these in 0 groups. **What happens?** There is always at least 1 group. Division by 0 can't be done.

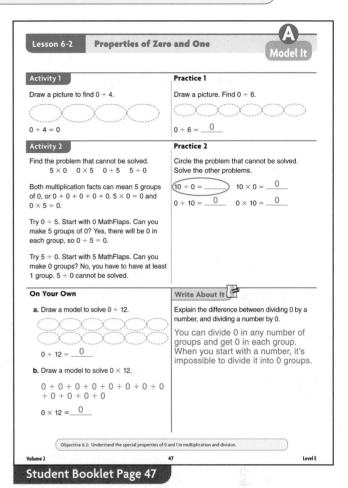

Student Booklet Page 47

 Progress Monitoring

Draw a picture to show 0 ÷ 9 and find the product. 0 Then draw a picture to show 5 × 0 and find the product. 0

Error Analysis

Have students write out an example of each use of 0. Help them see that for division and multiplication, the presence of 0 in a problem means the answer will be 0, except when the problem asks them to divide by 0. Those problems can't be solved.

If students struggle with the division, help them write the related multiplication problems.

Lesson 6-2 — Properties of Zero and One

B Understand It

Objective 6.2: Understand the special properties of 0 and 1 in multiplication and division.

Facilitate Student Understanding

Develop Academic Language
Tell students that *compute* means to calculate or solve.

Demonstrate the Examples

Example 1 Remind students how to show division by representing the first number with stars or counters and showing how to break the number into equal groups. Point out that if we divide a number of stars into 1 group, the group will contain all the stars. After they complete Practice 1 see whether students can draw a conclusion about dividing by 1. **A number divided by 1 is the number itself.**

Example 2 Point out that multiplication by 1 is a simple step that can be "saved" for the end of a long multiplication problem. Remind students that order is not important when multiplication is the only operation. Present similar problems with 1 in different places, such as $3 \times 1 \times 5$. You can also have students work through several problems such as $3 \times 0 \times 5$ and $6 \times 7 \times 2 \times 0$.

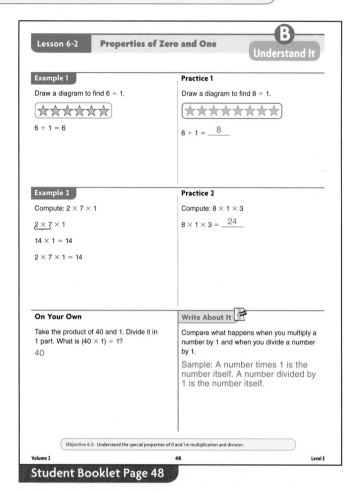

Student Booklet Page 48

⭐ Progress Monitoring
Draw a picture to help find $5 \div 1$ and find the product. **5** Then draw a picture to help find 5×1 and find the quotient. **5**

Error Analysis
Students might want to write 1 as the answer when they see 1 in a multiplication or division problem. Use repeated addition for multiplication and models for both operations to help students understand rather than simply to memorize the rules.

| Lesson 6-2 | **Properties of Zero and One** | |

Objective 6.2: Understand the special properties of 0 and 1 in multiplication and division.

Observe Student Progress

Computer Tutorial

Some students may benefit from completing a computer tutorial before they attempt the Try It page. A list of the tutorials for each lesson can be found beginning on page x in the front of this book.

Error Analysis

Exercise 3 Before students actually do the computations, have them review the rules for multiplying and dividing with 0 and 1. Students can use related facts to test each rule.

Remind students that the parentheses indicate the operation to be performed first.

Exercise 4 Remind students that they can model the problem or put it in word form: divide 4 items among 4 equal groups.

Exercise 8 Help students decide whether to refer to multiplication or division in their explanations.

Student Booklet Page 49

Lesson 6-2   Properties of Zero and One   Try It

1. Write the multiplication problems shown by this diagram.

   $5 \times 1 = 5$ and $1 \times 5 = 5$

2. Write *true* if the statement is correct. Write *false* if the statement is not correct.
   a. $6 \times 1 = 1$ __false__
   b. $0 \times 9 = 0$ __true__
   c. $15 \div 1 = 15$ __true__
   d. $0 \div 24 = 24$ __false__

3. Multiply or divide.
   a. $12 \times 9 \times 0 \times 3 = $ __0__
   b. $(8 \times 3) \div 1 = $ __24__
   c. $0 \div (7 \times 3) = $ __0__
   d. $18 \times (0 \div 12) = $ __0__

4. What is $4 \div 1$?
   A 0
   B 4
   C 3
   D 1

5. Circle the problem that cannot be solved. Solve the other problems.
   $0 \times 3 = $ __0__ $3 \times 0 = $ __0__
   $0 \div 3 = $ __0__   (($3 \div 0 = $ ____))

6. Explain why $0 \div 6 = 0$.
   Sample: Nothing divided into equal parts is nothing.

7. Write a rule to show what happens when you multiply a number by 1.
   The product of any number and 1 is the number itself.

8. Jon says he watches 0 hours of television each day. How much television will he watch in 7 days? A month? A year? Explain.
   Sample: The answer will always be 0 hours, because 0 times any number is always 0.

Objective 6.2: Understand the special properties of 0 and 1 in multiplication and division.

Lesson 6-3: Divide by 2, 3, 4, or 5

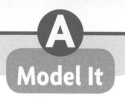

Objective 6.3: Find the basic facts involving division by 2, 3, 4, or 5.

Teach the Lesson

Materials ☐ 30 MathFlaps per student

Activate Prior Knowledge
Review skip counting and multiplication facts for 2, 3, 4, and 5.

Develop Academic Language
Emphasize that the "starting numbers" in students' division models are *dividends*.

ENGLISH LEARNERS Pass objects out to students to model a sharing situation in which some are left over. Write the division on the board, and ask what to do with the leftover objects. **We call the leftover number the *remainder*, because that number *remains* after we have made as many equal groups as possible.** Show the R that stands for *remainder*.

Model the Activities

Activity 1 Start with the total, the *dividend*. **What is the dividend in 8 ÷ 2?** 8 Connect 8 MathFlaps. **How many groups of 2 should you make?** 4 Show how to fold so 2 MathFlaps are in each group. **Do you have any counters remaining?** No. You have modeled the basic fact 8 ÷ 2. **What is the answer, the *quotient*?** 4 Repeat for 9 ÷ 2. **One MathFlap is left over. It can't be part of one of the equal groups. 9 ÷ 2 is not a basic fact for you to memorize, because it has a remainder of 1.** Show how to write the answer.

Activity 2 Use the same process as for Activity 1. Make sure students model the division correctly and identify the remainder in 16 ÷ 3.

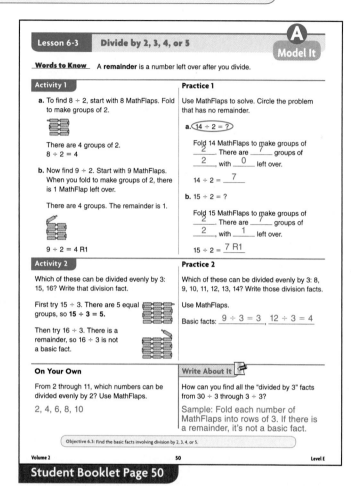

Student Booklet Page 50

Progress Monitoring
Write 17 ÷ 3 on the board. **Could we divide 17 into equal groups of 3?** No. **How many would be left over?** 2

Error Analysis
If students need more practice, have them work with all numbers from 2 through 20 for division by 2, and from 3 through 30 for division by 3. Students can also model by separating unconnected MathFlaps into equal groups if they find the "put one here, put one there" method of distributing items among groups easier.

Lesson 6-3 Divide by 2, 3, 4, or 5

Objective 6.3: Find the basic facts involving division by 2, 3, 4, or 5.

Facilitate Student Understanding

Develop Academic Language

We can divide any number by any other number. But we only call a number *divisible* by another number when there is no remainder.

ENGLISH LEARNERS Review *dividend, divisor,* and *quotient*. Have students label division problems on the board for reference.

Demonstrate the Examples

Example 1 Start with the number itself, since any number divided by itself is 1. What is the first fact? 4 ÷ 4 = 1 Try 5 as the dividend. Divide by 4. Are any left over? Yes, 1. Is 5 divisible by 4? No. 5 ÷ 4 = 1 R1 is not a basic fact. Guide students to continue for dividends through 20. Check that they find the correct remainders.

Example 2 Help students see that each dividend given in the first column should be divided by 5. The quotient and remainder are recorded in the next two columns. Start with 5. Divide by 5. What is the quotient? 1 Are any left over? No. Now start with 6. Continue through the dividend 12. What pattern can you see in the remainders when we divide by 5? They are 0, 1, 2, 3, 4 and then 0 again. What is the greatest remainder you will have when dividing by 5? 4 Why? When there are 5 left, you can make another group of 4.

Student Booklet Page 51

Progress Monitoring

If we had 21 books and divided them equally among 5 friends, would each friend get the same number of books? No. How do you know? There would be a remainder; 21 is not divisible by 5.

Error Analysis

Allow students to continue using MathFlaps. Make sure students test numbers in an orderly way to help them see the patterns. Help students see that On Your Own asks students to find the number that is divisible by 4.

Lesson 6-3 Divide by 2, 3, 4, or 5

Objective 6.3: Find the basic facts involving division by 2, 3, 4, or 5.

Observe Student Progress

Computer Tutorial

Some students may benefit from completing a computer tutorial before they attempt the Try It page. A list of the tutorials for each lesson can be found beginning on page x in the front of this book.

Error Analysis

Exercise 1 Allow students to duplicate each model with MathFlaps. **What is the total in the first model? 18 How many are in each group? 3 How many equal groups are there? 6**

Exercises 3 and 4 If necessary, allow students to use the Multiplication Facts Table. Remind students of the pattern of remainders (none, 1, 2, etc. until 1 less than the divisor).

Exercise 6 Explain that students should try to put each boy's seeds in 4 equal groups. If the division leaves a remainder, either the rows won't be equal or all the seeds won't be planted.

Exercise 7 Provide a few sample numbers for students to test. **Could he have had 3 pencils? Yes. 4? No.** Write ___ ÷ 3 = ___, and have students fill in the blanks to show known facts.

Exercise 8 Clarify that the greatest possible remainder is different for each divisor, and give students a few sample divisors to help them get started.

Student Booklet Page 52

Lesson 6-3   Divide by 2, 3, 4, or 5   Try It

1. What division problem does each set of MathFlaps show?

   18 ÷ 3 = 6   20 ÷ 3 = 6 R1

2. Find all the numbers in the list that are divisible by 2. Write the facts.

   11 12 13 14 15 16 17 18 19 20
   12 ÷ 2 = 6   14 ÷ 2 = 7
   16 ÷ 2 = 8
   18 ÷ 2 = 9   20 ÷ 2 = 10

3. Divide by 4. Fill in the table.

   Dividing by 4

Dividend	Quotient	Remainder
20	5	None
21	5	1
22	5	2
23	5	3
24	6	None

4. Divide by 3. Fill in the table.

   Dividing by 3.

Dividend	Quotient	Remainder
12	4	None
13	4	1
14	4	2
15	5	None

5. Solve.
   a. 30 ÷ 5 = __6__
   b. 40 ÷ 5 = __8__
   c. 25 ÷ 5 = __5__
   d. 15 ÷ 5 = __3__

6. Jack has 32 tomato seeds. Lucien has 33 pumpkin seeds. Who will be able to plant all his seeds in 4 equal rows?
   (A) Jack   B Lucien   C Both   D Neither

7. Mr. Washington had some pencils, and he gave equal numbers to 3 students. How many pencils might Mr. Washington have had?
   3, 6, 9 ... any multiple of 3

8. When you are dividing, what is the greatest possible remainder? Explain.
   Sample: The greatest remainder would be one less than the number you're dividing by. If you had more than that number remaining, you could make another group.

ENGLISH LEARNERS You can modify Exercise 8 to ask a specific question: **What is wrong with these answers: 36 ÷ 9 = 3 R9, 37 ÷ 9 = 3 R10, 38 ÷ 9 = 3 R11?**

Lesson 6-4 Divide by 6, 7, 8, or 9

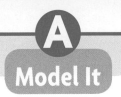

Objective 6.4: Find the basic facts involving division by 6, 7, 8, or 9.

Teach the Lesson

Materials ☐ MathFlaps

Activate Prior Knowledge

Review skip counting and multiplication facts through 10. You may want to discuss the relationship between 3s, 6s, and 9s, or between 2s, 4s, and 8s.

Develop Academic Language

Model 10 ÷ 2 with MathFlaps, and ask students to name each part of the problem—*dividend, divisor, quotient*—as you represent it in your model.

ENGLISH LEARNERS Point out that the problems without remainders in this lesson are some of the problems you refer to when you ask students to work with *basic facts*.

Model the Activities

Activity 1 Start with the total, the *dividend*. **What is the dividend in 18 ÷ 6? 8** Connect 18 MathFlaps. You'll fold the MathFlaps to show the division. **How many should be in each row? 6** Show how to fold so 6 MathFlaps are in each group. **Do you have any MathFlaps remaining? No.** You have modeled 18 ÷ 6. **What is the answer, or the *quotient*? 3** Repeat for 19 ÷ 6. One MathFlap is left over. It can't be part of one of the equal groups. 19 ÷ 6 is not a basic fact for you to memorize, because it has a remainder of 1. Show how to write the answer.

Activity 2 Use the same process as for Activity 1. Make sure students model the division correctly and identify the remainder in 20 ÷ 7.

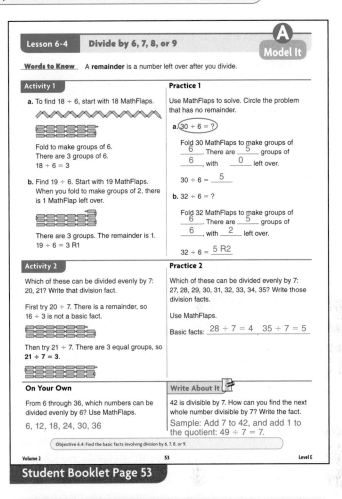

Student Booklet Page 53

Progress Monitoring

Write 17 ÷ 3 on the board. **Could we divide 17 into equal groups of 3? No. How many would be left over? 1 What "divided by 3" facts are close to 17 ÷ 3? 15 ÷ 3 = 5 and 18 ÷ 3 = 6**

Error Analysis

Students can also model the problems by separating unconnected MathFlaps into equal groups if they find the "put one here, put one there" method of distributing items among groups easier.

Lesson 6-4: Divide by 6, 7, 8, or 9

Objective 6.4: Find the basic facts involving division by 6, 7, 8, or 9.

Facilitate Student Understanding

Develop Academic Language

8 can be divided by 3, but 8 is not *divisible* by 3. What does *divisible* mean? **Able to be divided without leaving a remainder**

ENGLISH LEARNERS Review *dividend, divisor,* and *quotient.* Have students label division problems on the board for reference.

Demonstrate the Examples

Example 1 Start with the number itself, since any number divided by itself is 1. What is the first fact? $8 \div 8 = 1$ You may want to have students model this fact with 1 row of 8 MathFlaps. Then they can add more MathFlaps to see how many it takes to make a second row, or a quotient of 2. **Try 9 as the dividend. Divide by 8. Are any left over? Yes, 1. Is 9 divisible by 8? No. Try 10 as the dividend. Divide by 8. Are any left over? Yes, 2. Is 10 divisible by 8? No.** Continue until students see the pattern; they must add 8 to find another number divisible by 8. When students grasp the rule, help them generalize to other divisors.

Example 2 Explain that each dividend given in the first column should be divided by 9. The quotient and remainder are recorded in the next two columns. **Start with 18. Divide by 9. What is the quotient? 2 Are any left over? No. Now start with 19.** Continue through the dividend 27. **What pattern is starting to appear in the remainders when we divide by 9? They are 0 through 9 and then 0 again.** Point out that students will continue the pattern in Practice 1. They can make a similar chart on another piece of paper.

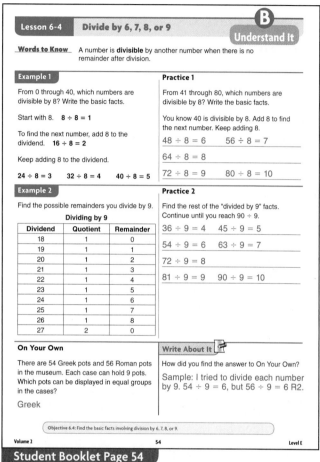

Student Booklet Page 54

Progress Monitoring

What is the greatest remainder you will have when dividing by 9? 8 Why? When there are 9 left, you can make another group of 9.

Error Analysis

Allow students to continue using MathFlaps. Make sure students test numbers in an orderly way to help them see the patterns. Help students see that On Your Own asks students to find the number that is divisible by 9.

| Lesson 6-4 | Divide by 6, 7, 8, or 9 |

Objective 6.4: Find the basic facts involving division by 6, 7, 8, or 9.

Observe Student Progress

Computer Tutorial

Some students may benefit from completing a computer tutorial before they attempt the Try It page. A list of the tutorials for each lesson can be found beginning on page x in the front of this book.

 Error Analysis

Exercise 1 What is the total in the first model? **48** How many are in each group? **8** How many equal groups are there? **6** 48 ÷ 8 = 6 and 48 ÷ 6 = 8 are both correct answers for the first picture. For the second picture, 55 ÷ 8 = 6 R7 and 55 ÷ 7 = 7 R6 are both correct answers.

Exercises 3 and 4 Point out that the given dividends are not consecutive. Each table is designed to produce a pattern of remainders, but the patterns will not be the same as those on the previous page.

Exercise 6 Explain that students should try to put each girl's comic strips in 8 equal groups. If the division leaves a remainder, either the number of comic strips on each won't be equal or all the comic strips can't be included.

Exercise 7 Provide a few sample numbers for students to test. **Could he have had 7 T-shirts? Yes. 8? No.** Write ___ ÷ 7 = ___, and have students fill in the blanks to show known facts.

Exercise 8 You may want to have students fill out a table to show division of the same dividends by 10, and then compare the results. You can then have students compare the table in Exercise 4 to a table in which they divide each dividend by 3.

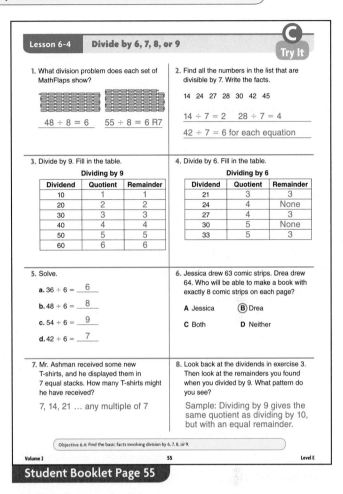

Student Booklet Page 55

ENGLISH LEARNERS For Exercise 8, allow students to use MathFlaps or a number line model to show their explanation.

Lesson 6-5 Relate Multiplication and Division

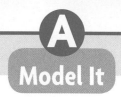

Objective 6.5: Write multiplication and division fact families.

Teach the Lesson

Materials ☐ MathFlaps

Activate Prior Knowledge

Have students connect and fold MathFlaps to model multiplication. **How can you model the fact 3 × 5 = 15?** Make 3 rows of 5, or 5 rows of 3. Or, since you know the total, you can start with 15 and fold it to make 5 rows with 3 in each row.

Develop Academic Language

Draw this model to help students identify rows and columns.

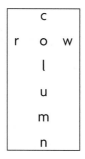

Model the Activities

Activity 1 Have students model 4 × 3 as shown. **If each row is a group, how many groups are there?** 4 **How many are in each group?** 3 **What is the total?** 12 4 × 3 = 12 Repeat for columns to find 3 × 4 = 12. **Now start with the total, 12. How many rows is it divided into?** 4 **How many are in each row?** 3 12 ÷ 4 = 3. Repeat for columns to find 12 ÷ 3 = 4. A *fact family* is a set of related number sentences. The model shows the fact family of 4, 3, and 12.

Activity 2 Guide students through the process of using tallies or MathFlaps to make the model. **When you start with 21, how many groups of 7 can you make?** 3 Guide students to understand the relationship of each fact to the question above it. After each question, ask students to explain how the model shows the answer.

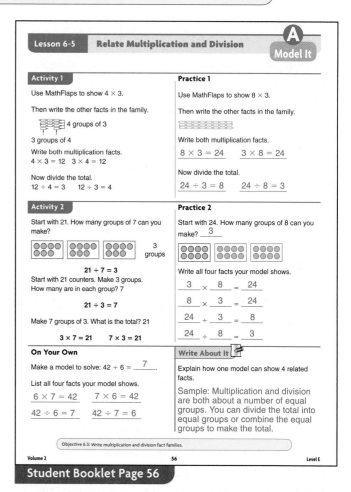

Student Booklet Page 56

Progress Monitoring

Model 40 ÷ 5. What four facts does your model show? 5 × 8 = 40, 8 × 5 = 40, 40 ÷ 8 = 5, 40 ÷ 5 = 8

Error Analysis

Explain fact families in terms of totals, equal groups, and the number in each group. Emphasize that multiplication results in a total, while division begins with a total.

Lesson 6-5: Relate Multiplication and Division

Objective 6.5: Write multiplication and division fact families.

Facilitate Student Understanding

Materials ☐ MathFlaps

Develop Academic Language

Display labeled multiplication and division facts. *Factors* are multiplied to get a *product*. A product is a total. Division begins with a total, called the *dividend*.

Demonstrate the Examples

Example 1 What are the factors in this fact family? **4 and 6** Explain that this could mean 4 groups with 6 in each group, or vice versa. **Either 4 or 6 could be the number of groups. Multiply. What is the product? 24** Now divide. What number will you start with? **24** Divide by a number of groups. Repeat for the other division fact. Explain that on the type of fact family flash card shown, the total belongs in the circle at the top. The total can be divided by either number below to equal the other number. The numbers below can be multiplied to equal the total.

Example 2 32 divided by 4 equals the missing number. What is another way to divide using the 2 known numbers shown in the triangle and the one missing number? **32 divided by the missing number equals 4.** Have students refer to the triangle to write 2 multiplication problems. **Which of these blanks is it easy to fill in, based on the facts you have memorized?** Answers will vary. **If you know that 4 × 8 = 32, then you know that those three numbers belong together in a family. If 4 and the total 32 are in any multiplication or division facts, the other number must be 8.**

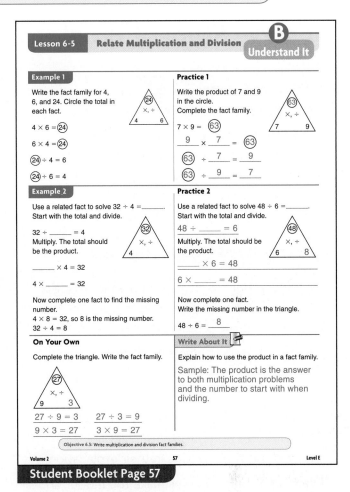

Student Booklet Page 57

Progress Monitoring

Make a fact triangle to help you solve 54 ÷ 6. What are the facts in this family? **9 × 6 = 54, 6 × 9 = 54, 54 ÷ 9 = 6, 54 ÷ 6 = 9**

Error Analysis

Students might think that 8 × 6 = 48 and 6 × 8 = 48 are the same fact. Explain that 8 × 6 can represent 8 groups of 6, and 6 × 8 can represent 6 groups of 8.

Lesson 6-5: Relate Multiplication and Division

Objective 6.5: Write multiplication and division fact families.

Observe Student Progress

Computer Tutorial

Some students may benefit from completing the computer tutorial before they attempt the Try It page of each lesson. A list of the tutorials for each lesson can be found beginning on page x in the front of this book.

 Error Analysis

Exercise 1 Help students set up a fact triangle. Since none of the answer choices is greater than 18, 18 must be the product, dividend, or total.

Exercise 2 Help students recall that factor × factor = product, so one factor is missing, and help them fill in the given numbers.

Exercise 3 Using MathFlaps to make a model that serves for both problems might help students use *equal groups* in their explanation.

Exercise 4 Help students make the model, and remind them to separately consider rows and columns as groups.

Exercise 5 Have students label the total in the given problem or place the given numbers in a fact triangle.

Exercise 6 If students forget the meaning of *dividend*, they might write the fact family that contains 16 ÷ 8 = 2. Remind them that the dividend is the total, the number we start with when we divide.

Exercise 7 Students can write each fact with a blank for the missing number first, or fill in the first blank they are able to and use that number to complete the rest of the facts.

Exercise 8 Provide examples such as the fact family 5, 5, and 25.

Student Booklet Page 58

ENGLISH LEARNERS Remind students that in any fact family, the dividend is the same as the product. Use *total* or circle the places for matching numbers to help students struggling with the vocabulary.

Volume 2 — Basic Facts
Topic 6: Division Facts

Topic Summary

Objective: Review division skills.

Have students complete the student summary page.

You may want to have students work in groups of four with each student analyzing a different choice.

Ask students to share their ideas about each answer choice. Be sure they confirm the correct answer for each problem at the end of the discussion.

Answer Evaluation

1. **A** Students gave the total distance instead of the distance per day.
   B This answer is correct.
   C Students multiplied 42 and 7 rather than dividing.
   D Students divided incorrectly.

2. **A** Students chose a fact that belongs in the same fact family.
   B Students chose a fact that belongs in the same fact family.
   C This answer is correct.
   D Students chose a fact that belongs in the same fact family.

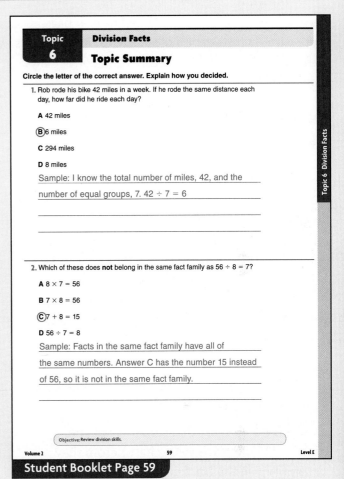

Student Booklet Page 59

 Progress Monitoring

When all assignments for this topic have been completed, assign the corresponding Progress Monitoring page for this topic (Assessment Resources Book, page 6). Be sure students complete the Progress Monitoring page before you administer the final assessment for this volume.

Volume 2 — Basic Facts
Topic 6: Division Facts
Mixed Review

Objective: Maintain concepts and skills.

Have students complete the Mixed Review page. Work with each student individually to review results. Identify strengths and weaknesses and correct any misunderstandings.

Develop Academic Language

Exercise 3 In each problem, ask students to identify the addends and the sum.

Error Analysis

Exercise 2 If students cannot fill in the blanks based on their knowledge of basic addition facts, suggest that they use subtraction to find the missing addend. Ask why this strategy works. **Because addition and subtraction are inverse operations**

Exercise 4 Students having difficulty doing this problem may benefit from placing the numbers on a number line.

Exercise 5 Suggest that students put the number into a place-value chart before they start filling in the blanks. Make sure that students understand that they are being asked for different information in each blank.

Exercise 6 Review rounding rules and the places through hundred thousands.

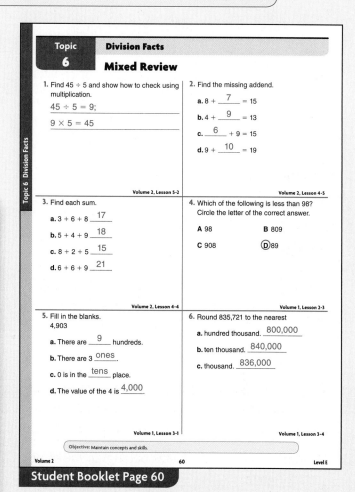

Student Booklet Page 60

Volume 3: Add and Subtract

Topic 7: Add or Subtract 1- and 2-Digit Numbers

Topic Introduction

Lesson 7-1 teaches students to add 1-digit to 2-digit numbers. Lessons 7-2 and 7-3 provide students with practice in adding and subtracting 2-digit numbers.

Lesson	Objective	Student Pages	Teacher Pages	Tutorials
Topic 7 Introduction	7.1 Solve problems involving one-digit numbers added to two-digit numbers. 7.2 Add two 2-digit numbers with and without regrouping. 7.3 Subtract two 2-digit numbers with and without regrouping.	1	110	
7-1 1-Digit and 2-Digit Numbers	7.1 Solve problems involving one-digit numbers added to two-digit numbers.	2–4	111–113	7a
7-2 Add 2-Digit Numbers	7.2 Add two 2-digit numbers with and without regrouping.	5–7	114–116	7b, 7c, 7a
7-3 Subtract 2-Digit Numbers	7.3 Subtract two 2-digit numbers with and without regrouping.	8–10	117–119	7d
Topic 7 Summary	Review adding and subtracting 1- and 2-digit numbers.	11	120	
Topic 7 Mixed Review	Maintain concepts and skills.	12	121	

Computer Tutorial

Some students may benefit from completing the computer tutorial before they attempt the Try It page of each lesson. If you are using the electronic components of *Pinpoint Math,* you will find a complete listing of the Tutorial codes and titles when you access them either online or via CD-ROM.

Volume 3: Add and Subtract

Topic 7: Add or Subtract 1- and 2-Digit Numbers

Topic Introduction

Objectives: 7.1 Solve problems involving one-digit numbers added to two-digit numbers. **7.2** Add two two-digit numbers with and without regrouping. **7.3** Subtract two two-digit numbers with and without regrouping.

Write one-digit and two-digit numbers on the board. Have students identify the number of tens and ones in each number. For example, 63 has 6 tens and 3 ones.

Informal Assessment

1. **Do you need to regroup in Part a? No. In Part b? Yes. Why? There are only 2 ones that need 5 ones subtracted.**

2. **What is the first number shown with the base ten blocks? 29 How did you know? There are 2 tens and 9 ones. What is the second number shown? 32 How do we write that addition problem? 29 + 32** Have students fill in the space provided in Problem 2.

3. **What does the picture show? Trading 1 ten for 10 ones Why do you think it shows that? Because there were only 2 ones, and 9 ones needed to be subtracted.** Have students fill in the spaces provided in Problem 3.

   How could you show subtracting 19? Cross out 1 ten and 9 ones. Have students cross out 1 ten and 9 ones on the base ten blocks shown in Problem 3. **How many tens and ones are left? 2 tens and 3 ones** Have students fill in the spaces.

4. **Which of these requires regrouping? B Why? There are more than 10 ones in the ones place. How do you regroup? Trade for 1 ten and 4 ones.**

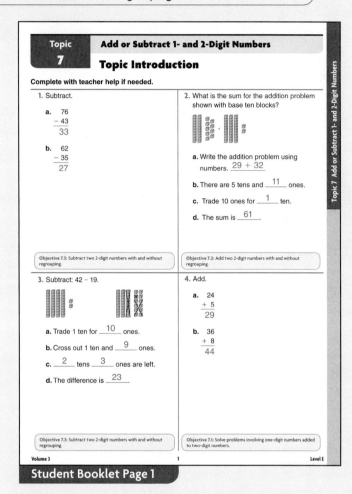

Student Booklet Page 1

Another Way To help students to see if their answers are reasonable, have them estimate the sums and then compare to their answers.

Lesson 7-1 1-Digit and 2-Digit Numbers

Model It

Objective 7.1: Solve addition problems with one- and two-digit numbers.

Teach the Lesson

Materials
☐ Base ten blocks
☐ Teaching Aid 1 Number Lines

Activate Prior Knowledge
Review basic addition and subtraction facts. Ask volunteers to give the sums and differences for the following basic facts: 4 + 5 **9**, 7 + 8 **15**, 17 − 9 **8**, 13 + 5 **18**, 20 − 9 **11**, 12 − 7 **5**. Have students model these problems with base ten blocks.

Model the Activities

Activity 1 Have students model the problem with base ten blocks. **Which blocks do you need to model 23?** 2 tens, 3 ones **To model 9?** 9 ones Always add the ones first. **How many ones are there?** 12 **Can you trade any of those blocks for a ten?** Yes, 10. Have students model the trade. **Now how many ones are there?** 2 Add the tens. **How many are there?** 3 There were only 2 tens in the original model. **Where did the other ten come from?** There were 10 ones that we traded for 1 ten. So the answer has 3 tens and 2 ones. **What is the sum of 23 + 9?** 32

Activity 2 Guide students through the same procedure they used in Activity 1. Alongside each step, show how to record the work in the vertical addition problem. **Add the ones.** 12 **Can you trade?** Yes. Have students show the trade. Your models now show the total. Point to the vertical addition problem. **What can I write as the total number of ones?** 2 If students suggest 12, show that this results in a recorded sum of 412. **What did we do with the other ones?** Traded for a ten Since we made 1 ten, I'll write 1 in the tens column. **What can I write as the total number of tens?** 5 Point to the addition problem as you say 4 tens from the original problem plus 1 ten that we made equals 5 tens. **What is the sum of 45 + 7?** 52

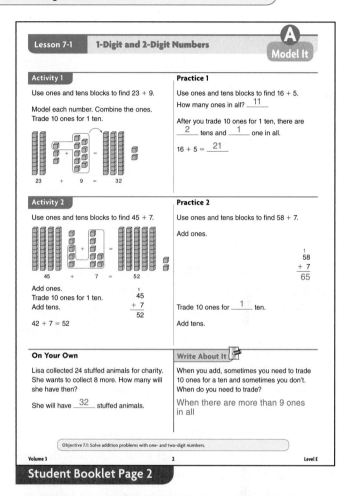

Student Booklet Page 2

Progress Monitoring
Show 25 + 9 with tens and ones blocks. How many ones do you have before you trade for a ten? 14 How many after you trade? 4 What is the sum of 25 + 9? 34

Error Analysis
Continue using block models to support the standard addition algorithm. Emphasize that the block model should use as few blocks as possible to show the model. Practice trading different amounts of ones for a ten and some ones.

Lesson 7-1: 1-Digit and 2-Digit Numbers

Objective 7.1: Solve addition problems with one- and two-digit numbers.

Facilitate Student Understanding

Materials ☐ Teaching Aid 1 Number Lines

Develop Academic Language

ENGLISH LEARNERS Point out that *carrying* and *regrouping* can also be used to name the process of trading ones for a ten.

Demonstrate the Examples

Example 1 Where should you start on the number line? **37** When you add to this number, will the answer be greater or less than 37? **Greater** Which direction on the number line leads to greater numbers? **Right** In order to add, move right on the number line. To add 6, how far right should you move? **6 spaces** Count on with students. Where do you stop? **43**

Example 2 What is the sum of 3 and 9? **12** How many tens and ones are in 12? **1 ten, 2 ones** Write the 1 over the tens column, and write the 2 as the sum of the ones column. Now there are two numbers to add in the tens column. What will you do? **Add 1 ten and 2 tens.** Guide students to complete the sum.

Practice 2 **ENGLISH LEARNERS** Students might benefit from underlining the verbs in word problems. The verbs often suggest the operation.

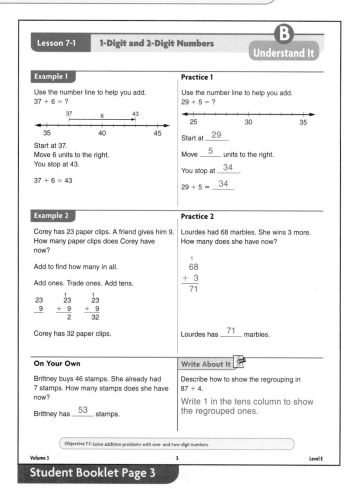

Student Booklet Page 3

Progress Monitoring

Show how to add 24 and 8 on a number line. **Start at 24. Move 8 to the right. The sum is 32.** Have students show the addition in vertical form.

Error Analysis

If students forget to record the regrouped numbers, have them work the problem in a place-value chart to help relate the numbers to the block models.

Lesson 7-1 | 1-Digit and 2-Digit Numbers

Objective 7.1: Solve addition problems with one- and two-digit numbers.

Observe Student Progress

Computer Tutorial

Some students may benefit from completing a computer tutorial before they attempt the Try It page. A list of the tutorials for each lesson can be found beginning on page x in the front of this book.

Error Analysis

Exercise 3 If students make errors finding the sum or difference, encourage them to go back and make sure they recorded any regrouping. You may want to have them work through the problems using place-value materials.

Exercise 4 If students select B, they may have made an error by neglecting to keep track of having regrouped 10 ones.

Exercise 7 The number at the start, rather than the result, is unknown. Even though it sounds like a problem that requires subtraction, students must add to find the solution. Read the word problem with the class. Guide students to see that they need to find a total number of magazines. They will need to add the number of magazines he has left and the number of magazines he gave away.

Exercise 8 If students have trouble knowing how to approach the problem, suggest that they draw a picture or use counters to help them understand the situation. Although "checked out" may seem to indicate subtraction, it is one part of a total here.

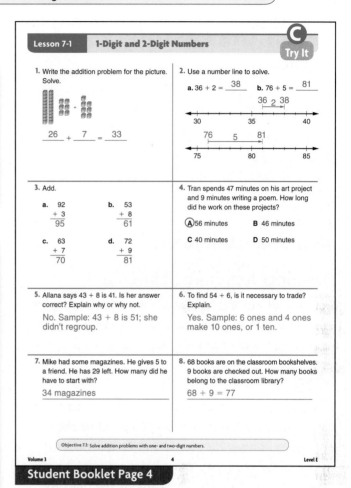

Student Booklet Page 4

Lesson 7-2   Add 2-Digit Numbers

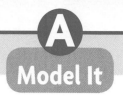

Objective 7.2: Add 2-digit numbers with and without regrouping.

Teach the Lesson

Materials ☐ Base ten blocks

Activate Prior Knowledge

Have students name each block's value. Ask them to show how many ones make a ten and how many tens make a hundred. Practice trading ones for tens and tens for hundreds.

ENGLISH LEARNERS Label each block so students learn the blocks' and values' names.

Develop Academic Language

Define *regroup* and have students model examples. **We know that 10 ones equal 10. I can regroup 10 ones as ten. How can I regroup 15 ones?** 1 ten, 5 ones

Model the Activities

Activity 1 Have students start with tens to model each addend. **What blocks do you need to show 38?** 3 tens, 8 ones **25?** 2 tens, 5 ones Tell students to combine the blocks. **What blocks do you have?** 5 tens, 13 ones **Can you trade any of the ones for a ten?** Yes, 10 ones for 1 ten. Instruct students to regroup the blocks. **What blocks do you have now?** 6 tens, 3 ones **Can you trade again?** No. **What is 38 + 25?** 63

Activity 2 Have students proceed in the same way as for Activity 1. After they have traded ones for tens, ask, **Can you trade any of the tens for anything?** Yes, 1 hundred **How many tens do you have then?** 1 ten **What is 18 + 96?** 114

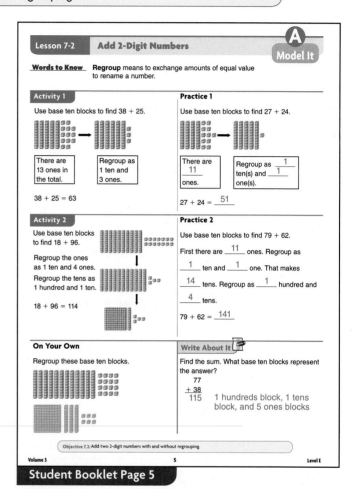

Student Booklet Page 5

Progress Monitoring

Add 64 + 28. **Do you need to regroup?** Yes. **What is the sum?** 92

Error Analysis

Point out that a sum of 10 in any place indicates a need for regrouping. Continue using base ten blocks to reinforce that 10 ones is the same as 1 ten and 0 ones.

Lesson 7-2 — Add 2-Digit Numbers

Objective 7.2: Add 2-digit numbers with and without regrouping.

Facilitate Student Understanding

Develop Academic Language

Students may know regrouping as *trading* or *carrying* in addition, and as *trading* or *borrowing* in subtraction. Use base ten blocks to talk about why *regrouping* is a good term.

Demonstrate the Examples

Example 1 Look at the ones column. What is 6 + 8? 14 How can we regroup? For 1 ten and 4 ones To make sure we remember the 1 ten, we write a small 1 above the tens column in the problem. Now we'll add all the numbers in the tens column. Don't forget about the 1 that we added at the top. 1 + 5 + 1 is what? 7 So what is 56 + 18? 74

Example 2 Look at the first problem. What digits are in the ones place? 2 and 5 Do we need to regroup? No, because 2 + 5 is less than 10. What digits are in the tens place? 3 and 6 Do we need to regroup? No, because 3 + 6 is less than 10. What is 32 + 65? 97 Look at the second problem. Do we need to regroup the digits in the ones column? Yes, because 13 equals 1 ten and 3 ones. What digits are in the tens column now? 1, 1, and 4 What is 1 + 1 + 4? 6 What is 18 + 45? 63

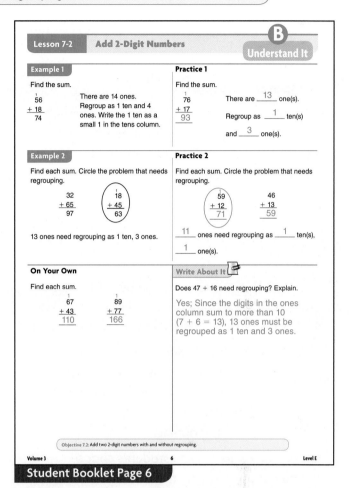

Student Booklet Page 6

Progress Monitoring

Explain the steps to add 58 and 24. First add the digits in the ones column: 8 + 4 = 12. Since 12 is greater than 10, we need to regroup. Keep the 2 in the ones column regroup 10 tens as 1, and write 1 in the tens column. Now add all the digits in the tens column: 1 + 5 + 2 = 8, so 58 + 24 = 82.

Error Analysis

Students often forget to write the 1 at the top of the next column when regrouping. Point out that showing their work as they regroup helps them keep track of all the addends.

Lesson 7-2 Add 2-Digit Numbers

Objective 7.2: Add 2-digit numbers with and without regrouping.

Observe Student Progress

Computer Tutorial

Some students may benefit from completing a computer tutorial before they attempt the Try It page. A list of the tutorials for each lesson can be found beginning on page x in the front of this book.

 Error Analysis

Exercise 1 Students can begin by solving each problem and checking whether their answer has the same number of tens and ones as the model.

Exercises 1 and 2 Students who choose answer A for Exercise 1 or answer C for Exercise 2 forgot to write the 1 in the tens column when they regrouped. Remind students to show all of their work when regrouping.

Exercises 3 and 4 Help students choose the operation, addition, needed to solve each problem.

Exercises 5 and 6 Remind students that if the digits in the ones column or tens column sum to 10 or more, regrouping is needed.

Exercise 7 Remind students that regrouping occurs when a column sums to 10, not just more than 10.

Exercise 8 Remind students that they need to add both the ones column and the tens column in order for their answers to be complete.

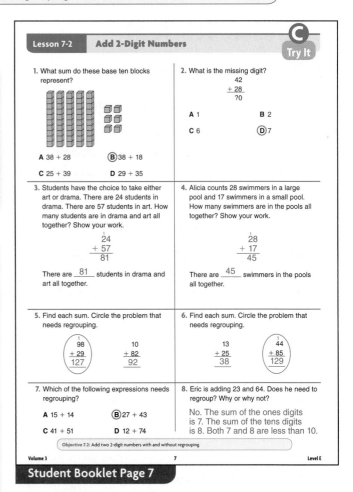

Student Booklet Page 7

ENGLISH LEARNERS Explain that the prefix *re-* in *regroup* means to do again. Words like *redo*, *reteach*, and *reset* indicate an action must be done again. So although numbers are already grouped in addition problems, they must be regrouped for the addition to take place.

Lesson 7-3: Subtract 2-Digit Numbers

Objective 7.3: Subtract two 2-digit numbers with and without regrouping.

Teach the Lesson

Materials ☐ Base ten blocks

Activate Prior Knowledge

Give students 1 ten and 5 ones. Ask students to take away 5 ones. **How much is left? 1 ten and 0 ones** Explain to students that the only way to take away more ones is to exchange the ten for 10 ones.

Develop Academic Language

Remind students that in subtraction, just like in addition, regrouping means to make an equivalent number in a problem.

Model the Activities

Activity 1 What blocks are needed to model 42? **4 tens and 2 ones** Instruct students to model 42. **What number are we subtracting? 5 Can we take away 5 ones from this model? Why or why not? No, because there are only 2 ones.** Let's exchange a ten for 10 ones. **Now can we subtract 5 ones? Yes.** Instruct students to take away 5 ones. **What is left? 3 tens and 7 ones What is 42 − 5? 37**

Activity 2 Talk students through the same steps as with Activity 1, but this time they will need to exchange 1 hundred for 10 tens. After they've regrouped and subtracted the ones, ask students how many tens they have. **0 How can you subtract 3? Regroup 1 hundred as 10 tens. Subtract 3 tens from 10 tens. Now how many hundreds do we have left? 0** There's nothing left to subtract, so we're done.

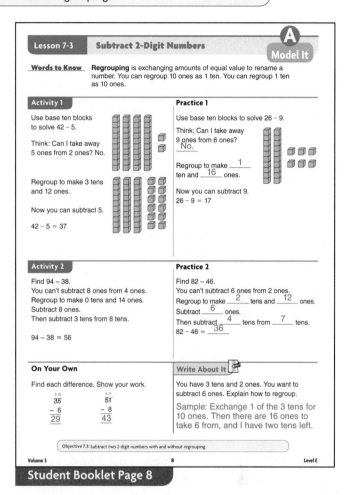

Student Booklet Page 8

Progress Monitoring

Use base ten blocks to subtract 78 from 105. Look at students' blocks to be sure they are regrouping correctly as they model the problem.

Error Analysis

Students might only add enough ones to make a total of 10 ones rather than add an additional 10 ones. Remind students they are performing an equal exchange, so if a ten is taken away, 10 ones must be added.

Lesson 7-3: Subtract 2-Digit Numbers

Objective 7.3: Subtract two 2-digit numbers with and without regrouping.

Facilitate Student Understanding

Develop Academic Language

Review the definition of difference. **What is 8 − 5?** 3 The answer to a subtraction problem is called the difference. 3 is the difference between 8 and 5.

Demonstrate the Examples

Example 1 What are we subtracting? 9 Can we take away 9 without regrouping? Explain. No, because there are only 4 ones. What should we do? Regroup by exchanging a ten for 10 ones. How much do you have now? 7 tens and 14 ones To help us remember what we regrouped, we cross off the 8 and write a 7 above it, and we cross off the 4 and write a 14 above it. Now subtract. How much do you have now? 7 tens and 5 ones What is 84 − 9? 75

Example 2 What is subtracted in the ones column? 2 is subtracted from 8. Do you need to regroup? No. What is subtracted in the tens column? 5 is subtracted from 3. Do you need to regroup? Yes. We can regroup 1 hundred from the hundreds column as 10 tens. What number is now in the tens column? 13 What number is now in the hundreds column? 0 What is the difference in the ones column? 6 Tens column? 8 Hundreds column? 0 What is 138 − 52? 86

Student Booklet Page 9

Progress Monitoring

When ten is regrouped, 1 is subtracted from the tens column and 10 is added to the ones column. Why is this? Sample: Regrouping is an equal exchange. Since ten equals 10 ones, I can take away 1 ten from the tens column and add 10 ones to the ones column.

Error Analysis

Students might write 10 above the columns rather than add 10 to the digits already in place. Remind students they are adding 10 to the digits in each column.

Lesson 7-3 Subtract 2-Digit Numbers

Objective 7.3: Subtract two 2-digit numbers with and without regrouping.

Observe Student Progress

Computer Tutorial

Some students may benefit from completing a computer tutorial before they attempt the Try It page. A list of the tutorials for each lesson can be found beginning on page x in the front of this book.

 Error Analysis

Exercise 2 Students may be confused when subtracting a digit from zero. Remind students they can regroup so 0 becomes 10.

Exercise 3 Remind students they might need to regroup twice. If they regroup twice, make sure they add 10 to the most current digit in the tens column. For example, in the first problem, when students regroup the first time, the digit in the tens column is changed from 3 to 2. So when students regroup the second time, they need to make sure they change 2 to 12, not 3 to 13.

Exercise 4 Students that chose answer A forgot to take away a ten when they added 10 ones. Remind students that when they regroup, something must be added AND taken away to create an equal exchange.

Exercise 5 Remind students to look for key words in a problem to indicate the operation. In this problem, point out that the phrase "how many more" indicates subtraction.

Exercise 6 Students who think the missing number is 4 forgot to change the 8 in the tens column to 7 when they regrouped. Remind students to write each step so they can keep track of how the numbers change.

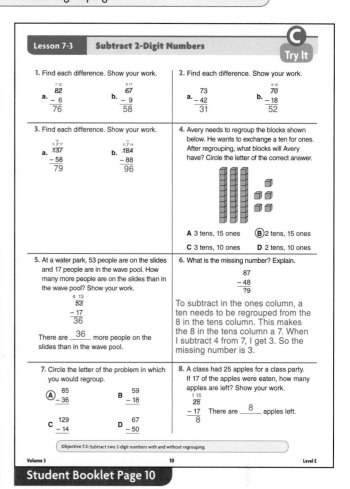

Student Booklet Page 10

ENGLISH LEARNERS Explain that regrouping can be used in both addition and subtraction problems. In each case an exchange is made of one form of a number for another. Allow students to work with base ten blocks to practice exchanging tens for 10 ones and hundreds for 10 tens, then vice versa.

Volume 3   119   Level E

Volume 3: Add and Subtract

Topic 7: Add or Subtract 1- and 2-Digit Numbers

Topic Summary

Objective: Review adding and subtracting 1- and 2-digit numbers.

Have students complete the student summary page.

You may want to have students work in groups of four with each student analyzing a different choice.

Ask students to share their ideas about each answer choice. Be sure they confirm the correct answer for each problem at the end of the discussion.

Answer Evaluation

1. **A** The student added instead of subtracting.
   B The student subtracted the lesser digit from the greater digit, not regrouping at all.
   C The student did not trade correctly in the tens place. He or she did not record the regrouping from 5 tens to 4 tens.
   D This choice is correct.

2. **A** The student rounded the numbers before adding.
   B The student subtracted instead of adding.
   C The student added but forgot to regroup the ones in the tens place.
   D This answer is correct.

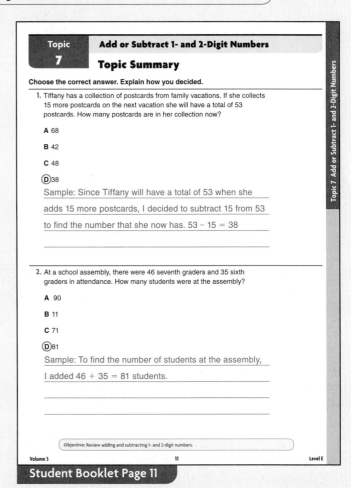

Student Booklet Page 11

 ## Error Analysis

Exercise 1 Students might choose the incorrect operation when solving this problem. Have students identify the part, part, whole in order to set the number sentence up correctly.

Exercise 2 Students might make addition errors regrouping in the ones and the tens. Encourage students to write the number they are regrouping above each place.

Progress Monitoring

When all assignments for this topic have been completed, assign the corresponding Progress Monitoring page for this topic (Assessment Resources Book, page 7). Be sure students complete the Progress Monitoring page before you administer the final assessment for this volume.

Volume 3: Add and Subtract
Topic 7: Add or Subtract 1- and 2-Digit Numbers
Mixed Review

Objective: Maintain concepts and skills.

 Error Analysis

Exercise 2 Encourage students to read the number aloud as they write the number in word form.

Exercise 3 Review the multiplication properties of 0 and 1. If students have difficulty, provide counters or MathFlaps and have them make, for example, 7 groups of 0, or 1 group of 18.

Exercise 4 Suggest that students write the problem vertically. Be sure they are writing the regrouping correctly.

Exercise 5 Suggest that students look at the relationship between the three numbers to decide if the fact family is made up of addition/subtraction facts or multiplication/division facts.

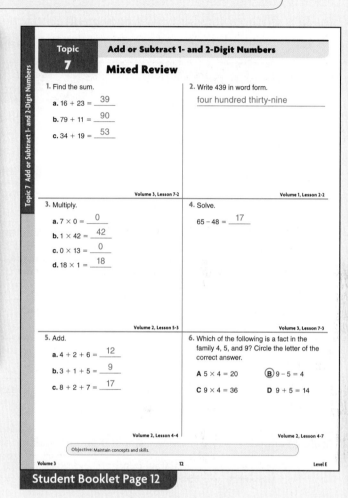

Student Booklet Page 12

Volume 3: Add and Subtract

Topic 8: Add or Subtract Multidigit Numbers

Topic Introduction

Lesson 8-1 teaches students mental math techniques for adding and subtracting 2-digit numbers. Lesson 8-2 broadens students' familiarity with adding and subtracting 2-digit numbers to include estimation. Lessons 8-3, 8-4, and 8-5 extend students' work with addition and subtraction into 3-digit and multidigit numbers.

Lesson	Objective	Student Pages	Teacher Pages	Tutorials
Topic 8 Introduction	8.1 Use mental arithmetic to find the sum or difference of two two-digit numbers. 8.3 Add two 2-digit numbers with and without regrouping. 8.4 Subtract two 2-digit numbers with and without regrouping. 8.5 Find the sum or difference of two whole numbers up to three digits long.	13	123	
8-1 Add or Subtract Mentally	8.1 Use mental arithmetic to find the sum or difference of two two-digit numbers.	14–16	124–126	
8-2 Add and Subtract with Estimation	8.2 Estimate sums and differences of 2-digit numbers.	17–19	127–129	8a, 8b
8-3 Add and Subtract 3-Digit Numbers	8.3 Find the sum or difference of two whole numbers up to three digits long.	20–22	130–132	8c, 8d
8-4 Whole Numbers to 10,000	8.4 Find the sum or difference of two whole numbers between 0 and 10,000.	23–25	133–135	8e, 8f
8-5 Multidigit Numbers	8.5 Demonstrate an understanding of, and the ability to use, standard algorithms for the addition and subtraction of multidigit numbers.	26–28	136–138	8d, 8d, 8e, 8f
Topic 8 Summary	Review computation with addition and subtraction.	29	139	
Topic 8 Mixed Review	Maintain concepts and skills.	30	140	

Computer Tutorial

Some students may benefit from completing the computer tutorial before they attempt the Try It page of each lesson. If you are using the electronic components of *Pinpoint Math*, you will find a complete listing of the Tutorial codes and titles when you access them either online or via CD-ROM.

Volume 3: Add and Subtract

Topic 8: Add or Subtract Multidigit Numbers

Topic Introduction

Objectives: 8.1 Use mental arithmetic to find the sum or difference of two two-digit numbers. **8.2** Estimate sums and differences. **8.3** Find the sum or difference of two whole numbers up to three digits long. **8.4** Find the sum or difference of two whole numbers between 1 and 10,000.

Write two-digit and three-digit numbers on the board. Have students identify the number of hundreds, tens, and ones in each number. For example, 63 would have 6 tens and 3 ones.

Informal Assessment

1. **What does it mean to use mental math to add?** To solve the problem in your head, without paper and pencil **What places do two-digit numbers have?** Tens and ones **Are there any ones in either number?** No. **Can you add the tens in your head?** Yes.

2. **When do you need to regroup in subtraction?** When the digit you're subtracting from is less than the digit you want to subtract

3. **What should we round to in this problem?** Tens **How do you know what to round 42 to?** 2 is less than 5, so we round down to 40. **How do you round 19?** 9 is greater than 5, so we round up to 20.

4. **What places do the base ten blocks represent?** Hundreds, tens, and ones **What can you trade 10 tens for?** 1 hundred **What can you trade 10 ones for?** 1 ten

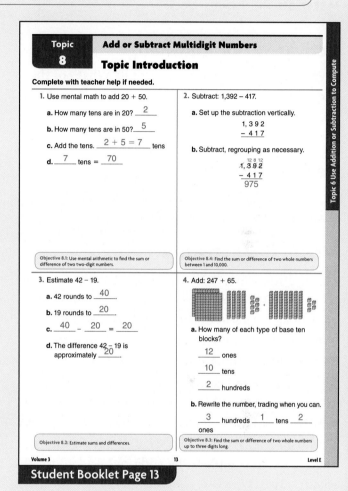

Student Booklet Page 13

Another Way To help students to see if their answers are reasonable for Exercises 2 and 4, have them estimate the sums and then compare to their answers.

| Lesson 8-1 | Add or Subtract Mentally |

Objective 8.1: Use mental arithmetic to find the sum or difference of two two-digit numbers.

Teach the Lesson

Materials ☐ Base ten blocks

Activate Prior Knowledge

Review addition facts with sums of 10. **What are the addition facts that have sums of 10?** 0 + 10 = 10, 1 + 9 = 10, 2 + 8 = 10, 3 + 7 = 10, 4 + 6 = 10, 5 + 5 = 10

Develop Academic Language

Be sure that students understand that "make 10" relates to finding sums of 10 in the ones place.

Model the Activities

Activity 1 Write on the board: *Find a number that is ten more than 34.* Show the number 34 with base ten blocks. **How can I show "ten more" with the base ten blocks?** Use 1 more ten. **What number is 10 more than 34?** 44

Activity 2 Write on the board: 32 + 38. **What fact with sums of 10 could we use to help us find the sum?** 2 + 8 = 10 On the board, show breaking apart 38 into 30 + 8. **What is 32 + 8?** 40 **What do we need to do to complete the addition?** Add 40 to 30, so 40 + 30 = 70.

ENGLISH LEARNERS You may need to explain the term *break apart*.

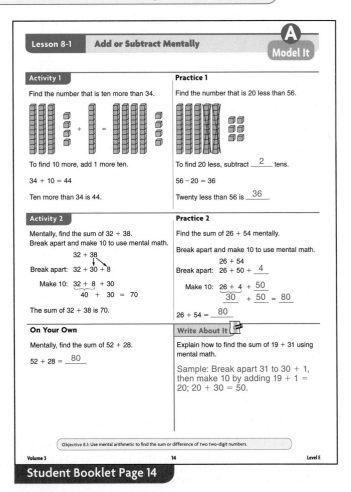

Student Booklet Page 14

Progress Monitoring

Ask students to explain how to find the number that is ten more than 35. **Add 1 ten to 35, 35 + 10 = 45.** Then ask students to add 34 and 16 mentally. **16 + 4 = 20, and 20 + 30 = 50**

Error Analysis

Some students may have difficulty with the concept of making 10 because they do not know their basic facts with confidence. Have them use addition flash cards to review and practice the basic facts with sums of 10.

Lesson 8-1: Add or Subtract Mentally

Objective 8.1: Use mental arithmetic to find the sum or difference of two two-digit numbers.

Facilitate Student Understanding

Develop Academic Language

Be sure students understand the idea of "break apart." Often we break numbers apart by the place value, for example, breaking 62 apart into 60 + 2, but sometimes we break apart to regroup as in Example 1, 60 = 50 + 10.

Demonstrate the Examples

Example 1 Review regrouping. On the board write: 60 − 25. Break 25 apart into 20 + 5. Remind students that to subtract 5, they need to regroup 10 ones from the 60. Guide students to describe regrouping 1 ten to 10 ones. Once they are thinking of 60 as 50 + 10, it is easy to subtract the 5 and the difference 60 − 25 is now 30 + 5 = 35. Be sure students know that *mentally* means by thinking, not by writing.

Example 2 As you discuss this example, point out that they are doing the subtraction mentally in two steps: first subtracting the tens, then subtracting the ones. Point out that this is the reverse of how problems are done on paper.

Computer Tutorial

Some students may benefit from completing a computer tutorial before they attempt the Try It page. A list of the tutorials for each lesson can be found beginning on page x in the front of this book.

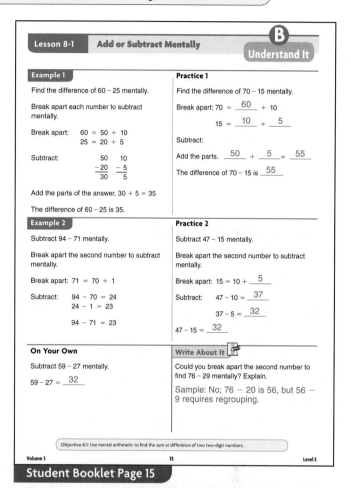

Student Booklet Page 15

Progress Monitoring

Ask students mentally to find the difference of 80 − 35. **45**

Error Analysis

When subtracting from multiples of 10 such as 60, students may forget that they had to regroup a ten, and so they give an answer that is 10 greater than the actual difference. For example, for 60 − 15, they might give the difference as 55. Encourage students to give a quick mental check by adding the difference and the number subtracted to make sure they get back to the original number.

Volume 3 125 Level E

Lesson 8-1: Add or Subtract Mentally

Objective 8.1: Use mental arithmetic to find the sum or difference of two two-digit numbers.

Observe Student Progress

Develop Academic Language

Exercise 4 Be sure that students understand the meaning of the term *compute*. To compute means to do the operation, in this case, either addition or subtraction. Also make sure to explain "mental arithmetic" so that students know what that means.

★ Error Analysis

Exercises 2 and 3 If students have difficulty finding the sums or differences mentally, encourage them to first find the sum or difference mentally, record that answer, and then check the answer using paper and pencil. If the answers do not agree, they need to compare the mental process with the paper and pencil process to determine where the error was made.

Exercise 7 Suggest that students do this problem on paper to find the error that Kyle made.

Exercise 8 If students have difficulty understanding this problem, read through the problem slowly with them. You may want to explain what it means to "make two 2-digit numbers using each of the digits 1, 2, 3, and 4 once." Help them find some of the numbers that meet those criteria, for example, 12 and 34. Point out that 33 would not be a possible number, because the instruction is to use the digits only once.

Student Booklet Page 16

Lesson 8-1 Add or Subtract Mentally — Try It

1. Find the number.
   a. ten more than 35 __45__
   b. ten less than 67 __57__
   c. twenty more than 38 __58__
   d. thirty less than 96 __66__

2. Find the sum mentally.
   a. 68 + 12 = __80__
   b. 52 + 35 = __87__
   c. 14 + 65 = __79__
   d. 55 + 15 = __70__

3. Find the difference mentally.
   a. 54 − 13 __41__
   b. 60 − 35 __25__
   c. 97 − 45 __52__
   d. 43 − 13 __30__

4. Use mental arithmetic to compute. Which is greater than 56? Circle the letter of the correct answer.
   (A) 32 + 26   B 27 + 20
   C 83 − 50   D 94 − 40

5. Dan biked 25 miles one day and 15 miles the next. How far did he bike in all? Circle the letter of the correct answer.
   A 30 miles   (B) 40 miles
   C 35 miles   D 50 miles

6. Dorothy says that it is easy to mentally add two numbers that have 5 in the ones place. Why do you think she says this?
   Since 5 + 5 = 10, the numbers make 10.

7. To add 27 + 17 mentally Kyle used 7 + 7 = 14, 20 + 10 = 30, 30 + 4 = 34. Did he get the right answer? Explain.
   No, he forgot to add in one of the tens.

8. The goal of this game is to create two numbers with the greatest possible sum. Each player makes two 2-digit numbers using each of the digits 1, 2, 3, and 4 once. Joshua makes 32 and 14. Lindsay makes 24 and 31. Who wins the game? Explain.
   Lindsay; 24 + 31 = 55; 32 + 14 = 46. Lindsay's sum is greater.

Lesson 8-2: Add and Subtract with Estimation

Objective 8.2: Estimate sums and differences of 2-digit numbers.

Teach the Lesson

Materials ☐ Teaching Aid 1 Number Lines

Activate Prior Knowledge

Have students count by tens to 100. Then name two-digit numbers and have students identify the ten just before and the ten just after.

Develop Academic Language

If we replace the numbers in a problem with numbers that are close but easier to work with, we get an answer called an *estimate*. The estimate will be close to the exact answer. Sometimes all we need to find is an estimate, and sometimes we estimate to help check our exact answer.

Model the Activities

Activity 1 Find 41 on the number line. 41 is between what two tens? **40 and 50** Is 41 closer to 40 or 50? **40** 41 rounds to 40. Find 18 on the number line. 18 is between what two tens? **10 and 20** Is 18 closer to 10 or 20? **20** 18 rounds to 20. What is 40 + 20? **60** Our estimate for 41 + 18 is 60. The exact answer to 41 + 18 will be close to 60.

Activity 2 55 is between what two tens? Students can count by tens until they hear the first digit of 55. **50 and 60** Is it closer to 50 or to 60? **Neither** The rule for numbers ending in 5 is to always round up. What does 55 round to? **60** 28 is between what two tens? **20 and 30** Is 28 closer to 20 or 30? **30** 28 rounds to 30. What is 60 − 30? **30** The estimate for 55 − 28 is 30.

Student Booklet Page 17

Progress Monitoring

What numbers round to 50? Generate an ordered list on the board. **45, 46, 47, 48, 49, 51, 52, 53, 54**

Error Analysis

Students might try to solve the actual problem and then round the answer. Explain that estimation is supposed to shorten and simplify the work. Therefore, estimation is done before calculations, not after.

Lesson 8-2: Add and Subtract with Estimation

Objective 8.2: Estimate sums and differences of 2-digit numbers.

Facilitate Student Understanding

Develop Academic Language

Clarify that in *front-end estimation,* students will pay attention to only the first digit, or "front end" of a number. This is different from rounding, in which we can round to any place greater than ones, and the digit following that place is important in indicating the closest ten.

ENGLISH LEARNERS Tell students that *estimate* can be a noun or a verb. **The verb *estimate* describes the process of finding about how much or how many. For example,** *Estimate how much the groceries will cost.*

Demonstrate the Examples

Example 1 Tell students they will be using front-end estimation. **In front-end estimation, the first digit of the number remains in place and the digits following it become zeros. Using front-end estimation, what number do we use in place of 67?** 60 **How do you know?** The 6 stays the same and the 7 becomes 0. **What do we use in place of 22?** 20 **How do you know?** The first 2 stays the same and the other 2 becomes 0. **What is 60 + 20?** 80 **Using front-end estimation, the estimate for 67 + 22 is 80.**

Example 2 **What does 94 become using front-end estimation?** 90 **45?** 40 **What is the estimate for 94 − 45 using front-end estimation?** 90 − 40 = 50

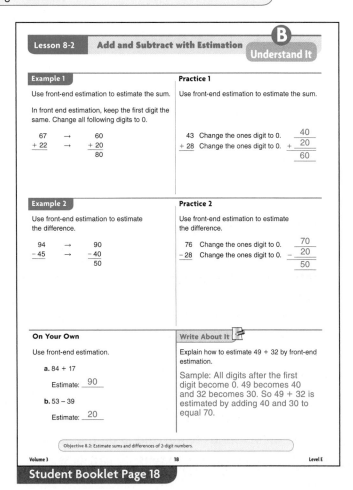

Student Booklet Page 18

Progress Monitoring

Ask students to estimate 42 + 29 using each method of estimation. **What is the estimate for 42 + 29 using front-end estimation?** 40 + 20 = 60 **What is the estimate for 42 + 29 using rounding?** 40 + 30 = 70 **Which is the better estimate?** Rounding is better because we changed the numbers less than with front-end estimation.

Error Analysis

Show students confused by front-end estimation how to cover up the ones digit and use only the tens digit.

Lesson 8-2: Add and Subtract with Estimation

Objective 8.2: Estimate sums and differences of 2-digit numbers.

Observe Student Progress

Computer Tutorial

Some students may benefit from completing a computer tutorial before they attempt the Try It page. A list of the tutorials for each lesson can be found beginning on page x in the front of this book.

 Error Analysis

Exercise 1 If students find estimates of 80 and 40, they have used front-end estimation instead of rounding. Explain that both methods are correct, but the question specifies that rounding should be used.

Exercise 2 If students find estimates of 140 and 40, they have used rounding instead of front-end estimation.

Exercise 3 Remind students of the rule for rounding numbers that end in 5.

Exercise 6 Students might find the incorrect answer 120 by rounding or doing the actual addition and then applying front-end estimation.

ENGLISH LEARNERS Point out the word *about* and explain that this word tells us we should estimate. Provide examples to help students understand. **64 is *about* 60; 64 is an *estimate* of 60. 98 is *about* 100; 98 is an *estimate* of 100.**

Exercise 7 Clarify the question. Students should estimate twice for each problem. They'll use rounding once and front-end estimation once on each. They should record the estimates and then find the problem for which the estimates are the same.

Student Booklet Page 19

Lesson 8-2: Add and Subtract with Estimation — Try It

1. Round to estimate.
   a. 67 + 21 Estimate: 90
   b. 53 − 18 Estimate: 30

2. Use front-end estimation to estimate.
   a. 91 + 48 Estimate: 130
   b. 72 − 26 Estimate: 50

3. Round to estimate.
   a. 88 + 63 Estimate: 150
   b. 55 − 19 Estimate: 40

4. Use front-end estimation to estimate.
   a. 67 + 38 Estimate: 90
   b. 54 − 22 Estimate: 30

5. What is the estimate for 68 + 35 when the addends are rounded? Circle the letter of the correct answer.
   A 80   B 90   C 100   (D) 110

6. A pair of shoes costs $48 and a jacket costs $74. About how much do the shoes and jacket cost all together? Use front-end estimation. Show your work.
   48 becomes 40 and 74 becomes 70. 40 + 70 = 110
   The shoes and jacket cost about $110.

7. If each sum is estimated by rounding and then by front-end estimation, which sum will have the same estimate from both methods?
   A 12 + 19   B 25 + 48   (C) 62 + 84   D 36 + 91

8. What is the advantage of using rounding? What is the advantage of using front-end estimation?
   Sample: When you round, you change the numbers to a close-but-easier number. When you use front-end estimation, it's very fast to figure out what numbers to use, because you just change the ones digit to zero.

Lesson 8-3: Add and Subtract 3-Digit Numbers

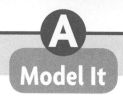

Objective 8.3: Find the sum or difference of two whole numbers up to three digits long.

Teach the Lesson

Materials
- ☐ Base ten blocks
- ☐ Teaching Aid 6 Place-Value Charts

Activate Prior Knowledge

Review place value for 3-digit numbers. Direct students to model 258. **How many hundreds, tens, and ones do you need to show 258?** 2 hundreds, 5 tens, and 8 ones

Display 12 ones, 9 tens, and 3 hundreds. **How can I show this same amount using the least number of base ten blocks?** Trade 10 ones for 1 tens. Now there are 10 tens. Trade the 10 tens for 1 hundreds. So there are 2 ones and 4 hundreds. **What number does that represent?** 402

Model the Activities

Activity 1 Write this problem on the board: 143 + 65 = ? Have students model the addends with hundreds, tens, and ones. **How many ones in all?** 8 **How many tens in all?** 10 **How many tens do you trade for 1 hundreds?** 10 tens

Demonstrate trading 10 tens for 1 hundreds. **How many hundreds do you have now?** 2 hundreds **How many tens?** 0 **How many ones?** 8 Record the sum on the board: 143 + 65 = 208.

Activity 2 Write this subtraction problem on the board: 100 − 57 = ? **How do you show 100 with base ten blocks?** 1 hundreds **How do you subtract 57?** Take away 5 tens and 7 ones. **There are no ones or tens. What can you do?** Trade 1 hundred for 10 tens. Then trade 1 ten for 10 ones. Direct students to make the trades and complete the subtraction. Then write the problem on the board in vertical form.

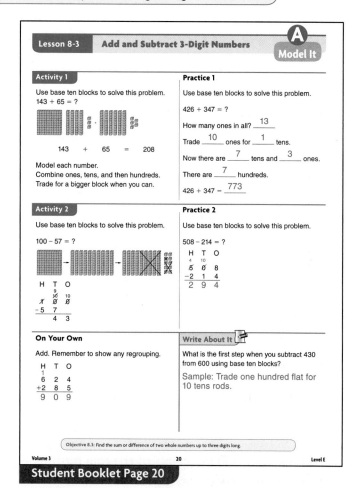

Student Booklet Page 20

⭐ Progress Monitoring

Show 312 − 62 with hundreds, tens, and ones blocks. 250 Check students' blocks.

Error Analysis

Some students may have difficulty keeping columns aligned when working with the standard algorithm. You may want to give them Teaching Aid 4 Centimeter Grid Paper to help them keep the place-value columns aligned.

Lesson 8-3 Add and Subtract 3-Digit Numbers

Understand It

Objective 8.3: Find the sum or difference of two whole numbers up to three digits long.

Facilitate Student Understanding

Develop Academic Language

As you present number lines with different scales, such as marked in tens or hundreds, have the students name the numbers that are not labeled on the number lines to be sure that they understand the number line representations.

Demonstrate the Examples

Example 1 Draw the number line on the board. Discuss with students how to choose the numbers to use on the number line in order to show this subtraction. Sample: Since both numbers in the problem have ones, use intervals of 1 on the number line from 350 to 360. Demonstrate the subtraction on the number line. Record the difference, $357 - 4 = 353$.

Example 2 Record the addition problem on the board in vertical format and work through the addition emphasizing recording the regrouping.

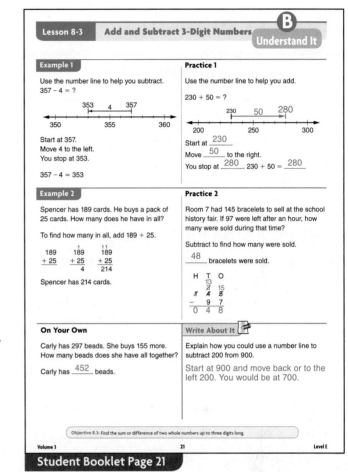
Student Booklet Page 21

Progress Monitoring

Show how to subtract 40 from 260 on a number line. Move 40 to the left. The difference is 220.

Error Analysis

Some students may simply subtract the greater digit from the lesser digit in each place rather than regroup. Have these students work more with concrete models to reinforce the meaning of the operation.

Lesson 8-3 Add and Subtract 3-Digit Numbers

Objective 8.3: Find the sum or difference of two whole numbers up to three digits long.

Observe Student Progress

Computer Tutorial

Some students may benefit from completing a computer tutorial before they attempt the Try It page. A list of the tutorials for each lesson can be found beginning on page x in the front of this book.

Develop Academic Language

Exercise 7 Be sure that students understand the meaning of the term "3-digit number" and that students understand the difference between digits and the values of the digits. For example, in 419, the digits are 4, 1, and 9. The values of the digits are 400, 10, and 9.

 Error Analysis

Exercise 3 Encourage students to go back and check their work using the inverse operation. In other words, use addition to check subtraction and subtraction to check addition.

Exercise 4 If students have difficulty seeing the answer to this problem when the subtractions are set up horizontally, suggest they rewrite them vertically so it is easier to compare the tens places.

Exercise 6 If students do not see the relationship, have them write each addend of 265 + 43 in expanded form and have them match the numbers with those give in the second addition expression.

Exercise 8 You may wish to show students the relationship in words and then in numbers:

Miles on Monday + Miles on Tuesday = 240

$\quad$ 93 $\quad$ + $\quad$? $\quad$ = 240

Lesson 8-4 — Whole Numbers to 10,000

Objective 8.4: Find the sum or difference of two whole numbers between 0 and 10,000.

Teach the Lesson

Materials ☐ Base ten blocks

Activate Prior Knowledge
Review place value for 3- and 4-digit numbers. **Look at 381. How many hundreds, tens, and ones?** 3 hundreds, 8 tens, 1 one **Look at 1,290. How many thousands, hundreds, tens, and ones?** 1, 2, 9, 0

Develop Academic Language
Discuss *trading*. **What other words can you use to describe this process?** Sample: carrying, borrowing

Model the Activities

Activity 1 Add ones first. **Can you regroup?** Yes. **What is the result?** There is 1 one and 1 more ten. Add tens. **Can you regroup?** No. **What is the sum of the tens?** 7 tens Add hundreds. **Can you regroup?** No. **What is the sum of the hundreds?** 4 hundreds Keep your model as you work the addition in vertical form. **What do you do first?** Add the ones: 8 and 3. **Do you need to regroup?** Yes. **What will you write as the sum of the ones?** 1 **What else will you write?** 1 in the tens column **What do you do next?** Add the tens: 1, 4, and 2. **Do you need to regroup?** No. **What will you write as the sum of the tens?** 7 **What do you do next?** Add the hundreds: 3 and 1. **Do you need to regroup?** No. **What will you write as the sum of the hundreds?** 4 **What number have you written as the sum of 343 + 128?** 471 Use your model to check. **Does your model show 471?** Yes.

Activity 2 Have students model 417. Tell them to show each step on their models and in the subtraction problem. **Can you take 4 ones away from 7 ones?** Yes. **What do you ask yourself next?** Can I subtract 3 tens from 1 ten? **Can you subtract 3 tens from 1 ten?** No. **What do you need to do?** Trade 1 hundred for 10 tens. **Now that you have more tens, what can you do?** Subtract 3 tens from 11 tens. **What do you ask yourself next?** Can I subtract 2 hundreds from 3 hundreds? **Can you subtract 2 hundreds from 3 hundreds?** Yes. **What is the result?** 183

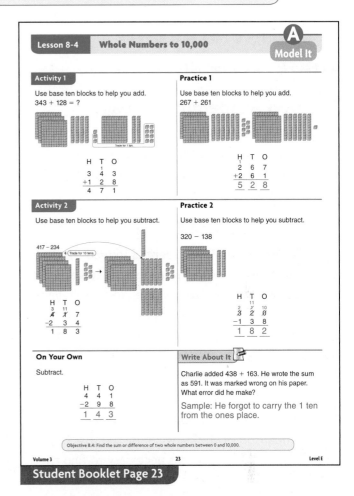

Student Booklet Page 23

Progress Monitoring

Show how you would align 27 + 459 to perform the addition. What is the sum? 2 would line up with the 5 and the 7 would line up with the 9; 486.

Error Analysis
Some students might make errors because they do not know their basic facts. Before students begin work, have them practice basic facts and write the ones that they have not mastered. Students may keep the list handy for reference while they work.

Lesson 8-4 Whole Numbers to 10,000

Objective 8.4: Find the sum or difference of two whole numbers between 0 and 10,000.

Facilitate Student Understanding

Develop Academic Language

Review place-value terms. Create a thousands model by stacking 10 hundreds blocks together. Have students say the names of the places and touch the appropriate models.

Demonstrate the Examples

Example 1 Make sure students can read the numbers correctly. **Why is it important to keep the places lined up correctly?** Sample: You don't want to add thousands to hundreds by mistake. Guide students to add in each place. Discuss the regrouping, and support students' work with models if necessary.

Example 2 Ask a student to read the word problem aloud. Discuss why subtraction is used to solve this problem. Mention that the problem asks for a comparison of the sales during the first week and the second week, so it is necessary to find the difference. Guide students to subtract. Discuss the regrouping with special attention to the hundreds column, and support students' work with models if necessary.

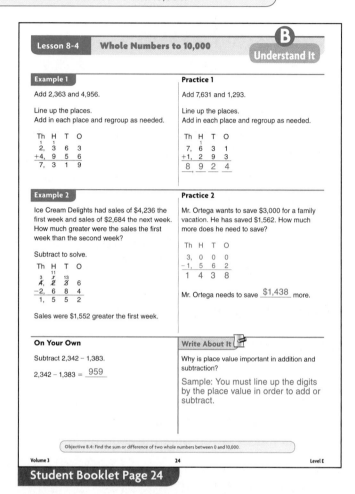

Student Booklet Page 24

Progress Monitoring

Subtract: 1,630 − 273. 1,357

Error Analysis

Encourage students to write neatly and leave enough space to see the place-value columns easily. Working on grid paper may help students align the numbers.

| Lesson 8-4 | **Whole Numbers to 10,000** | |

Objective 8.4: Find the sum or difference of two whole numbers between 0 and 10,000.

Observe Student Progress

Computer Tutorial

Some students may benefit from completing a computer tutorial before they attempt the Try It page. A list of the tutorials for each lesson can be found beginning on page x in the front of this book.

Develop Academic Language

Exercise 4 Focus on the phrase *more than*. Help students to see that the operation required is addition. Point out that they could replace the phrase *more than* with *added to*. Ask students to make a simpler problem by replacing the multidigit numbers with 1-digit numbers. For example, "What is 2 more than 1?"

 Error Analysis

Exercise 2 Make sure students write 2c and 2d correctly in vertical form. You may want to provide graph paper to help students align the digits.

Exercise 5 Have students draw a diagram of the park with the pond inside. Help them see that the problem requires subtraction.

Exercise 6 Students might benefit from looking at a simpler problem. Rewrite the problem for them this way: The sum of two numbers is 5. If one of the addends is 3, what is the other addend? Ask them what they did to solve this problem. **Find 5 – 3.** Then have students read the original problem.

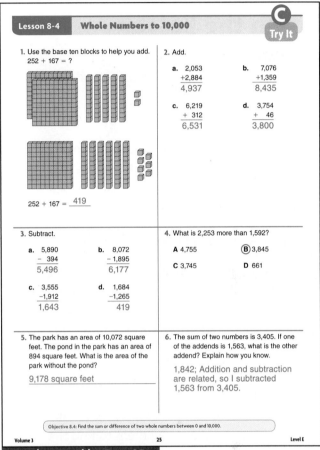

Student Booklet Page 25

Lesson 8-5 — Multidigit Numbers

Objective 8.5: Demonstrate an understanding of, and the ability to use, standard algorithms for the addition and subtraction of multidigit numbers.

Teach the Lesson

Materials ☐ Base ten blocks
☐ Teaching Aid 7 Place-Value Chart

Activate Prior Knowledge

Review place value for 5- and 6-digit numbers. Write on the board: *42,709.* **What digit is in the hundreds place?** 7 **What digit is in the thousands place?** 2 **What is the value of the digit 4?** 40,000

Develop Academic Language

Discuss with students the importance of using mathematical language when writing explanations. **What are some of the math words that you might use in explaining addition work?** Sample: places, sum, trading, estimating, rounding

ENGLISH LEARNERS Make sure that students understand the terminology on this page. For example, when we say "Find 627 + 393," make sure students know we are asking them to find the answer, or, in this case, add.

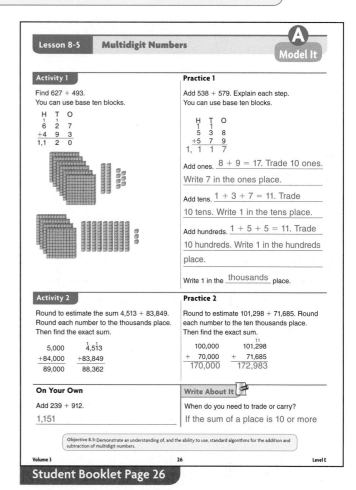

Student Booklet Page 26

Model the Activities

Activity 1 Write the addition problem vertically on the board. Have students model the addends with base ten blocks to perform the addition as you record the algorithm on the board. **Add ones.** 7 + 3 = 10 **Do you need to trade?** Yes. **How many?** 10 ones for 1 ten **There are no more ones left. Write 0 in the ones place.** Repeat this procedure for each place. **What is the answer?** 1,120

Activity 2 Ask students to round the addends to the nearest thousand. **Underline the digit in the thousands place and circle the digit to its right. To what numbers did you round 4,513 and 83,849?** 5,000 and 84,000 Write the addition problem vertically on the board. Guide students to add in each place. Record the algorithm on the board. Then have students add using the exact numbers. **Is your answer reasonable?** Yes, because the estimated sum is close to the exact sum. If students do not mention it, point out that you rounded to the thousands, and the exact sum is within 1 thousand of the estimate.

Progress Monitoring

Show how you would align 15,948 + 223 to perform the addition. Line up the places. **What is the sum?** 16,171

Error Analysis

Students might have trouble writing good explanations of their work. Encourage them to use complete sentences and to think of it as describing what they do with the base ten blocks.

Lesson 8-5: Multidigit Numbers

Understand It

Objective 8.5: Demonstrate an understanding of, and the ability to use, standard algorithms for the addition and subtraction of multidigit numbers.

Facilitate Student Understanding

Materials
- ☐ Base ten blocks
- ☐ Teaching Aid 7 Place-Value Chart

Activate Prior Knowledge

Invite a student to demonstrate this subtraction on the board: 658 − 431. **227** Discuss that no trading is necessary, because in each place the number being subtracted is less than the number it is being subtracted from.

Develop Academic Language

Review place-value terms used with 4-, 5-, and 6-digit numbers: thousand, ten thousand, and hundred thousand. Point out the pattern of ones, tens, and hundreds in the thousands period.

Demonstrate the Examples

Example 1 Write the subtraction problem vertically on the board. Have students use base ten blocks to perform the subtraction. Have students subtract in each place, as you record the algorithm on the board. **Subtract ones.** 4 − 1 = 3 **Write 3 in ones place. Subtract tens. Do you need to regroup?** Yes, regroup 1 hundred as 10 tens, which gives 13 − 5 = 8. **Write 8 in the tens place. Subtract hundreds.** 1 − 1 = 0 **What is the answer?** 83

Example 2 Ask students to round the numbers to the nearest ten thousand. **810,000 and 50,000 Write the numbers in the place-value chart. Perform the subtraction for each place.** Guide students to explain the work. Repeat by using the exact numbers. **Is your answer reasonable? Why?** Yes, because we rounded to ten thousands and the exact difference is within ten thousand of the estimate.

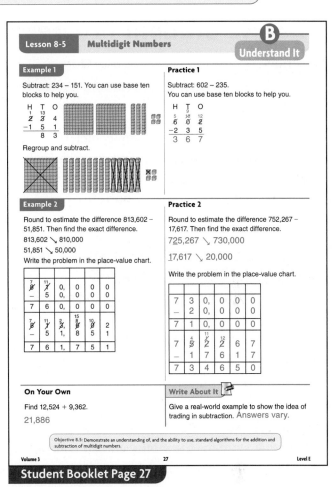

Student Booklet Page 27

Progress Monitoring

To subtract 2,087 − 145, is any trading necessary? Explain. Yes, to subtract hundreds you need to trade 1 thousand for 10 hundreds.

Error Analysis

Students might have difficulty recording all the regrouping correctly. Suggest that students think of regrouping as a pair of changes that affects two places, and the changes should compensate each other. For example, 1 less in the tens place should pair with 10 more in the ones place.

Lesson 8-5: Multidigit Numbers

Objective 8.5: Demonstrate an understanding of, and the ability to use, standard algorithms for the addition and subtraction of multidigit numbers.

Observe Student Progress

Materials ☐ Teaching Aid 7 Place-Value Chart

Computer Tutorial

Some students may benefit from completing a computer tutorial before they attempt the Try It page. A list of the tutorials for each lesson can be found beginning on page x in the front of this book.

Develop Academic Language

Exercise 4 Point out the abbreviation, sq mi. Tell students it means square miles and that a square mile would be a square of land that is 1 mile on each side

Exercise 5 Read through the word problem and check that students understand that taxes are taken from total income.

ENGLISH LEARNERS Make sure students understand that another word for *the money earned* is *income*.

Error Analysis

Exercise 2 Have students use Teaching Aid 11 Place-Value Charts to perform the addition and subtraction. Make sure they show all regrouping.

Exercise 3 Remind students how to round the numbers to the nearest ten thousand. Write a few multidigit numbers on the board and have students round the numbers to different places.

Exercise 5 Make sure the students identify the correct operation to solve the problem. **Subtraction** After paying taxes, Mrs. Rodriquez's income must be less than the original amount, $32,565. Ask which answer choices can be eliminated. **C and D**

Exercise 7 You may want to point out that if they are adding more than 2 numbers, then it is possible to trade more than 10 ones. For example, 24 + 156 + 39 + 128; in this case there are 27 ones, so you would need to trade 20 ones for 2 tens.

Exercise 8 Tell the students that the word *about* indicates estimation and the phrase *how much taller* indicates comparison by subtraction.

Volume 3: Whole Number Operations

Topic 8: Add or Subtract Multidigit Numbers

Topic Summary

Objective: Review computation with addition and subtraction.

Have students complete the student summary page.

You may want to have students work in groups of four with each student analyzing a different choice.

Ask students to share their ideas about each answer choice. Be sure they confirm the correct answer for each problem at the end of the discussion.

Answer Evaluation

1. **A** The student added instead of subtracting.

   B The student subtracted the lesser digit from the greater digit, not regrouping at all.

   C The student did not trade correctly in the tens place. He or she did not record the second regrouping from 10 tens to 9 tens.

   D This choice is correct.

2. **A** The student subtracted and regrouped incorrectly.

   B The student subtracted instead of adding.

   C The student added but forgot to regroup the ones in the tens place.

   D This answer is correct.

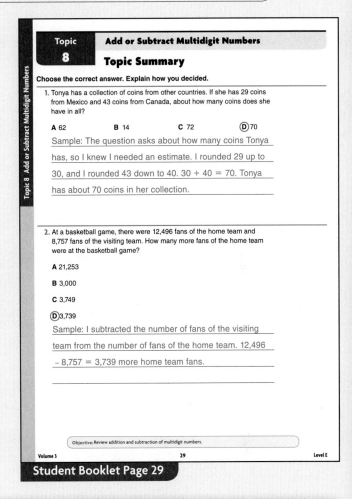

Student Booklet Page 29

 ## Error Analysis

Exercise 1 Students might choose the incorrect operation when solving this problem. Have students identify the part, part, whole in order to set the number sentence up correctly.

Exercise 2 Students might make addition errors regrouping in the ones and the tens. Encourage students to write the number they are regrouping above each place.

Progress Monitoring

When all assignments for this topic have been completed, assign the corresponding Progress Monitoring page for this topic (Assessment Resources Book, page 8). Be sure students complete the Progress Monitoring page before you administer the final assessment for this volume.

Volume 3 — Topic 8
Whole Number Operations
Add or Subtract Multidigit Numbers
Mixed Review

Objective: Maintain concepts and skills.

Error Analysis

Exercise 2 Encourage students to read the word name aloud as they write the number in standard form.

Exercise 4 Although all of the answer choices are factors of 24, only the third choice includes all of the factors. Make sure students identify all of the factors before making a choice.

Exercise 6 Suggest that students look at the relationship between the three numbers to decide if the fact family is made up of addition/subtraction facts or multiplication/division facts.

Exercise 7 Review the terms *prime* and *composite* with any student who wrote that 17 is a composite number.

Exercise 8 Have students identify what place Kayla is rounding to. Once students have established it is the nearest ten, review the rules for rounding.

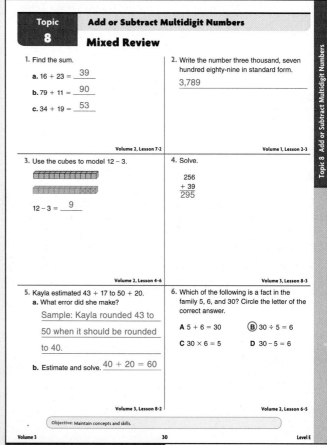

Student Booklet Page 30

Volume 4: Multiply and Divide

Topic 9: Use Multiplication to Compute

Topic Introduction

Lesson 9-1 expands students' knowledge of basic multiplication facts to multiplying multiples of 10. Lesson 9-2 gives students practice in estimating products. Lesson 9-3 builds on students' knowledge of basic facts and extended facts to explore multiplying with multidigit numbers. In Lesson 9-4 students choose from the methods they learned in previous lessons to solve multiplication problems.

Lesson	Objective	Student Pages	Teacher Pages	Tutorials
Topic 9 Introduction	9.1 Multiply multiples of 10.	1	142	
	9.2 Estimate products by rounding factors and using mental math techniques.			
	9.3 Solve simple problems involving multiplication of multidigit numbers by one-digit numbers.			
	9.4 Multiply 2-digit numbers by 2-digit numbers.			
9-1 Multiply by Multiples of 10	9.1 Multiply multiples of 10.	2–4	143–145	9a, 9b
9-2 Estimate Products	9.2 Estimate products by rounding factors and using mental math techniques.	5–7	146–148	9a, 9c
9-3 Multiply: Four Digits by One Digit	9.3 Solve simple problems involving multiplication of multidigit numbers by 1-digit numbers.	8–10	149–151	9d, 9a
9-4 Choose a Method and Multiply	9.4 Solve problems by choosing between using estimation, mental math, or pencil and paper.	11–13	152–154	9e
Topic 9 Summary	Review computations with multiplication.	14	155	
Topic 9 Mixed Review	Maintain concepts and skills.	15	156	

Computer Tutorial

Some students may benefit from completing the computer tutorial before they attempt the Try It page of each lesson. If you are using the electronic components of *Pinpoint Math,* you will find a complete listing of the Tutorial codes and titles when you access them either online or via CD-ROM.

Volume 4 — Multiply and Divide
Topic 9: Use Multiplication to Compute
Topic Introduction

Objectives: 9.1 Multiply multiples of 10. **9.2** Estimate products by rounding factors and using mental math techniques. **9.3** Multiply multidigit numbers by 1-digit numbers. **9.4** Solve problems by choosing between using estimation, mental math, or pencil and paper.

Materials ☐ Base ten blocks

Have students use base ten blocks to create multiplication facts using ten as a factor. Start with 3 × 10. Have pairs of students make 3 groups of 10 to find the product. Repeat using other factors times 10. Discuss any patterns students may see.

Informal Assessment

1. **What basic fact do you see?** 5 × 3 **How can you find the product?** Add 5 three times. **What do you notice about the pattern?** Each problem has the basic fact 5 × 3 and adds a zero at the end each time.

2. **Which number will you round?** 78 **How can you use the number line to round 78?** Make a mark at 78 and see if it is closer to 70 or 80. **What number will we multiply 4 by?** 80

3. **Why do you break 14 into tens and ones?** It makes it easier to multiply. **How many tens does 14 have?** 1 **How many ones?** 4 **Why do you multiply the tens and ones by 8?** In two-digit multiplication you have to multiply the one-digit number by the tens and ones.

4. If students have not memorized facts through 12, post them. **How can you break apart 15 into numbers that are easy to multiply by 12?** 10 and 5 **How does that help you multiply?** You can multiply 10 by 12 and 5 by 12, and then add the products.

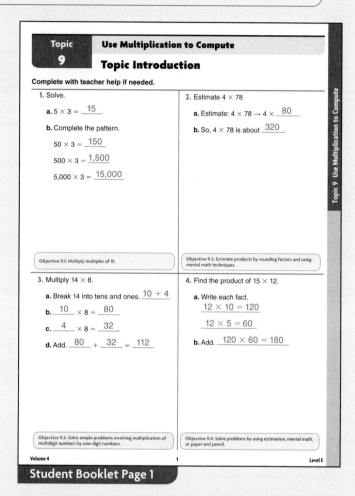

Student Booklet Page 1

Another Way To help students understand multiplying by multiples of ten, have students box in the basic fact and underline each zero.

Lesson 9-1: Multiply by Multiples of 10

Objective 9.1: Multiply multiples of 10.

Teach the Lesson

Materials ☐ Numeral tiles

Activate Prior Knowledge
Have students take turns picking two numeral tiles and finding the product.

Develop Academic Language
Use skip-counting to show the meaning of *multiples*. Address multiples of 10, 100, and 1,000. Have students find patterns in the products of 1-digit numbers and 10, 100, 1,000, or 10,000.

Model the Activities

Activity 1 Make sure each group of students has seven 0 tiles or slips of paper labeled 0. Have students copy this frame four times in a column, with plenty of space after each blank: __ × __ = __. Have students write 3 × 4 = 12. **What is the product?** 12 **We can use this basic fact to find greater products.** Have students copy the equation in the next frame and place a 0 tile next to the 4 to produce 3 × 40. **We multiplied a factor by 10. Will this increase or decrease the product?** increase **How will we increase the product?** Multiply by 10; put 0 at the end. Repeat for the other problems shown. Have students write the zeros before removing the tiles.

Activity 2 **What basic fact do you see when you ignore the end zeros in 5,000 × 8?** 5 × 8, which equals 40. **We can build this product if we start with 5 × 8.** Have students write the 40 without using a tile.

Refer to multiplying the product by 10 in the next problem to avoid confusion over the number of zeros.

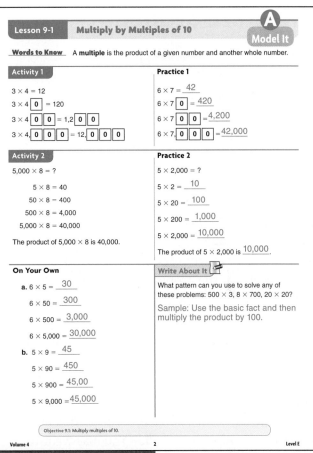

Student Booklet Page 2

Progress Monitoring
What basic fact do you see in 6 × 300?
6 × 3 = 18 **Since we multiplied one factor of the basic fact by 100, what will we multiply the product by?** 100 **What will it be?** 1,800

Error Analysis
Have students use repeated addition or base ten blocks to check their answers.

Lesson 9-1: Multiply by Multiples of 10

Objective 9.1: Multiply multiples of 10.

Facilitate Student Understanding

Develop Academic Language

ENGLISH LEARNERS Have students practice labeling the parts of a multiplication sentence using the terms *factor* and *product*.

Demonstrate the Examples

Example 1 Ask students to cover the zeros. **What basic fact do you see?** $3 \times 4 = 12$ If students know the commutative property, you can introduce it here by showing that the problem is equal to $3 \times 10 \times 4 \times 100$. Otherwise, discuss each number separately. **30 is the same as 3×10. We multiplied the first factor by 10, so we'll have to do the same to the product. How is 4 related to 400?** 4 is multiplied by 100. **We'll have to multiply the product by 100 as well. Multiplying by 10 and then by 100 is the same as multiplying by 1,000. What happens to a number when you multiply it by 1,000?** You may want to help students find the pattern in a few examples. 3 zeros are put at the end of the number. **What is the product?** 12,000

Example 2 Have students circle the leading non-zero digits and then underline the zeros. **What basic fact do you see?** $4 \times 5 = 20$ Point out that the product of the basic fact ends in zero. You may want to show a few examples of multiplying numbers ending in zero by 10, 100, and 1,000. Guide students through the same process as for Example 1.

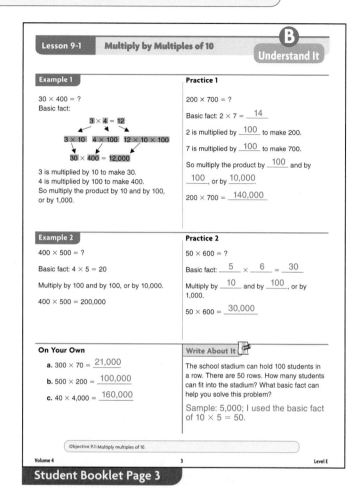

Student Booklet Page 3

 Progress Monitoring

Will the product of 200×50 have 3 or 4 zeros? How do you know? 4; the product of the basic fact has a zero, and then we multiply by 1,000.

Error Analysis

If students write an incorrect number of zeros, suggest they record the basic fact in black and the number of zeros in red. The number of red zeros should match the number of zeros in the factors.

Lesson 9-1: Multiply by Multiples of 10

Objective 9.1: Multiply multiples of 10.

Observe Student Progress

Computer Tutorial

Some students may benefit from completing a computer tutorial before they attempt the Try It page. A list of the tutorials for each lesson can be found beginning on page x in the front of this book.

ENGLISH LEARNERS Make sure students understand that *basic facts* are the multiplication facts to 10 × 10 or 12 × 12. *Basic* means what you start with before moving on to harder problems.

 Error Analysis

Exercises 1–8 Allow students to use MathFlaps to model the basic facts. Students who make basic fact errors may need to review with multiplication flash cards. Students can also use a multiplication fact table to check their answers.

Exercises 2 and 5 Students who make basic fact errors may need to review with multiplication flash cards. Students can also use a multiplication fact table to check their answers.

Exercise 3 The product of the basic fact has a zero. Remind students to first record the product of the basic fact and then multiply by 10, 100, or the appropriate power of 10. Using two different colors to record the two components can be helpful.

Exercise 7 Students should solve the basic facts first.

Exercise 8 Have students write the problem first. Ask why this problem requires multiplication; each page is an equal group of words.

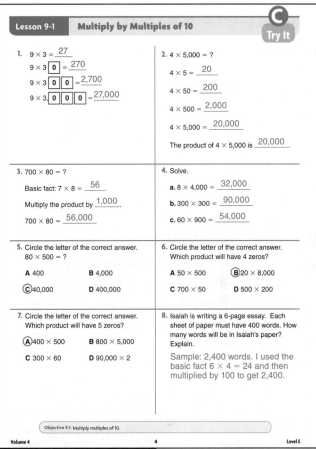

Student Booklet Page 4

Lesson 9-2: Estimate Products

Objective 9.2: Estimate products by rounding factors and using mental math techniques.

Teach the Lesson

Materials ☐ Teaching Aid 1 Number Lines

Activate Prior Knowledge

Remind students that when they move right on a number line, the numbers increase, and when they move left, the numbers decrease.

Write 4,537 in a place-value chart. Have students practice rounding the number to different places.

Develop Academic Language

Write 56 + 21 on the board. **We will *round* these numbers to find an *estimate*. What is 56 rounded to the nearest ten?** 60 **What is 21 rounded to the nearest ten?** 20 **What is the sum?** 80 Explain that 80 is an *estimate* of the sum of 56 and 21. 80 is close to, but not exactly, the actual sum.

Write several multidigit numbers on the board. Ask students to name the greatest place in each number. For example, write 36. **What is the greatest place?** Tens

Model the Activities

Activity 1 Let's make this problem easier. We can change the 2-digit factor to a multiple of 10 by rounding to the tens place. Provide copies of Teaching Aid 1. Have students locate 52. **What does 52 round to?** 50 **How do you know?** On the number line, 52 is closer to 50 than 60. Now, use what we know about multiplying multiples of 10 to multiply 3 × 50. **What basic fact will you use?** 3 × 5 = 15 Then what is 3 × 50? 150 Our estimate is 150.

Activity 2 Have students locate both numbers on the number lines. **To what number do you round 24?** 20 **To what number do you round 28?** 30 **How many zeros will be in the product?** 2

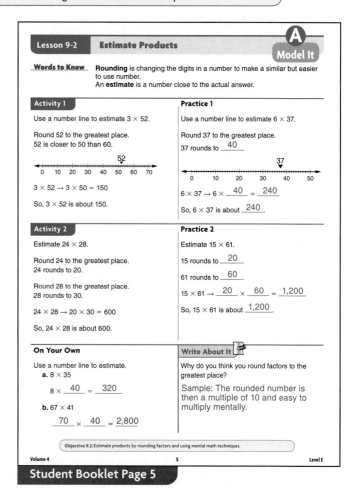

Student Booklet Page 5

 Progress Monitoring

What numbers will you use to estimate the product of 42 × 31? 40 and 30

Error Analysis

If estimates are incorrect, you may need to review rounding rules. Point out that the 1-digit numbers don't need to be rounded, since they are already part of basic facts. You can have students color-code the basic fact and the zeros in each problem to see whether the zeros on both sides of the equal symbol match.

Lesson 9-2: Estimate Products

Objective 9.2: Estimate products by rounding factors and using mental math techniques.

Facilitate Student Understanding

Develop Academic Language

Discuss terms such as *about* and *approximate*, which signify that a value is an estimate, and compare them to *exact*.

ENGLISH LEARNERS Describe items in the classroom using the words *about*, *estimate*, and *approximate*. For example, **The pen is *about* the length of my hand.** Explain that the pen may be longer or shorter than your hand, but the word *about* refers to an approximation, or an estimate.

Demonstrate the Examples

Example 1 Point to 183. Ask students to underline the digit in the hundreds place and circle the digit to its right. **What digit did you circle? 8 Is 8 less than 5, or greater or equal to 5?** Greater than **What happens to the 1 in the hundreds place?** It increases by one. **To what number did you round 183? 200 The new problem is 200 × 5. What is the product of 200 × 5? 1,000 How do you know?** I used the basic fact 2 × 5 = 10 to get 200 × 5 = 1,000.

Example 2 **Why do you round both factors?** So that both factors are multiples of 10 or 100. **To what numbers did you round 45 and 239? 50 and 200 What basic fact do you use? 5 × 2 = 10 How many zeros do you add to find the product of 50 × 200? 3**

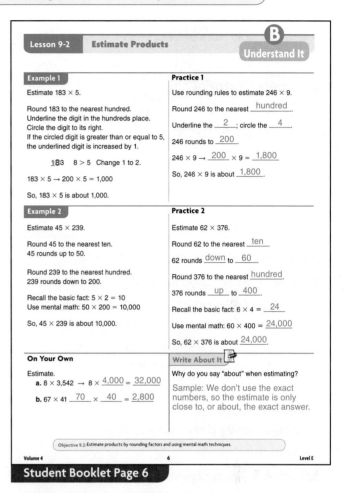

Student Booklet Page 6

Progress Monitoring

How do you estimate 34 × 582? Round the factors, then multiply. 30 × 600 = 18,000

Error Analysis

Continue to provide practice in rounding to the greatest place. Help students create a graphic as a reminder of the "5 or more, round up" rule.

You can help students use numeral tiles or slips of paper marked 0 to multiply multiples of 10.

Lesson 9-2: Estimate Products

Objective 9.2: Estimate products by rounding factors and using mental math techniques.

Observe Student Progress

Computer Tutorial

Some students may benefit from completing a computer tutorial before they attempt the Try It page. A list of the tutorials for each lesson can be found beginning on page x in the front of this book.

Error Analysis

Exercise 2 If students make rounding errors, allow them to plot the factors on Teaching Aid 1 Number Lines. Check that students plot the numbers correctly and then chose the ten that is closest to their plotted number.

Exercise 4 Be sure students round the factors and then multiply to find the estimates. Students may make errors if they only look at the first digits of the factors.

Exercise 5 Encourage students to use number lines to round and numeral tiles to work through the multiplication. Remind students that if the basic fact contains a zero, the product will have 1 more zero than the factors.

Exercise 6 Students may incorrectly round 85 to 80. Remind students that since 5 is no closer to 0 than it is to 10 on a number line, there is a rule stating that 5 means to round up.

Exercise 8 Make sure students work through each step logically. They need to round and use basic facts. If they miss part of the procedure, have them work the problem and match up the steps they do with their written description.

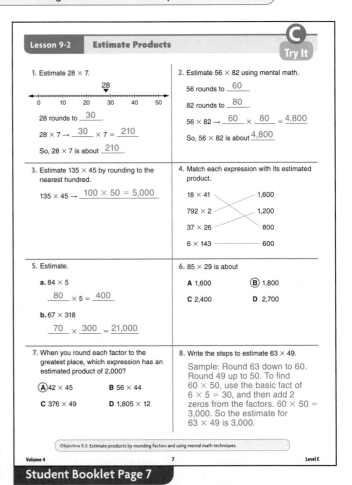

Student Booklet Page 7

Lesson 9-3 — Multiply: Four Digits by One Digit

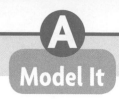
Model It

Objective 9.3: Solve simple problems involving multiplication of multidigit numbers by one-digit numbers.

Teach the Lesson

Materials ☐ Base ten blocks

Activate Prior Knowledge

Illustrate the number 423 using base ten blocks. **How many hundreds are there?** 4 **How many tens?** 2 **How many ones?** 3 Write 541 on the board. **How can we use base ten blocks to show 541?** Use 5 hundreds, 4 tens, and 1 one.

Develop Academic Language

Ask students to give terms that are used to write 12 ones as 1 ten 2 ones. **Sample: rename, regroup, carry the 1, carry a 10** Remind students of the meanings of the words *renaming*, *regrouping*, and *trading* as used in this lesson.

Model the Activities

Activity 1 Write 213 on the board. Have students use base ten blocks to model the number as you draw the corresponding blocks on the board. **What problem have we modeled?** 213 × 2 **How many hundreds, tens, and ones are in the product?** 4 hundreds, 2 tens, 6 ones Show how to count the total number of hundreds, tens, and ones in the product. Write 213 × 2 = 426 on the board.

Activity 2 Use base ten blocks to model 136. **How many time will we model 136 to solve the problem?** twice **Why?** Each of two people sold 136 tickets. **What problem does the model on the page show?** 2 × 136 **How many hundreds, tens, and ones are in the product?** 2 hundreds, 6 tens, and 12 ones **Since there are more than 9 ones, what do you need to do to write the product as a number in standard form?** First regroup 12 ones as 1 ten and 2 ones. Model the regrouping with the students and help them find the total.

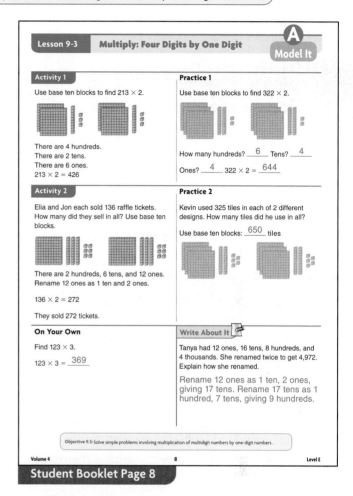

Student Booklet Page 8

Progress Monitoring

Show how to find 235 × 3 using base ten blocks. Show 2 hundreds, 3 tens, 5 ones three times. Regroup 15 ones as 1 ten and 5 ones. Rename 10 tens as 1 hundred. 235 × 3 = 705

Error Analysis

You may need to help students see that there should be 3 groups of blocks in On Your Own. Modeling Write About It with students may be helpful. Make sure they explain each step of the process.

Lesson 9-3 — Multiply: Four Digits by One Digit

Understand It

Objective 9.3: Solve simple problems involving multiplication of multidigit numbers by one-digit numbers.

Facilitate Student Understanding

Develop Academic Language

For a number such as 3,418, have students tell how many thousands, **3**, how many hundreds, **4**, how many tens, **1**, and how many ones, **8**, there are.

ENGLISH LEARNERS Relate *partial* in *partial products* to *part*. Each partial product is part of the final product, so they must be combined by addition.

Demonstrate the Examples

Example 1 Remind students that any number can be written in expanded notation by using the place values in the number. **How do you write 1,324 in expanded notation?** 1,000 + 300 + 20 + 4 **When writing a number in expanded form, what are we doing to the number?** Writing out the value for each number **Multiply each of those values by the other factor, 4.** Remind students that it is very important to align the places properly before adding the partial products.

Practice 1 Have students describe how to write 2,153 in expanded form 2,000 + 100 + 50 + 3 and then how to use place value to find the product. Show how the product 4 × 5 tens is written as 200 in this method, which automatically takes care of the renaming.

Example 2 Work the problem on the board as a volunteer reads the steps aloud. Remind students to align the places. **What is 7 × 8?** 56 **How many ones and tens are there?** 6 ones and 5 tens Write 6 in the ones place and write 5 above the 2 in 28. **2 is in the tens place. What is 2 tens × 7?** 14 **What is the next step?** Add 5 tens to 14 tens. **How many tens?** 19 Point out that since the multiplication is complete, writing 19 in the product shows the regrouping, since the 1 falls in the hundreds place. **What is the answer?** 196

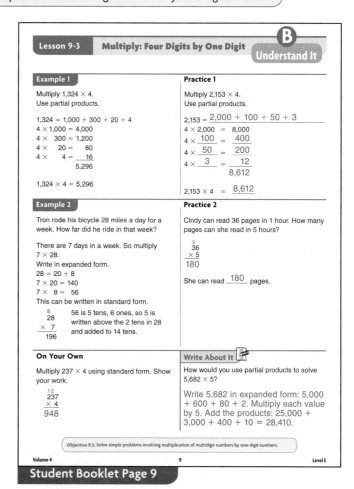

Student Booklet Page 9

Progress Monitoring

What is the product of 3,147 × 5? **15,735** Explain the regrouping. **35 ones is 3 tens and 5 ones; 23 tens is 2 hundreds and 3 tens.**

Error Analysis

If regrouping occurs two or three times in a problem, have students work with base ten blocks until they feel comfortable with the shortcut in the algorithm.

Lesson 9-3 Multiply: Four Digits by One Digit

Objective 9.3: Solve simple problems involving multiplication of multidigit numbers by one-digit numbers.

Observe Student Progress

Computer Tutorial

Some students may benefit from completing a computer tutorial before they attempt the Try It page. A list of the tutorials for each lesson can be found beginning on page x in the front of this book.

Develop Academic Language

Exercise 3 To enhance understanding of the term *rename*, have students explain the renaming used in each answer choice of this exercise.

Error Analysis

Exercise 3 Remind students that the meaning of finding partial products is to multiply the value of each digit by the one-digit factor. Students should begin by writing the two- or three-digit factor in expanded notation. Make sure students write the expanded form correctly and understand how to align the places before adding.

Exercise 4 If students continue to make errors in regrouping, suggest they find the solution using partial products. Check the expanded form to be sure they are interpreting place value correctly.

Exercise 5 If students do not see the pattern, have them solve simpler problems such as 2×1, 2×10, and 2×100 to discover the pattern.

Exercise 7 Make sure students recall that there are 31 days in July.

Exercise 8 Make sure the students write the expanded form correctly and understand how to align the places. Remind them that they can also solve the problem by writing in standard form.

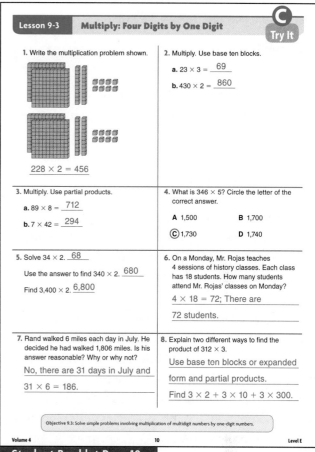

Student Booklet Page 10

Lesson 9-4 — Choose a Method and Multiply

Objective 9.4: Solve problems by choosing a multiplication method.

Teach the Lesson

Materials ☐ Number cubes

Activate Prior Knowledge

Have pairs of students practice multiplication facts by tossing number cubes or drawing numbered slips of paper to find two factors. Both players multiply; the greater product earns a point. Then have students roll multiple times and use the digits to form multidigit factors. Review the partial-products method and the standard algorithm. For this page, allow students to use whichever method they are more comfortable with.

Develop Academic Language

Ask students to solve simple math problems such as 5 × 2, 3 + 6 and 10 − 4. Explain that the phrase *mental math* refers to solving math problems without manipulatives, paper and pencil, or a calculator. Discuss some mental math strategies with emphasis on patterns and place value.

Model the Activities

Activity 1 Review how to round factors to the greatest place and estimate products. **Is 87 closer to 80 or 90?** 90 **Is 64 closer to 60 or 70?** 60 **How do you find the estimate?** Multiply 9 by 6 by 100: 5,400. **The estimate can help you check whether the exact answer you find is reasonable.** Guide students to multiple and check their work.

Activity 2 Have a volunteer read the problem aloud. Discuss why multiplication can be used to solve the problem. **The factors are not easy to multiply mentally. Use pencil and paper to find the product. How can you check whether the product is reasonable?** Estimate, and compare the exact answer to the estimate. Guide students to estimate by rounding to the greatest place before calculating the exact product and checking the answer against the estimate.

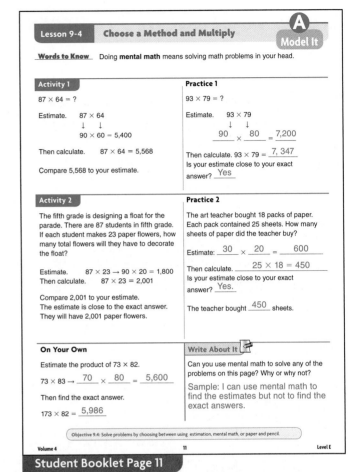

Student Booklet Page 11

Progress Monitoring

What is the product of 328 × 4? 1,312 **Estimate.** 1,200 **Is your answer reasonable?** Yes.

Error Analysis

You may need to give students various numbers to round for practice. Point out that estimation is especially useful for catching errors in place value; an estimate of 2,000 and a product of 213 indicate an error. Review methods of multiplication.

ENGLISH LEARNERS Provide visuals to aid in understanding.

Volume 4 — 152 — Level E

Lesson 9-4: Choose a Method and Multiply

B Understand It

Objective 9.4: Solve problems by choosing a multiplication method.

Facilitate Student Understanding

Develop Academic Language

Work the *partial-products method* and the *standard algorithm* for the same multiplication problem side-by-side on the board. Show how the meaning of the steps in each method is the same.

Demonstrate the Examples

Example 1 What are you asked to find? The cost of 5 tickets What are you given? The cost of 1 ticket Since one factor is 5, and you know the fives facts, and since there are only two digits to multiply by 5, you can use the partial-products method to multiply mentally. Point out that students can jot down the partial products as they find them in order not to forget any. What is the value of the 3? 30 Multiply by 5 mentally. $3 \times 5 = 15$, so $30 \times 5 = 150$. Jot down 150 as the first partial product. Look back at the problem. What is the value of the 6? 6 Multiply by 5 mentally. $6 \times 5 = 30$ Jot down 30 as the second partial product. Add partial products. $30 + 150 = 180$ What is the total cost? $180

Example 2 Do you need to find an exact answer? Yes. The problem doesn't ask for an estimate. What problem must you solve to find the exact answer? 38.95×2 Why would it be difficult to use partial products to solve this problem mentally? There are 4 partial products to keep track of, and there will be a lot of regrouping in the addition. Guide students to solve.

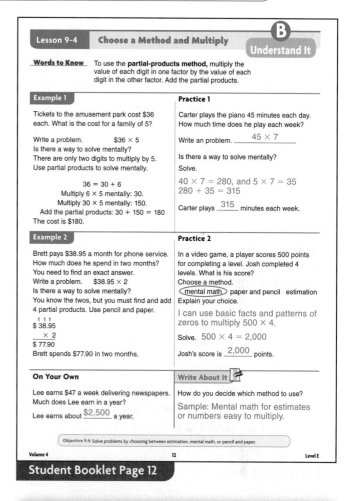

Student Booklet Page 12

⭐ Progress Monitoring

Which method would you use to find 322×4? Why? Sample: Mental math; I know the fours, and there will be no regrouping when I add.

Error Analysis

If students make basic fact errors, encourage them to use a multiplication table to check each step of a problem. Also encourage students to use estimation to check every answer.

Lesson 9-4: Choose a Method and Multiply

Objective 9.4: Solve problems by choosing a multiplication method.

Observe Student Progress

Computer Tutorial

Some students may benefit from completing a computer tutorial before they attempt the Try It page. A list of the tutorials for each lesson can be found beginning on page x in the front of this book.

Error Analysis

Exercise 1 If students estimate incorrectly, have them use a number line to help them round the factors, and then use one color to record the basic fact and another color to record the zeros.

Exercise 2 Guide students to see that the problem does not require estimation unless it is to check the exact answer. Students who see that they can multiply by 10 and then subtract 429 can multiply mentally.

Exercises 3 and 4 Be sure students tell why they chose a particular method. Accept all reasonable explanations.

Exercise 5 Choices B and D are estimates. The question asks for an exact answer. Remind students to read the exercises carefully and answer the question that is being asked.

Exercise 6 Guide students to find each answer mentally and check their answers using paper and pencil or a calculator. **Were you able to find both products mentally? Which product did you find quicker?**

Exercise 8 Students should write a problem with numbers that are easy to multiply or that asks for an estimate. If they make a problem that is too complex for mental math, ask them to try to do the problem without paper and pencil or any manipulatives.

Student Booklet Page 13

Lesson 9-4 Choose a Method and Multiply — C Try It

1. $71 \times 86 = ?$
   Estimate. 71×86
   $\underline{600} \times \underline{90} = \underline{6,300}$
   Calculate.
   Check.
   The product of 71×86 is $\underline{6,106}$.

2. The drama club sold 429 tickets for $9 each. How much did the drama club make in ticket sales?
   The drama club made $\underline{\$3,861}$.

3. Mr. Jones drove for 6 hours at 58 miles per hour. How many miles did Mr. Jones drive?
   Choose a method.
   (mental math) paper and pencil estimate
   Explain your choice.
   I can multiply 50 and 8 by 6 mentally.
   Mr. Jones drove $\underline{348}$ miles.

4. Each camper uses 36 beads to make a necklace. There are 41 campers. About how many beads are needed?
   Choose a method.
   mental math paper and pencil (estimate)
   Explain your choice.
   "About" tells me to estimate. I can round and multiply mentally.
   About $\underline{1,600}$ beads are needed.

5. A photo album has 72 pages. Each page holds 6 photos. How many photos are in the album?
   A 412   B 420
   (C) 432   D 700

6. Which problem is easier to solve mentally: 203×4 or 7×36? Explain your choice.
   Sample: 203×4; I can break it apart, multiply $200 \times 4 = 800$ and $3 \times 4 = 12$ and then add $800 + 12 = 812$.

7. Write a multiplication problem that can be solved using paper and pencil. Show the solution.
   Sample: $23 \times 8 = 184$

8. Write a multiplication problem that you can solve mentally. Show the solution.
   Sample: $250 \times 17 = 4,250$

Objective 9.4: Solve problems by choosing between using estimation, mental math, or pencil and paper.

ENGLISH LEARNERS Students might need help when explaining why they chose a method. Have students make a chart listing each method and when to use each. They can reference this chart when completing the problems.

Volume 4 — Multiply and Divide
Topic 9: Use Multiplication to Compute

Topic Summary

Objective: Review computations with multiplication.

Have students complete the student summary page.

You may want to have students work in groups of four with each student analyzing a different choice.

Ask students to share their ideas about each answer choice. Be sure they confirm the correct answer for each problem at the end of the discussion.

Answer Evaluation

1. **A** Students rounded 134 to 100 and $8.75 to $10.
   B Students did not estimate.
   C This is the correct answer.
   D Students rounded 134 to 140 and multiplied by 10.

2. **A** Students added 45 and 8.
   B Students estimated 40 × 8.
   C This is the correct answer.
   D Students estimated 45 × 10.

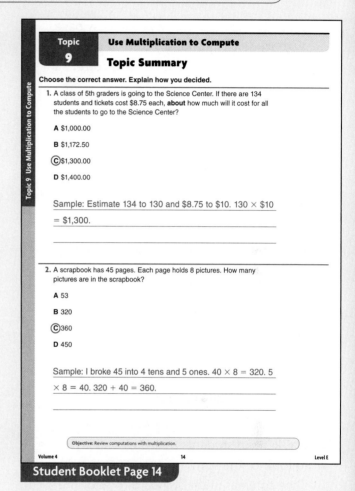

Student Booklet Page 14

 ## Error Analysis

Exercise 1 Make sure students read the problem carefully and underline the key phrase that tells them to estimate.

Exercise 2 Students may solve this problem by breaking the factors apart or by using the algorithm.

Progress Monitoring

When all assignments for this topic have been completed, assign the corresponding Progress Monitoring page for this topic (Assessment Resources Book, page 9). Be sure students complete the Progress Monitoring page before you administer the final assessment for this volume.

Volume 4 — Multiply and Divide
Topic 9 — Use Multiplication to Compute
Mixed Review

Objective: Maintain skills and concepts.

Have students complete the Mixed Review page. Work with students individually to review results. Identify strengths and weaknesses and correct any misunderstandings.

 Error Analysis

Exercise 1 Encourage students to draw a box around the basic fact and underline the zeros in each factor.

Exercise 2 Suggest students use a place-value chart. Also, make sure students understand that they are being asked for the place of the digit, not its value.

Exercise 3 Have students write out the problem and their solutions. Ask them to identify the error that Dylan probably made.

Exercise 4 It may help students to act out this problem with smaller numbers.

Exercise 5 Provide Teaching Aid 7 Place-Value Charts. Remind students that the commas in the word form of the number separate the periods, that the commas in the standard form of the number will fall in the same places, and that each comma should be followed by 3 digits.

Exercise 6 Help students rewrite each division problem as a missing-factor multiplication problem. Making fact-family triangles with the total at the peak and the factors at the lower corners may also be helpful.

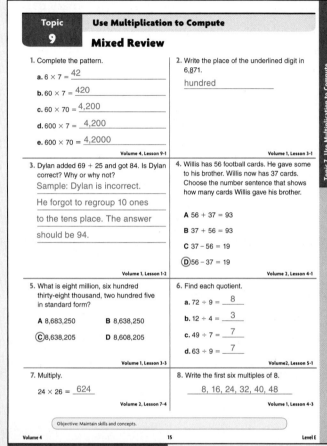

Student Booklet Page 15

Volume 4 — Multiply and Divide
Topic 10 — Use Division to Compute

Topic Introduction

Lesson 10-1 extends students' knowledge of basic division facts into dividing multiples of 10, 100, and 1,000. Lesson 10-2 gives students practice with estimating quotients and using mental math techniques. Lesson 10-3 expands upon students' previous work with division into dividing multidigit numbers. Lessons 10-4 and 10-5 give students opportunities to practice the various division techniques they've learned and to solve more problems with multiplication and division.

Lesson	Objective	Student Pages	Teacher Pages	Tutorials
Topic 10 Introduction	10.1 Solve problems dividing multiples of 10, 100, and 1,000.	16	158	
	10.2 Estimate quotients by rounding numbers and using mental math techniques.			
	10.3 Solve simple problems involving division of multidigit numbers by one-digit numbers with and without remainders.			
	10.5 Solve problems by multiplying or dividing.			
10-1 Divide Multiples of 10	10.1 Solve problems dividing multiples of 10, 100, and 1,000.	17–19	159–161	10a
10-2 Estimate Quotients	10.2 Estimate quotients by rounding numbers and using mental math techniques.	20–22	162–164	10b
10-3 Divide by 1-Digit Numbers	10.3 Solve simple problems involving division of multidigit numbers by 1-digit numbers with and without remainders.	23–25	165–167	10c, 10d, 10e, 10f, 10g
10-4 Choose a Method for Division	10.4 Solve division problems by choosing between using estimation, mental math, or pencil and paper to find the quotients.	26–28	168–170	10h
10-5 Multiplication and Division Problems	10.5 Solve problems by multiplying or dividing.	29–31	171–173	10i, 10d, 10j
Topic 10 Summary	Review computation using division.	32	174	
Topic 10 Mixed Review	Maintain concepts and skills.	33	175	

Computer Tutorial

Some students may benefit from completing the computer tutorial before they attempt the Try It page of each lesson. If you are using the electronic components of *Pinpoint Math,* you will find a complete listing of the Tutorial codes and titles when you access them either online or via CD-ROM.

Volume 4: Multiply and Divide
Topic 10: Use Division to Compute
Topic Introduction

Objectives: 10.1 Solve problems involving dividing multiples of 10, 100, and 1,000. **10.2** Estimate quotients by rounding factors and using mental math techniques. **10.3** Solve simple problems involving division of multidigit numbers by 1-digit numbers with and without remainders. **10.5** Solve problems by multiplying or dividing.

Materials ☐ Counters

Distribute counters to each pair of students. Write basic division facts on the board and have students model using the counters. Discuss the concept that in division you take a whole and separate it into equal groups.

Informal Assessment

1. **What basic fact do you see?** 28 ÷ 4 **What is the quotient of 28 divided by 4?** 7 **How will that fact help you with the pattern?** Each expression has the same division fact, so the quotient will be 7 and any zeros that could not be crossed out.

2. **What are compatible numbers?** Numbers that are easy to compute **Why did you choose 35 divided by 7 instead of 42 divided by 7?** 35 is closer to 37 than 42, which means my estimate will be closer to the exact answer.

3. **How many base ten blocks do you start with?** 112 **How do you know how many groups to make?** Use the divisor, which is 8. **Did you have to trade any tens?** Yes. **How many tens did you trade?** 3 **How many ones did you then have?** 32 **Is 32 a multiple of 8?** Yes. **Did you have any ones left over?** No.

4. **How did you know that you needed to divide?** The problem told the total number of jump ropes and how many could be packaged in a box, so I knew I needed to divide. **What is 486 divided by 9?** 54 **If you estimated to see if your answer was reasonable, what would you get?** 50 **Is 50 close to 54?** Yes.

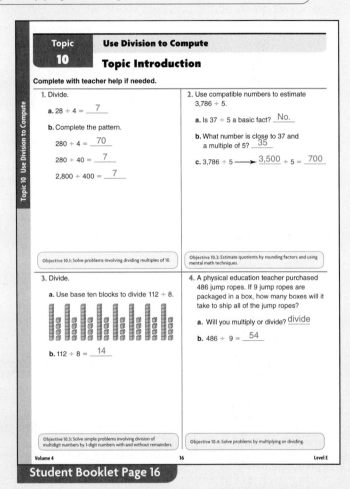

Student Booklet Page 16

Another Way Instead of base ten blocks, students could also draw arrays and separate the rows or columns into equal groups based on the divisor.

Lesson 10-1 | Divide Multiples of 10

Objective 10.1: Solve division problems with multiples of 10, 100, and 1,000.

Teach the Lesson

Materials ☐ Numeral Tiles
☐ Index cards

Activate Prior Knowledge

Have students solve and find patterns in 1 ÷ 1, 10 ÷ 10, 100 ÷ 100, 10 ÷ 1, 100 ÷ 1, and 100 ÷ 10.

Develop Academic Language

Review *dividend*, *divisor*, and *quotient*.

Model the Activities

Activity 1 Have students use numeral tiles to create the equation 6 ÷ 3 = 2 on the index card. On the index card, students should have ___ ÷ ___ = ___. **What is the quotient of 6 ÷ 3?** 2 **We can use this basic fact to find greater quotients.** Have students use a zero tile to turn the dividend 6 into 60.

We've multiplied the dividend by 10, and we're still dividing by the same divisor. What happens to the quotient? Multiply it by 10. Repeat for the other problems.

Activity 2 Have students model 60 ÷ 30. **We multiplied the dividend by 10. But then we multiplied the divisor by 10. The increases cancel each other.** Have students cross off one zero in 60 and the zero in 30 on their papers. **What happens to the quotient?** It stays the same. **Look at 600 ÷ 30. We multiplied the dividend by 10 and then by 10 again. We multiplied the divisor by 10. Which had a greater increase?** The dividend **How much greater?** 10 times **It is as if we only multiplied the dividend by ten. The other tens cancel each other. What happens to the quotient?** Multiply it by 10. Repeat for the other problems. Students may notice that the quotient has the number of zeros in the dividend minus the number of zeros in the divisor.

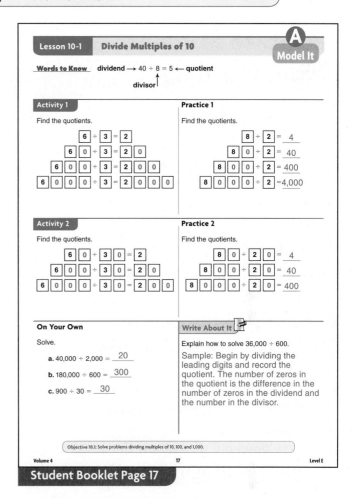

Student Booklet Page 17

 Progress Monitoring

What basic fact can you use to solve 120,000 ÷ 30? 12 ÷ 3 = 4 **How many zeros will be in the quotient?** 3 **What is the quotient?** 4,000

Error Analysis

Have students circle leading digits to find the basic fact at the root of each problem.

Volume 4   159   Level E

Lesson 10-1: Divide Multiples of 10

Objective 10.1: Solve division problems with multiples of 10, 100, and 1,000.

Facilitate Student Understanding

Develop Academic Language

Write multiplication equations with a factor of 10. Circle the products. **These are multiples of 10. A multiple of 10 is the product of 10 and a whole number.** Repeat with 100's and 1000's. Ask students to identify multiples of 10 on the page.

ENGLISH LEARNERS Distinguish between *division* and *divisor*. To divide is to perform the action 12 ÷ 4 = 3. In this division equation, 4 is the divisor.

Demonstrate the Examples

Example 1 Have students underline the basic fact, 42 ÷ 7, in each problem. Help students identify a related multiplication fact if necessary. **What is the quotient of 42 ÷ 7?** 6 Look at 4,200 ÷ 7. **How many times did we multiply the dividend by 10?** Twice **How many times did we multiply the divisor by 10?** none **What happens to the quotient?** Multiply by 10 and by 10 again, or by 100.

Example 2 Is 4 ÷ 8 a basic fact? **No.** We can use another digit in the dividend. Can we divide 40 by 8? **Yes.** What is the basic fact? 40 ÷ 8 = 5 Have students circle the basic fact in each problem. **How many times did we multiply 40 by 10 to get 4,000?** Twice

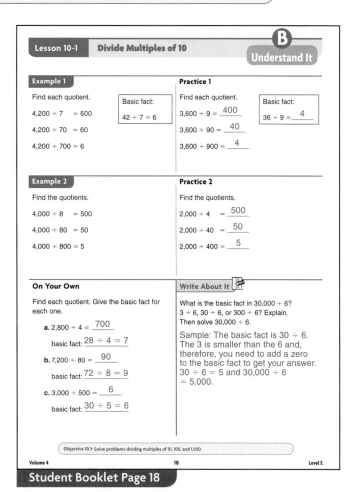

Student Booklet Page 18

 Progress Monitoring

What basic fact would you use to solve 200 ÷ 50? 20 ÷ 5 = 4 How many zeros will be in the quotient? None

Error Analysis

If quotients have an incorrect number of zeros, have students circle the basic fact, then count zeros.

Lesson 10-1 — Divide Multiples of 10

Objective 10.1: Solve division problems with multiples of 10, 100, and 1,000.

Observe Student Progress

Computer Tutorial

Some students may benefit from completing a computer tutorial before they attempt the Try It page. A list of the tutorials for each lesson can be found beginning on page x in the front of this book.

Error Analysis

Exercises 1–8 If students make basic fact errors, have them use flashcards to review multiplication facts through 10.

Exercise 1 Suggest that students use base ten blocks to model each expression and check their answers. Have them model 4 ones and divide them into 2 groups; do the same for 4 tens, 4 hundreds, and 4 thousands.

Exercise 3 Have students circle the basic fact so they exclude the zero of the basic fact when determining the number of zeros in the final quotient.

Exercise 6 Make sure students understand they are choosing the answer that is NOT equal to 5. Students may find it helpful to underline or circle the basic fact in each expression.

Exercise 7 Encourage students to use numeral tiles to model each expression. Remind students that they can pair zeros in the dividend and divisor to cancel them.

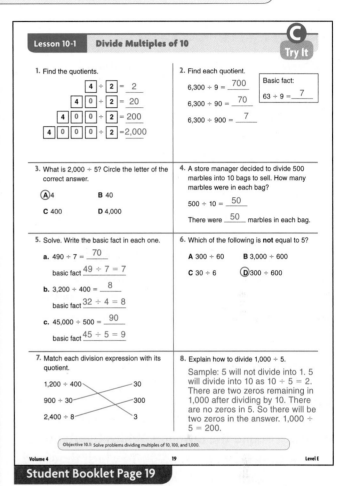

Student Booklet Page 19

ENGLISH LEARNERS The phrase *500 into 10* can also be interpreted as 10 ÷ 500. Remind students to model and understand the problem rather than focus on key words.

Lesson 10-2 Estimate Quotients

Objective 10.2: Estimate quotients by rounding numbers and using mental math techniques.

Teach the Lesson

Materials ☐ Teaching Aid 1 Number Lines

Activate Prior Knowledge

Write 4,500 ÷ 50 on the board. Underline 45 in 4,500 and 5 in 50. **What is the basic fact?** 45 ÷ 5 = 9 **How many zeros are in the final quotient?** One **How do you know?** Sample: 100 times the dividend divided by 10 times the divisor leaves 10 times the quotient.

Develop Academic Language

Write 247 → 200 and 4,821 → 5,000 on the board. *Rounded* numbers are close to the exact number. **To which place is each number rounded?** Hundreds, thousands

Model the Activities

Activity 1 Distribute Teaching Aid 1 Number Lines. **To estimate, we use numbers that are easy to divide mentally. Round the dividend. Which thousands is 7,615 between?** 7,000 and 8,000 **To which thousand is 7,615 closest?** 8,000 **Now, what basic fact can we use?** 8 ÷ 4 = 2 **How many zeros will be in the final quotient?** 3 Use the word *about* to show the quotient is an estimate.

Activity 2 Point out that both the divisor and dividend can be rounded. **To which thousand is 2,040 closest?** 2,000 **To which ten is 43 closest?** 40 **How do you find the quotient of 2,000 ÷ 40?** Use the basic fact 20 ÷ 4 = 5 and end with a 0. Point to the "≈" symbol. **This symbol means** *approximately equal to*.

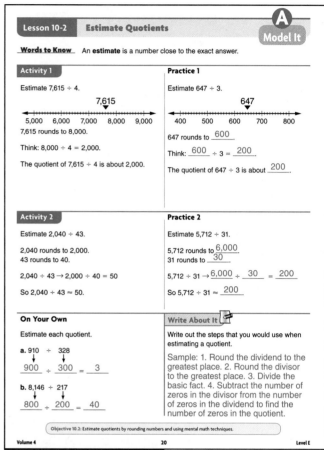

Student Booklet Page 20

Progress Monitoring

How would you round to estimate the quotient of 1,987 ÷ 11? Since each number has two or more digits, round each to the greatest place value. 2,000 ÷ 10 = 20, so 1,987 ÷ 11 ≈ 20.

Error Analysis

Help students use basic division facts with multiples of 10 to derive rules for division with zeros.

ENGLISH LEARNERS Help students brainstorm estimation words such as *equal, about, close to,* and *approximate*.

Lesson 10-2: Estimate Quotients

Objective 10.2: Estimate quotients by rounding numbers and using mental math techniques.

Facilitate Student Understanding

Develop Academic Language

On the board, show:

5,512 ÷ 28
round ↓   ↓ round
6,000 ÷ 30 ≈ ◯ estimate

Calculating with numbers that are close to the actual numbers is *estimating*. The answer is called an *estimate*. What is the estimate of 5,512 ÷ 28? **200**

ENGLISH LEARNERS Talk about the term *compatible*. Explain that when two things are compatible, they go well together.

Demonstrate the Examples

Example 1 If we use rounding rules, to what number does 4,415 round? **4,000** Can we use a basic fact to divide 4,000 by 7 mentally? **No.** We need to round the dividend to a number that is a multiple of 7. What basic fact with 7 is closest to 44 ÷ 7? **42 ÷ 7** 4,200 and 7 are compatible numbers. They are close to the original numbers and easy to divide mentally. What is 4,200 ÷ 7? **600**

Example 2 Which is the divisor? **612** Which hundred is closest to 612? **600** Should we round the dividend to the greatest place? **No.** Seeing this step done may help students test whether 5,200 ÷ 600 involves a basic fact. Why not? **600 does not divide evenly into 5,000; 52 ÷ 6 is not a basic fact.** Look at the first two digits in the dividend. What number is close to 52 and a multiple of 6? **54** Point to the final quotient. Why are there no zeros in the final quotient? **The dividend and divisor have the same number of zeros.**

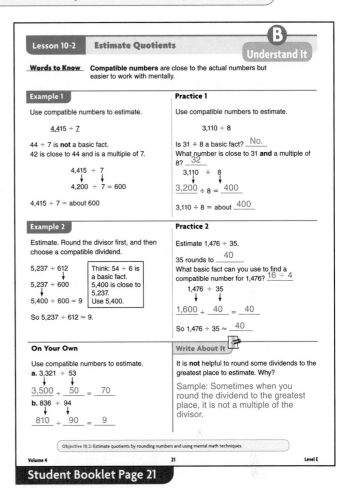

Student Booklet Page 21

Progress Monitoring

Would you use 200 ÷ 8 to estimate 236 ÷ 8? Why or why not? **No; 8 does not evenly divide into 200. I would round 238 to 240 because 240 is a multiple of 8.**

Error Analysis

Review basic facts to help students choose compatible numbers.

Lesson 10-2 — Estimate Quotients

Objective 10.2: Estimate quotients by rounding numbers and using mental math techniques.

Observe Student Progress

Computer Tutorial

Some students may benefit from completing a computer tutorial before they attempt the Try It page. A list of the tutorials for each lesson can be found beginning on page x in the front of this book.

 Error Analysis

Exercise 1 Point out that the single-digit number doesn't have to be rounded, since division by 5 can be done on the basis of basic facts. The less rounding is done, the closer the estimate will be to the exact answer.

Exercise 2 To help students find compatible numbers, have them first make a list of all the basic facts for 9. Then they can use the list to find a multiple of 9 that is close to the first two digits of the dividend. Another solution is to divide 3,400 by 10, yielding an estimate of 340.

Exercise 3 Explain that the closer the chosen number is to the original dividend, the closer the estimate will be to the exact answer.

Exercises 4–6 Errors may occur in rounding the divisor and dividend or in dividing multiples of 10. Provide students with a multiplication table to will help them find compatible numbers and to divide basic facts.

Exercise 7 You may want to provide a few choices. Have students check each other's work.

Student Booklet Page 22

Lesson 10-2 — Estimate Quotients — Try It

1. Round to estimate $429 \div 5$.
   429 rounds to __400__
   Think: __400__ $\div 5 =$ __80__
   $429 \div 5 =$ about __80__

2. Use compatible numbers to estimate.
   $3,408 \div 9$
   What basic fact can you use to round the dividend? __$36 \div 9 = 4$__
   $3,408 \div 9$
   $3,600 \div 9 =$ __400__
   $3,408 \div 9 \approx$ __400__

3. For the **best** estimate of $4,125 \div 60$, which number would you choose as a compatible number for the dividend? Circle the letter of the correct answer.
   A 6,000   B 4,000
   (C) 4,200   D 4,800

4. Round to estimate $7,833 \div 421$.
   7,833 rounds to __8,000__
   421 rounds to __400__
   So, $7,833 \div 421 \approx$ __20__

5. Use compatible numbers to estimate.
   $5,800 \div 655$
   $5,600 \div 700 =$ __80__

6. The tennis coach collected 174 tennis balls. If 3 balls fit in one container, about how many containers will the coach need for all the balls?
   __174__ $\div$ __3__ $=$ __60__

7. Write a division expression with an estimated quotient of 50.
   Sample: $247 \div 5 \rightarrow 250 \div 5 = 50$

8. Marcie and Collin both use compatible numbers to estimate the quotient of $3,020 \div 741$. Marcie rounds the problem to $3,500 \div 700$. Collin rounds the problem to $2,800 \div 700$. Whose estimate will be closest to the exact answer? How do you know?
   Sample: Collin's will be closer. 2,800 is closer to 3,020 than 3,500 is, so it is better to choose 2,800 for this problem.

Exercise 8 Explain that students do not need to find the exact answer to solve the problem. Point out that both rounded the divisor to 700 but each rounded the dividend to a different multiple of 700. They need to decide whose rounded dividend is closer to the original dividend. **What is the difference of 3,500 and 3,020?** 480 **What is the difference of 3,020 and 2,800?** 220 **Whose rounded dividend is closer to the original dividend?** Collin's

Lesson 10-3: Divide by 1-Digit Numbers

Objective 10.3: Solve simple problems involving division of multidigit numbers by 1-digit numbers with and without remainders.

Teach the Lesson

Materials ☐ Base ten blocks

Activate Prior Knowledge

Have students model 6 ÷ 2 by separating 6 ones into 2 groups. **How many blocks did you put in each group? 3 What is 6 ÷ 2? 3** Repeat with other basic facts.

Develop Academic Language

Discuss the place and value of the digits in 259. **In what place is the 2? Hundreds** The value of the 2 is 200. 5 is in the tens place. **What is the value of the 5? 50** 9 is in the ones place. The 9 has a value of 9.

Model the Activities

Activity 1 Have students represent 693 with 6 hundreds 9 tens 3 ones. Division separates into equal groups. **How many equal groups should we make? 3 Of 6 hundreds, how many will go into each group? 2 Of 9 tens, how many will go into each group? 3 Of 3 ones, how many will go into each group? 1 What number is shown in each group? 231** The quotient of 693 ÷ 3 is 231.

Activity 2 Have students represent 87 with 8 tens 7 ones. **Can 8 tens be divided equally into 4 groups? Yes. How many will go into each? 2 Can 7 ones be divided equally into 4 groups? No.** Put 1 one in each group. This is the most that will distribute equally. **How many ones are left over? 3** 3 ones is the remainder. Use an R to record the remainder. Write 87 ÷ 4 = 21 R3 on the board.

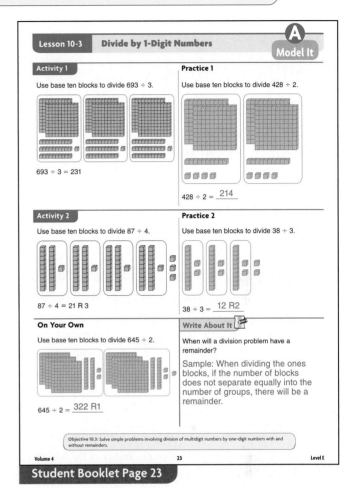

Student Booklet Page 23

Progress Monitoring

How can you check that division with base ten blocks has been done correctly? **Samples: The number of groups is the same as the divisor. The sum of all blocks is the same as the dividend. Each group is identical.**

Error Analysis

Point out to students that the remainder is never greater than the divisor.

ENGLISH LEARNERS Discuss the terms *equal* and *equally*. Show that dividing something equally means to make groups that are the same.

Lesson 10-3: Divide by 1-Digit Numbers

Objective 10.3: Solve simple problems involving division of multidigit numbers by 1-digit numbers with and without remainders.

Facilitate Student Understanding

Develop Academic Language

What are some words that could indicate that only an estimate is needed? **Estimate, approximation, about**

Demonstrate the Examples

Example 1 Set up the division problem on the board. Check to make sure students place the divisor and dividend in the appropriate places. **Will 7 divide into 9 tens? Yes. How many times? 1** Write 1 above the 9 in 96. Perform each division step on the board as it is discussed. **Multiply the 1 you just placed on top of the 9 by 7. What is 1 × 7? 7** Write the 7 below 9. Subtract. **What is 9 − 7? 2** Compare the numbers. The difference must be less than the divisor. Bring down the 6 ones. Divide the 26 ones. **What is 26 ÷ 7? 3** Place the 3 next to the 1. Multiply 3 by 7. **What is it? 21** Subtract 26 − 21. **What is it? 5** Can 7 divide into 5? **No.** 5 is the remainder. Write this number next to the quotient, on the top. **What is 96 ÷ 7? 13 R5**

Example 2 **What number does 12 divide into? 45** How many times? **3** What is 3 × 12? **36** Complete the division on the board as each step is discussed. **What is 45 − 36? 9** What is next? Bring down the 6. **How many times does 12 divide into 96? 8** What is 8 × 12? **96** What is 96 − 96? **0** What is 456 ÷ 12? **38**

Student Booklet Page 24

Lesson 10-3 Divide by 1-Digit Numbers Understand It

Example 1
Use paper and pencil to find 175 ÷ 7. Check your answer.

175 ÷ 7 = 25

```
    13 R5
  7)96
    7
    26
    21
    5
```

Practice 1
Use paper and pencil to find 276 ÷ 4. Check your answer.

276 ÷ 4 = __69__

```
    68 R3
  4)275
    24
    35
    32
    3
```

Example 2
Use paper and pencil to find 456 ÷ 12. Check your answer.

456 ÷ 12 = 38

```
      38
  12)456
     36
     96
     96
     0
```

Practice 2
Use paper and pencil to find 627 ÷ 11. Check your answer.

627 ÷ 11 = __57__

```
      57
  11)627
     55
     77
     77
     0
```

On Your Own
Use paper and pencil to find 3,740 ÷ 5. Check your answer.

3,740 ÷ 5 = __748__

```
      748
   5)3740
     35
     24
     20
     40
     40
      0
```

Write About It
How can an estimate using mental math help you when using pencil and paper to divide?

Sample: An estimate helps in providing you with an answer that is close to the exact one. If the estimation is way off from the pencil and paper answer, your division might be wrong on paper.

Objective 10.3: Solve simple problems involving division of multidigit numbers by one-digit numbers with and without remainders.

Progress Monitoring

Write 5,733 ÷ 91 on the board. **When using long division, which number goes on the inside of the division sign? 5,733 Does pencil and paper give an exact answer or an estimate? Exact answer**

Error Analysis

At each step in long division, the difference should be less than the divisor.

ENGLISH LEARNERS Students should be familiar with the terms dividend, divisor, and quotient. Have the students label the parts of a division expression.

Lesson 10-3: Divide by 1-Digit Numbers

Objective 10.3: Solve simple problems involving division of multidigit numbers by 1-digit numbers with and without remainders.

Observe Student Progress

Computer Tutorial

Some students may benefit from completing the computer tutorial before they attempt the Try It page of each lesson. A list of the tutorials for each lesson can be found beginning on page x in the front of this book.

Error Analysis

Exercise 1 Students may put the 6 underneath the 4 in 84 instead of under the 8. Remind students that the 2 over top of the 8 in the quotient represents 2 tens, not 2 ones.

Exercise 2 Students might have difficulty setting up this problem or in finding the solution. First, walk students through the process of correctly setting up the long division. Ask them to identify the divisor and the dividend. Next, guide them through each step until they have reached a quotient. It might be useful to have students come up to the board and fill out each part of the steps in solving the expression.

Exercise 4 Students should model each answer choice with base ten blocks. Remind them that the remainder cannot be greater than the divisor.

Exercise 5 Encourage students to use base ten blocks to model 37 ÷ 6. This should help them to visualize the problem.

Exercise 6 Have students work the long division of 746 ÷ 2. Encourage them to compare their work to the choices before completion. They may be able to choose the correct answer before completing the entire division problem.

ENGLISH LEARNERS Discuss *divide* and *divisible*. Division is an action of separating into groups; *divisible* means that there is no remainder from a division. For example, 16 can be divided by 3, 16 ÷ 3 = 5 R1. But 16 is *not* divisible by 3. Have students practice using the terms to describe problems on the page.

Lesson 10-4: Choose a Method for Division

Objective 10.4: Solve division problems by choosing between using estimation, mental math, or pencil and paper to find the quotients.

Teach the Lesson

Activate Prior Knowledge

Write 3,825 on the board. **Round 3,825 using front-end estimation.** 3,000 **Round 3,825 to the nearest hundred.** 3,800 **Round 3,825 to the nearest ten.** 3,830

Develop Academic Language

If we don't have access to paper and pencil, how can we solve problems? Estimate. What can we do to make numbers easier to work with? Round.

Model the Activities

Activity 1 What is 714 rounded to the nearest hundred? 700 Does 3 divide evenly into 700? No. 3 × 1 is 3, 3 × 2 is 6, and 3 × 3 is 9. We can use these facts to help us. What is 3 × 200? 600 This is too small. What is 3 × 300? 900 This is too big. The answer must be in the middle. What is a number in the middle of 200 and 300? 250

Activity 2 1,312 is about how many hundreds? 13 32 rounded to the nearest ten is what? 30 About how many 30s are in 100? 3 So 3 groups of 30 times 13 hundreds is how much? 39 Therefore, 1,312 ÷ 32 is about how much? 39

Student Booklet Page 26

Lesson 10-4 Choose a Method for Division — Model It

Activity 1
Estimate 714 ÷ 3.
Think: 714 rounded to the nearest hundred is 700.
Think: 3 × 200 = 600 and 3 × 300 = 900, so the answer is between 200 and 300.
714 ÷ 3 ≈ 250

Practice 1
Estimate 2,358 ÷ 6.
2,358 rounded to the nearest hundred is 2,400.
6 × 4 = 24, so 6 × 400 = 2,400
2,358 ÷ 6 ≈ 400

Activity 2
Estimate 1,312 ÷ 32.
Think: 1,312 rounded to the nearest hundred is 1,300, and 32 rounded to the nearest ten is 30.
Think: There are about 3 groups of 30 in one hundred. How many groups of 30 are in 13 hundreds?
3 groups of 30 × 13 hundreds = 39
1,312 ÷ 32 ≈ 39

Practice 2
Estimate 2,698 ÷ 21.
2,698 rounded to the nearest hundred is 2,700, and 21 rounded to the nearest ten is 20.
There are 5 groups of 20 in 1 hundred. How many groups of 20 are in 27 hundreds?
5 groups of 20 × 27 hundreds = 135
2,698 ÷ 22 ≈ 135

On Your Own
Use mental math to find 4,800 ÷ 60.
4,800 ÷ 60 = 80

Write About It
Does mental math always provide just an estimate or can it give an exact answer? Explain.
Sample: Mental math can provide an exact answer. This could happen if the division required no regrouping (example: 639 ÷ 3 = 213).

Progress Monitoring

Write 5,733 ÷ 91 on the board with 60 and 63 below it. **These are answers to 5,733 ÷ 91. Which is the exact answer?** 63 The other answer, 60, is an estimate. **Which is found using mental math?** The estimate

Error Analysis

Estimates should be reasonably close to the exact answer. Usually, the exact answer can be rounded to the estimate.

Lesson 10-4: Choose a Method for Division

Objective 10.4: Solve division problems by choosing between using estimation, mental math, or pencil and paper to find the quotients.

Facilitate Student Understanding

Develop Academic Language

What are some key words that would indicate an estimate is needed? **Estimate, approximation, about**

Demonstrate the Examples

Example 1 Which operation should be used to solve this problem? **Division** Should we divide 175 by 7, or 7 by 175? **Divide 175 by 7.** How do you know? **We're trying to find out how many groups of 7 are in 175.** Set up the division problem on your paper. How many sevens are in 17? **2** What is 2 × 7? **14** Subtract. What is 17 − 14? **3** Bring down the 5 next to the 3. How many 7s are in 35? **5** Subtract this from 35. What is 35 − 35? **0**

Explain to students that once there's a zero at the bottom, or a number less than the divisor, the problem is complete. If there's a remainder, put an R next to the quotient and next to the R, put the digit that remains at the bottom. So 175 ÷ 7 is how much? **25** Is this an estimate or exact answer? **Exact answer**

Example 2 Are we trying to find an exact answer or an estimate? **Estimate** How do you know? **The problem says *about*.** What is 523 rounded to the nearest ten? **520** What is 12 rounded to the nearest ten? **10** How many tens are in 520? **52; 520 ÷ 10 = 52** So about how many pages did Marcia read in one hour? **About 52 pages**

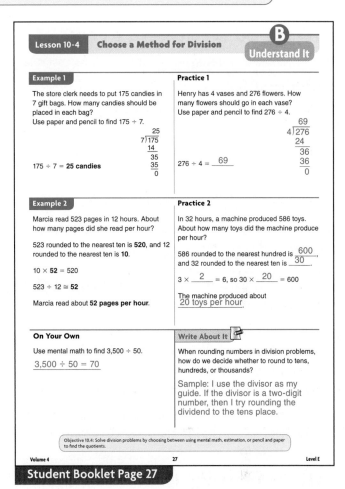

Student Booklet Page 27

Progress Monitoring

Solve this problem by estimation 5,633 ÷ 71.
5,600 ÷ 70 = 80

Error Analysis

At each step in long division, the difference should be less than the divisor.

ENGLISH LEARNERS Be sure that students understand that the word *estimate* is not an exact answer, but an educated guess.

Lesson 10-4: Choose a Method for Division

Objective 10.4: Solve division problems by choosing between using estimation, mental math, or pencil and paper to find the quotients.

Observe Student Progress

Computer Tutorial

Some students may benefit from completing the computer tutorial before they attempt the Try It page of each lesson. A list of the tutorials for each lesson can be found beginning on page x in the front of this book.

Error Analysis

Exercise 1 If students write 6 or 600 instead of 60, ask them to use multiplication to check their estimate.

Exercise 2 All of the answer choices will provide good estimates. Be sure to encourage students to find the *closest* estimate.

Exercise 4 If students have great difficulty with division, encourage them to try repeated subtraction.

Exercise 5 Encourage students to show work to prove their answers.

Exercise 7 If students provide an estimated answer, ask them to read the problem again.

Exercise 8 Students who choose to estimate do not realize that the problem has to have an exact answer. The pencil and paper method should be used. Help students to set up the division problems on their paper. Walk them through part a if they are having trouble, and have them solve Part b on their own.

Student Booklet Page 28

Lesson 10-4 — Choose a Method for Division — Try It

1. Estimate 2,510 ÷ 39.
   2,510 ÷ 39 ≈ __60__

2. Which calculation will provide an estimation that is closest to the exact answer?
   A 3,000 ÷ 50 = 60   **B 3,000 ÷ 60 = 50**
   C 2,800 ÷ 70 = 40   D 2,400 ÷ 60 = 40

3. After working for 90 hours, Harrison earned $4,500. How much did he earn per hour?
   __$50__

4. Toni opened 13 bags of snacks and counted 166 total snacks. About how many were in each bag?
   Sample: 17 snacks

5. Which is greater: 137 ÷ 8, or 232 ÷ 12?
   __232 ÷ 12__
   How did you decide?
   Sample: I estimated both quotients by rounding.

6. A series of 30 books has a total of 3,250 pages. If each book is the same length, about how many pages are in each book?
   Sample: 110 pages

7. Find an exact answer for 2,001 ÷ 23.
   2,001 ÷ 23 = __87__
   Use estimation to show that your answer is reasonable.
   Sample: 90 × 20 = 1,800

8. Tameca and Selena want to build a tree fort. They need 153 feet of lumber. If the lumber comes in 8-foot lengths and 9-foot lengths, how many pieces of lumber will they need if they use:
   a. 8-foot lengths of lumber? __19 R1__
   b. 9-foot lengths of lumber? __17__

ENGLISH LEARNERS Distinguish between estimating and rounding numbers. To estimate an answer, students use mental math and rounding skills, along with division and multiplication basic facts.

Lesson 10-5 Multiplication and Division

Objective 10.5: Solve problems by multiplying or dividing.

Teach the Lesson

Activate Prior Knowledge

Write 58 × 31 on the board. **Estimate the product.** 1,800 **Use paper and pencil. What is the exact product?** 1,798

Develop Academic Language

Write 3 × 7 = 21 on the board. **Explain what this problem represents.** There are 3 groups with 7 items in each group, with a total number of 21 items. **What is the answer to a multiplication problem called?** Product

Model the Activities

Activity 1 Read the problem aloud. **How many eggs are in each omelet?** 3 **How many omelets are made each day?** 78 **What operation is needed to find the number of eggs in 78 groups of 3 eggs?** Multiplication **To estimate 3 × 78, what numbers are mentally multiplied?** 3 × 80 **What is 3 × 80?** 240 The cook needs about 240 eggs.

Activity 2 Is an estimate or an exact answer needed? Exact **What operation will be used?** Multiplication **Why?** Sample: The total in 3 groups of 12 is needed. **What is 12 × 3?** 36 There are 36 movie channels in the three packages.

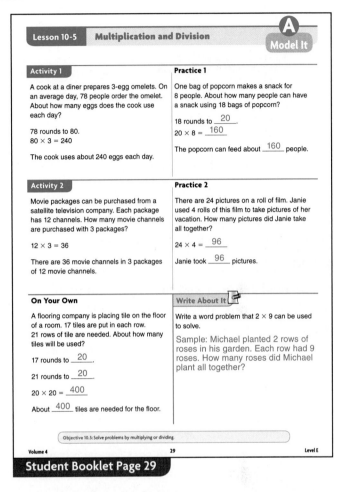

Student Booklet Page 29

ENGLISH LEARNERS Carefully define *each*. *Each* means one of two or more objects that are similar. In this context, *each* implies identical groups and usually indicates multiplication or division.

 Progress Monitoring

How can you decide if an exact answer or an estimate is needed? Sample: If an estimate is needed, the problem will use words like *estimate*, *approximate*, or *about*.

Error Analysis

To determine if an answer is correct, the student should check that (1) the operation was done correctly and (2) the calculation answers the question which was asked. The latter can be accomplished by re-reading the problem after the answer is found.

Lesson 10-5: Multiplication and Division

Objective 10.5: Solve problems by multiplying or dividing.

Facilitate Student Understanding

Develop Academic Language

Teach students to recognize verbal cues in word problems. For example, statements such as "What is the cost of each?" indicate division. Statements with the phrase "How much all together?" indicate multiplication.

Demonstrate the Examples

Example 1 We know the number of pages that can be read per hour and the total number of pages in the book. What are we trying to find? The number of hours needed to finish the book What operation is needed? Division Estimate 112 ÷ 8. 10 Use paper and pencil to find 112 ÷ 8. What is the result? 14 Check with a calculator.

Example 2 What operation will be used? Division Why? Because we are already given the total, and we only need to find the cost for each Is an estimate or exact answer needed? Estimate What is the mental division problem? 200 ÷ 20 Each calculator costs about $10. You can also use 240 ÷ 20 for an estimated cost of $12.

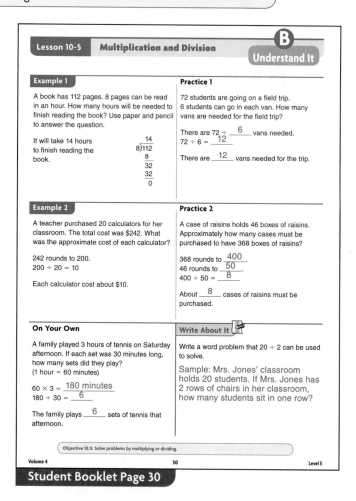

Student Booklet Page 30

Progress Monitoring

Think of a multiplication problem as

number of groups × number of items in a group = total items

Write this on the board. **Using these terms, how can you decide whether to multiply or divide when solving a word problem?** If the number of groups and the number of items in the group are both given, use multiplication to find the total items. If the total items are given, use division.

Error Analysis

Encourage students to estimate first when finding an exact answer, even if using the calculator. An incorrect key can easily be pushed on a calculator and give the wrong answer. Having first estimated the result will help students catch this mistake.

ENGLISH LEARNERS Carefully define *per*. Per means "for each." It generally means that multiplication or division is involved.

Lesson 10-5: Multiplication and Division

Objective 10.5: Solve problems by multiplying or dividing.

Observe Student Progress

Computer Tutorial

Some students may benefit from completing a computer tutorial before they attempt the Try It page. A list of the tutorials for each lesson can be found beginning on page x in the front of this book.

 Error Analysis

Exercise 1 Encourage students to label the word problem to make it easier to set up the division problem. The total earned is $135. The number of groups, or hours, is 9. The unknown is the number of items, or dollars, in one group. This indicates that division will be used.

Exercise 3 Because multiplication involves repeated addition, students may choose choice B. The problem asks for the number of eggs in 4 dozen. If the problem was an addition problem, the 12 would need to be added to itself 4 times (12 + 12 + 12 + 12), and not only to 4, as in choice B. The only choice that represents 12 added to itself 4 times is answer A, 12 × 4.

Exercise 4 Students may have difficulty with an abstract problem such as this. Encourage them to use any number to represent the various parts of the problem, to make it easier to solve. However, since the problem isn't asking for a number, make sure the students correctly identify what is being asked for in the exercise.

Student Booklet Page 31

Exercise 5 Encourage students to use labeling here, as it will help them identify what is being asked for in this problem. Those who choose answer A have not labeled the exercise correctly.

Exercise 7 A solution is another way to describe an answer. Make sure students understand that they should estimate the solution by rounding the factors, then multiplying.

Volume 4 — Multiply and Divide
Topic 10 — Use Division to Compute

Topic Summary

Objective: Review computation using division.

Have students complete the student summary page. You may want to have students work in groups of four with each student analyzing a different choice.

Ask students to share their ideas about each answer choice. Be sure they confirm the correct answer for each problem at the end of the discussion.

Answer Evaluation

1. **A** This is the correct answer.
   B Students counted the zero in 30 twice.
   C Students did not cancel any zeros.
   D Students added the number of zeros from both numbers, as they would in multiplication.

2. **A** Students estimated instead of dividing.
   B This is the correct answer.
   C Students added instead of dividing.
   D Students multiplied instead of dividing.

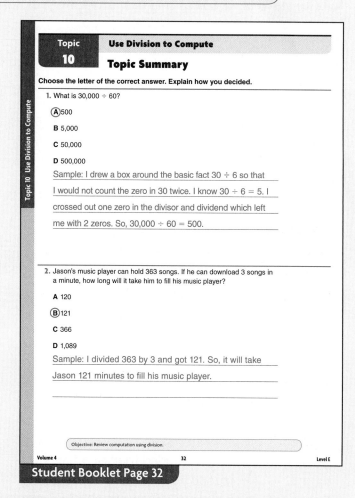

Student Booklet Page 32

 ## Error Analysis

Exercise 1 Make sure students draw a box around the basic fact of 30 and 6 so that they do not count the zero in 30 twice.

Exercise 2 For students who choose C or D, ask whether the answer should be larger than the number of songs that the music player can hold.

Progress Monitoring

When all assignments for this topic have been completed, assign the corresponding Progress Monitoring page for this topic (Assessment Resources Book, page 10). Be sure students complete the Progress Monitoring page before you administer the final assessment for this volume.

Volume 4: Multiply and Divide

Topic 10: Use Division to Compute

Mixed Review

Objective: Maintain skills and concepts.

Have students complete the Mixed Review page. Work with each student individually to review results. Identify strengths and weaknesses and correct any misunderstandings.

Error Analysis

Exercise 1 For students having difficulty determining the second number in the problem, review how many months are in a year.

Exercise 2 Encourage students to draw a diagram to determine what operation to use.

Exercise 3 Make sure students read the problem carefully so they know to order the numbers from least to greatest, not greatest to least.

Exercise 4 At first students may not see what is wrong with Amanda's estimate. Amanda did estimate correctly, it is just that 60 and 7 are not compatible numbers.

Exercise 5 Suggest that students use a place-value chart.

Exercise 7 Encourage students to use base ten blocks to visualize how to make ten for this sum.

Exercise 8 Encourage students to rewrite the problem vertically so that they can solve it more easily. Remind them that they need to regroup.

Student Booklet Page 33

Topic 10 — Use Division to Compute — Mixed Review

1. A car dealership sells 87 cars each month. How many cars will they sell in a year?
   A 870   B 957   **C 1,044**   D 1,131

2. At football practice, there are 22 players on the field and 19 more on the sidelines. How many players are at practice?
   A 31   **B 41**   C 418   D 51

3. Order 610, 61, and 601 from least to greatest.
   61, 601, 610

4. Amanda estimated 5,683 ÷ 7 as 6,000 ÷ 7. What would be a better estimate?
   Sample: Amanda should estimate 5,683 to 5,600 because 56 and 7 are compatible numbers. 60 and 7 are not compatible numbers.

5. Write the place of each underlined digit.
   a. 8,0<u>3</u>6   hundreds
   b. 27<u>4</u>   ones
   c. <u>1</u>,224   thousands
   d. 3,<u>5</u>08   hundreds

6. Find each quotient.
   a. 14 ÷ 7 = 2
   b. 12 ÷ 1 = 12
   c. 36 ÷ 4 = 9
   d. 24 ÷ 3 = 8

7. Use the make-ten strategy to find the sum 7 + 5.
   Sample: I added 3 to 7 to make 10 and 2 more is 12, so 7 + 5 = 12.

8. Subtract 572 − 37.
   535

Volume 4 — Multiply and Divide
Topic 11 — Expressions and Equations
Topic Introduction

Students will work with expressions and equations in Topic 11. In Lesson 11-1, students will represent patterns in number relationships as expressions. In Lesson 11-2, students will build on this concept to use various symbols such as letters and boxes to represent unknowns in expressions. Lesson 11-3 will introduce students to the writing equations. Finally, in Lesson 11-4, students will extend their work with equations to solving equations in one variable.

Lesson	Objective	Student Pages	Teacher Pages	Tutorials
Topic 11 Introduction	11.1 Record the rule for a pattern as an expression. 11.2 Write expressions for situations that include an unknown quantity. 11.3 Write equations for word problems that include an unknown quantity.	34	177	
11-1 Write Expressions for Patterns	11.1 Record the rule for a pattern as an expression.	35–37	178–180	11a
11-2 Write Expressions	11.2 Write expressions for situations that include an unknown quantity.	38–40	181–183	11b
11-3 Write Equations with Unknowns	11.3 Write equations for word problems that include an unknown quantity.	41–43	184–186	11c
11-4 Solve Equations with Unknowns	11.4 Write and solve simple equations for word problems that include an unknown quantity.	44–46	187–189	11d, 11e
Topic 11 Summary	Review writing and solving expressions and equations.	47	190	
Topic 11 Mixed Review	Maintain concepts and skills.	48	191	

Computer Tutorial

Some students may benefit from completing the computer tutorial before the attempt the Try It page of each lesson. If you are using the electronic components of *Pinpoint Math,* you will find a complete listing of Tutorial codes and titles when you access them either online or via CD-ROM.

Volume 4: Multiply and Divide
Topic 11: Expressions and Equations
Topic Introduction

Objectives: 11.1 Record the rule for a pattern as an expression. **11.2** Write expressions for situations that include an unknown quantity. **11.3** Write equations for word problems that include an unknown quantity.

A child's ticket is $3 less than an adult's ticket for the movies. **What is known in this statement and what is unknown?** A child's ticket is $3 less than an adult's ticket; the cost of an adult's ticket **What can you use to represent an unknown?** A symbol Tell students that ☐ will stand for the price of an adult's ticket. **What operation would you use to show the cost of a child's ticket in terms of an adult's ticket?** Subtraction Write the expression ☐ − 3 on the board. **Does this expression represent the statement above?** Yes. **Why?** ☐ stands for the price of an adult ticket and subtracting $3 will tell you the price of a child's ticket.

Informal Assessment

1. **What does the diamond represent?** The unknown value **Is that value added to 12 or subtracted from 12?** Subtracted Have students complete Exercise 1.

2. **What is the unknown?** x **What is the value of x?** 6 **What is put in place of x?** 6 **What is the new expression?** 6 ÷ 3 Have students complete Parts b and c of Exercise 2.

3. **If you know the price of one set of earrings, how do you find the price of more than one?** Multiply the price, $10, by the number of sets of earrings. Have students complete Exercise 3.

4. **What is known?** One box contains 30 cards, and the total number of cards is 95. **What is not known?** The number of cards in the second box Have students complete Parts a and b of Exercise 4.

An equation is another name for a number sentence with an equal sign in it. **When you write an equation, how will you represent the unknown amount of tokens in the second box?** Sample: with a variable Have students complete Part c of Exercise 4.

Student Booklet Page 34

Another Way Allow students to model these problems with MathFlaps or other objects.

Lesson 11-1 — Write Expressions for Patterns

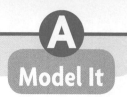

Objective 11.1: Record the rule for a pattern as an expression.

Teach the Lesson

Materials ☐ Counters

Activate Prior Knowledge

On the board, write number facts with missing values, such as 7 − ☐ = 2 and ☐ + 9 = 14. Have students provide the missing numbers.

Develop Academic Language

In the number facts on the board, change the box to various shapes and letters. Help students see that any symbol can represent an unknown value.

Explain that a numerical expression, such as 100 + 500, has at least two numbers and one operation. An algebraic expression, has at least one variable and one operation.

Model the Activities

Activity 1 What addition expression is shown? 3 + ■ How do you find the value if you know ■ = 6? Replace ■ with 6 in the expression. What is the value? 9

Activity 2 You may want to have students model the problem. Let each counter represent $10. For 1 hour, show 1 counter, for 2 hours, show 2 counters, and so on. Repeat with additional examples as needed. Point out that when a multiplication sign is not used in a multiplication expression, the variable is usually written after the number.

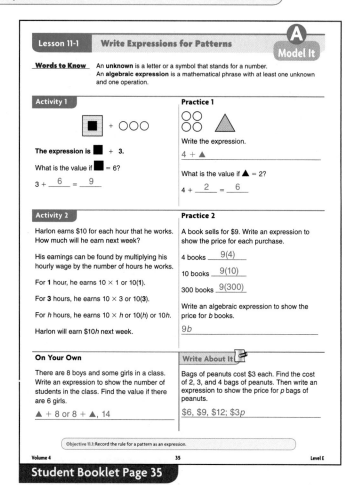

Student Booklet Page 35

⭐ Progress Monitoring

What expression would show 13 added to an unknown amount? **Sample:** 13 +

Error Analysis

Remind students that although they can use symbols or any letter for an unknown, they should try to use meaningful letters in expressions, such as *h* for hours.

| Lesson 11-1 | **Write Expressions for Patterns** | **B** Understand It |

Objective 11.1: Record the rule for a pattern as an expression.

Facilitate Student Understanding

Develop Academic Language

Explain the terms *Input* and *Output* as the value of an unknown in an expression and the value of the expression.

Demonstrate the Examples

Example 1 Help students see that the Input is the number Colin starts with and the Output is the result of his calculations. Ask students how they would find the Output number if the Input number was 12. **Multiply 12 by 2.** How could you show this relationship if the Input number was a star? **Show 2 times a star.**

Example 2 In this table, what is the input? **Number of yards** What is the output? **Number of feet** What is the rule in words? **Multiply by 3.** What is the rule as an expression? **3f**

Computer Tutorial

Some students may benefit from completing a computer tutorial before they attempt the Try It page. A list of the tutorials for each lesson can be found beginning on page x in the front of this book.

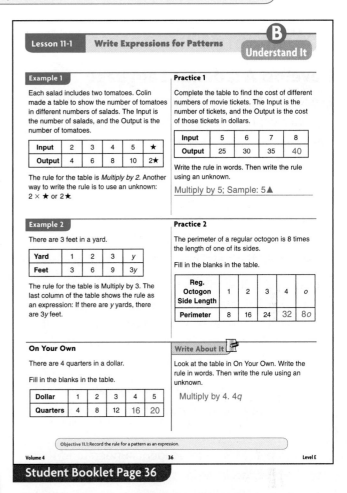

Student Booklet Page 36

⭐ Progress Monitoring

A rule in words is "Subtract 3." What is the expression? $x - 3$

Error Analysis

Remind students that for the expression to be a rule, it must be true for each pair of Input and Output numbers. Looking at just the first pair of numbers in Example 1, students might think the rule is "Add 2." But when they look at the other pairs of numbers, they will see this rule does not hold.

Lesson 11-1 Write Expressions for Patterns

Objective 11.1: Record the rule for a pattern as an expression.

Observe Student Progress

Develop Academic Language

Exercise 1 Be sure students understand that an algebraic expression has at least one unknown and at least one operation.

 Error Analysis

Exercise 1 Be sure students count the objects shown and then represent that quantity with a number in their expressions.

Exercise 2 If students have difficulty writing the rule as an algebraic expression, suggest that they add a column to the end of the table and write the unknown, *n*, as the input value. The expression for the output value will be the rule.

Exercise 3 Suggest that students make an input/output table of values to organize their work.

Exercise 4 Make sure students understand that this expression includes subtraction.

Exercise 6 Students may benefit from writing the rule for this table in words and as an expression to help them fill in the blanks.

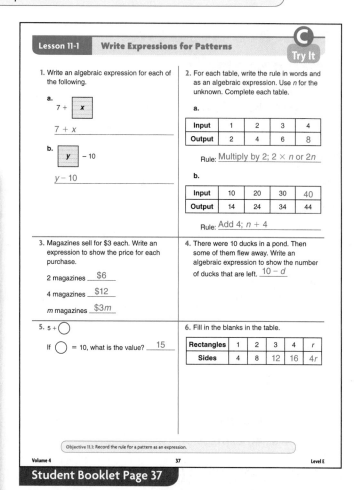

Student Booklet Page 37

Lesson 11-2: Write Expressions

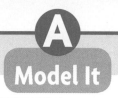

Objective 11.2: Write expressions for situations that include an unknown quantity.

Teach the Lesson

Activate Prior Knowledge

Write 2 × 2, 2 × 3, 3 × 3, and 4 × 5 on the board. Have students identify the factors in each expression. 2, 2; 2, 3; 3, 3; 4, 5

Develop Academic Language

Ask a student to tell the number of people in her family. Then ask how large a family twice that size would be. Discuss that *twice* means to multiply by 2.

Model the Activities

Activity 1 Do we know exactly how long the sign for the card shop is? *No.* How can we describe its length? *x feet* What do we know about the length of the sign for the furniture store? *It is twice as long as the sign for the card shop.* How can we write that as an expression? *2x feet*

Activity 2 Describe the cost of *x* cartons of oil paints in words. *Multiply 25 times x.* How can we write that as an expression? *25x* How do you find the cost of 20 cartons? *Substitute 20 for x.* What is the cost? *$500*

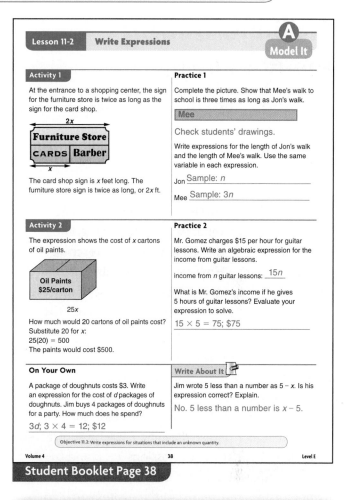

Student Booklet Page 38

Progress Monitoring

If lunch costs $8 give an expression for the price of *l* lunches. $8*l*

Error Analysis

If students have difficulty substituting values for variables correctly, suggest that they write the value on an index card and then cover the variable in the algebraic expression with the index card to see the number expression.

Lesson 11-2: Write Expressions

Objective 11.2: Write expressions for situations that include an unkown quantity.

Facilitate Student Understanding

Develop Academic Language

Have students describe situations using the words *fewer* and *greater*.

Demonstrate the Examples

Example 1 Tell students that they should not use two different variables to describe the two different kinds of shirts. **How can you describe the blue shirts?** *x* **How can you describe the white shirts using the same variable?** There are 3 fewer white shirts than blue shirts, so you can use the expression $x - 3$ to describe the white shirts.

Example 2 Emphasize that students should carefully read each expression. Many students immediately write the first number in the phrase as the first term in the expression. Point out that "7 less than" is a phrase meaning 7 will be subtracted, and point out that the expressions $7 - b$ and $b - 7$ are very different. Suggest that students substitute number values to check that the expressions make sense.

Computer Tutorial

Some students may benefit from completing a computer tutorial before they attempt the Try It page. A list of the tutorials for each lesson can be found beginning on page x in the front of this book.

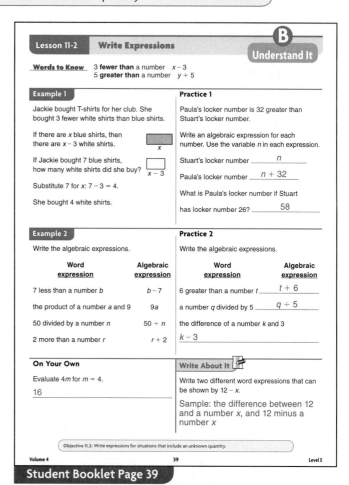

Student Booklet Page 39

Progress Monitoring

Write an algebraic expression for *12 more than a number.* $x + 12$

Error Analysis

Check that students are using the correct order of terms and the correct operation.

Lesson 11-2 | Write Expressions

Objective 11.2: Write expressions for situations that include an unkown quantity.

Observe Student Progress

Develop Academic Language

Exercise 6 Have students share their answers to illustrate that different word expressions can be used to describe the same algebraic expression.

Error Analysis

Exercise 1 Encourage students to use parentheses to show multiplication. They can write the number of items within the parentheses and then multiply by each respective price. This will help them place the variable in the correct position when called for.

Exercise 2 Watch for students who use two variables to write these expressions. Point out that the phrase *6 more points than* shows that the numbers are related and it is possible to use one variable to write both unknowns.

Exercise 3 Encourage students to rewrite each expression with an empty set of parentheses for the variable when evaluating.

Lesson 11-2 | Write Expressions — Try It

1. The band's T-shirts cost $12. Posters cost $6. Write an expression to represent the cost of each purchase.
   a. 4 shirts ___12(4)___
   b. 5 shirts and 9 posters
      ___12(5) + 6(9)___
   c. n shirts and 6 posters
      ___12(n) + 6(6)___
   d. 3 shirts and d posters
      ___12(3) + 6(d)___

2. Trina has 6 more points than Alex in a video game. Write an algebraic expression for the number of points each person has earned. Use the same variable in each expression.
   a. Alex ___Sample: x___
   b. Trina ___Sample: $x + 6$___
   c. Alex has 85 points. How many points does Trina have? Evaluate your expression to solve. Show your work.
      ___91; 85 + 6 = 91___

3. Write an algebraic expression for each of the following. Use n as your variable. Then evaluate each expression for $n = 8$.
   a. the sum of 12 and a number
      ___12 + n; 20___
   b. the quotient of 24 divided by a number
      ___24 ÷ n; 3___

4. Ribbon costs $3.00 per package. Which expression gives the total cost of p packages of ribbon? Circle the letter of the correct answer.
   A $p + 3$ B $3 - p$
   C $p ÷ 3$ **D** $3p$

5. Explain how to evaluate $4d$ for $d = 3$.
   Substitute 3 for d. Then multiply $4 × 3 = 12$.

6. Write a word expression for each algebraic expression.
   $b + 8$ ___8 more than b___
   $b ÷ 2$ ___half a number___

Student Booklet Page 40

Lesson 11-3: Write Equations with Unknowns

Objective 11.3: Write equations for word problems that include an unknown quantity.

Teach the Lesson

Materials
- ☐ Counters
- ☐ Index cards

Activate Prior Knowledge

Review using unknowns in expressions. Tell number stories such as: **Tina has some mystery books. Bella has 4 more mystery books than Tina.** Then have students describe with one unknown the number of mystery books that Tina and Bella each have. $t, t + 4$

Develop Academic Language

Direct students to look at the definition of *equation* at the top of the student page. Write *equation* on the board, and next to it, write *equal*. Underline *equa-* in each word. **The words equation and equal are related.**

Model the Activities

Activity 1 Provide counters so students can model this activity. **Put together 5 counters. Then write a ? on an index card and put that next to the counters. Then put together six counters. How could we write this as an equation?** $5 + ? = 6$

Activity 2 **What operation will be used in this equation?** Addition **How do you know?** We start with one number, and end up with a greater number. Write an addition frame on the board: ___ + ___ = ___. Fill in the frame as you work through the problem with students. **What number do we start with?** 3 If students say 2, make sure they know they should count Jake, too. **How many do we add?** It doesn't say. **Let's use *f* to stand for the number of friends that we add. What is the sum?** 5 **So the equation for this number story is $3 + f = 5$.**

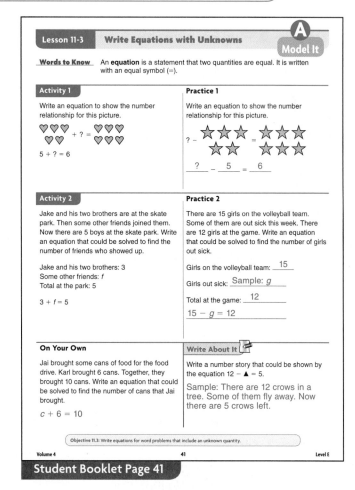

Student Booklet Page 41

Progress Monitoring

There are 10 tigers at the zoo. Some of them are adults. 3 of them are cubs. Write an equation that could be solved to find the number of adults. $10 - a = 3$

Error Analysis

If students have difficulty deciding what operation to use in an equation, have them think about if the numbers are increasing or decreasing.

Lesson 11-3: Write Equations with Unknowns

Objective 11.3: Write equations for word problems that include an unknown quantity.

Facilitate Student Understanding

Develop Academic Language

Review that multiplication is joining equal groups and that division is splitting something into equal groups. This will help students better understand what operation they should choose for their equations.

Demonstrate the Examples

Example 1 **What do we know in this problem?** The number of students, 12, and the total number of sandwiches, 36 **What is the unknown?** The number of sandwiches each student brought **Let's use s for the unknown. What operation should we use?** Multiplication **Why?** Each student brought the same number of sandwiches, so we are joining equal groups. **What equation can we write?** $12 \times s = 36$

Example 2 **What operation should we use in this equation?** Division **How do you know?** Scott gives an equal number of muffins to each of his aunts. **What is the unknown in this equation?** The number of muffins **How should we represent that number?** m **What equation can we write?** $20 \div m = 5$

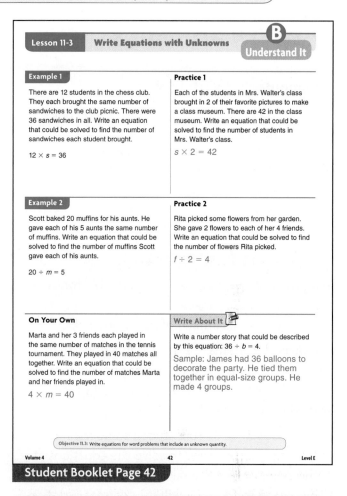

Student Booklet Page 42

Progress Monitoring

Sharee made some bracelets and gave them to her 4 friends. Each friend got 2 bracelets. If you wrote an equation for this story, what would the unknown be? The number of bracelets she made **What operation would you use?** Division

Error Analysis

Make sure students understand that, although they could use any letter or symbol to stand for an unknown, it makes the most sense to use a meaningful letter, such as the first letter of the word the unknown stands for, to make it easier to understand the equation.

Lesson 11-3: Write Equations with Unknowns

Objective 11.3: Write equations for word problems that include an unknown quantity.

Observe Student Progress

Computer Tutorial

Some students may benefit from completing a computer tutorial before they attempt the Try It page. A list of the tutorials for each lesson can be found beginning on page x in the front of this book.

 Error Analysis

Exercise 1 It would also be acceptable for students to use variables or other symbols in these equations instead of question marks.

Exercise 2 To make sure students don't think the unknown is the total number of clients that Kim has now, ask them if the story tells them the total. **Yes, 9.**

Exercises 3–6 Be sure students understand that they do not need to solve the equations at this point or find the value of the unknown. They only need to focus on writing a correct equation to describe the story.

Exercises 3 and 4 Because of the inverse relationship of addition and subtraction, students might write a subtraction equation for Exercise 3 and an addition equation for Exercise 4. Although this is not incorrect, explain that they should try to write the equation in the same order as the story. This will lead them to use addition for Exercise 3 and subtraction for Exercise 4. This is also true for Exercises 5 and 6, with division and multiplication.

Student Booklet Page 43

Lesson 11-3 Write Equations with Unknowns — Try It

1. Write a number sentence to show the number relationship for each picture.
   a. $3 + ? = 8$
   b. $? - 4 = 3$

2. Kim had 3 clients. She added more clients to her list. Now she has 9 clients. Which equation can be used to find the number of clients Kim added to her list? Circle the letter of the correct answer.
   - **(A)** $3 + c = 9$
   - **B** $3 - c = 9$
   - **C** $c + 9 = 3$
   - **D** $3 + 9 = c$

3. John has a collection of 9 photographs. Julie gives him some more photographs for his birthday. Now he has 13 photographs. Write an equation that could be solved to find the number of photographs Julie gave John.
   $9 + p = 13$

4. Kay had some students in her cooking class. Then 4 of them moved away. Now she has 10 students. Write an equation that could be solved to find the number of students that Kay had originally.
   $s - 4 = 10$

5. At Greta's party, an extra-large pizza was cut into 16 slices. Each person got 2 slices. Write an equation that could be solved to find the number of people at Greta's party.
   $16 \div p = 2$

6. Each member of the McGregor family brought 2 pies to the school bake sale. There were 8 pies all together. Write an equation that could be solved to find the number of people in the McGregor family.
   $p \times 2 = 8$

7. Donatella volunteered to deliver 7 lunches to senior citizens. She delivered some. Now she has 3 lunches left. Which equation could be used to find how many lunches Donatella has already delivered? Circle the letter of the correct answer.
   - **A** $3 + 7 = ?$
   - **(B)** $7 - ? = 3$
   - **C** $? - 3 = 7$
   - **D** $? - 7 = 3$

8. Write a number story that could go with this equation: $y \times 3 = 18$.
   Sample: There were some dancers in a showcase. They each performed 3 dances. All together, there were 18 dances.

Lesson 11-4 — Solve Equations with Unknowns

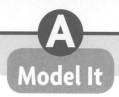

Objective 11.4: Write and solve simple equations for word problems that include an unknown quantity.

Teach the Lesson

Materials
- ☐ Counters
- ☐ Two containers, one with a lid

Activate Prior Knowledge

Review fact families. Give students one fact, such as $4 + 3 = 7$, and have them find the other facts in the family. Emphasize the fact that addition and subtraction facts belong in the same fact family because they are inverse operations.

Develop Academic Language

Write $5 + \square = 8$ on the board. **How is the unknown shown in the equation?** By an open square **How can you find the number that can replace the square to make the statement true?** Use mental math; subtract. **What is the missing number?** 3 Explain that 3 is called the *solution* to the equation because it makes the statement true when substituted for the unknown.

Model the Activities

Activity 1 **Is y the number we start with, the number we take away, or the number left?** The number we start with **What must be done to y in order to get 11?** Take 4 away. **Will the value of y be greater or less than 11?** Greater Students can think of the solution as starting with 11 and "putting back" the 4 that were taken away from y. Have students use the picture to determine the value of y. **How can you find the value of y?** Count the total number of counters; $y = 15$.

Activity 2 Students can use counters and two containers, one with a lid, to model the addition sentence. **How is the unknown shown in this number sentence?** By the letter x Point out that in algebraic expressions, a letter is usually used to represent the unknown. **How could you use a related fact to find the value of x?** Sample: 2, 8, and 10 are in one fact family, so $2 + 8 = 10$. Point out that the solution is

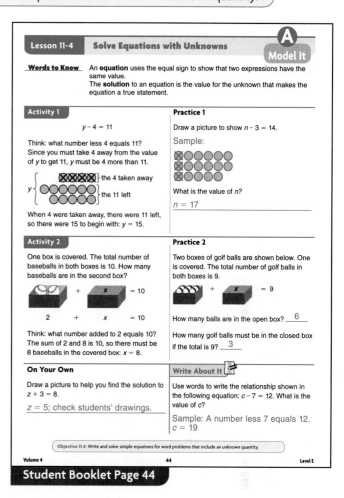

Student Booklet Page 44

written as an equation with the unknown on one side of the equal sign and the value of the unknown on the other: $x = 8$.

✓ Progress Monitoring

Have students draw pictures to show $3 + a = 6$ and $7 - x = 1$. Check students' drawings.

Error Analysis

If students try to solve an addition equation like $3 + a = 6$ by adding 3 and 6, ask them to check their solutions by substitute them back into the original equation.

Lesson 11-4 — Solve Equations with Unknowns

Objective 11.4: Write and solve simple equations for word problems that include an unknown quantity.

Facilitate Student Understanding

Develop Academic Language

Advise students that an addition story requires a setting where two or more values will be added to find a greater total. A subtraction story involves decreasing one number by another.

Demonstrate the Examples

Example 1 Help students see that the total number of photographs stored minus the number of photographs taken before the party will give the number of photographs taken at the party.

Example 2 What do we know in this exercise? They sold 62 doughnuts. They had 18 doughnuts left. What is the unknown? The number of doughnuts they baked What operation will we use? Subtraction Write the equation. $d - 62 = 18$ How can we find the value for d? Add 18 and 62. Why? $18 + 62 = d$ is in the same fact family as $d - 62 = 18$.

Write About It

ENGLISH LEARNERS Allow students to work with partners fluent in English to write stories.

Student Booklet Page 45

Progress Monitoring

Have students write the number relationship for the following problem and then solve.

Dr. Ruiz saw 12 patients before lunch. By the end of the day, she had seen 20 patients. How many patients did Dr. Ruiz see after lunch?
Sample: $12 + ? = 20$; 8

Error Analysis

Remind students that if an addend is missing, they can use related facts, mental math, or subtraction to find the missing value.

Lesson 11-4: Solve Equations with Unknowns

Objective 11.4: Write and solve simple equations for word problems that include an unknown quantity.

Observe Student Progress

Computer Tutorial

Some students may benefit from completing a computer tutorial before they attempt the Try It page. A list of the tutorials for each lesson can be found beginning on page x in the front of this book.

Error Analysis

Exercise 1 Review mental math techniques and basic addition and subtraction facts to help students solve these problems.

Exercise 3 Make sure students see the phrase *addition equation* in the directions. They might otherwise write a subtraction equation.

Exercise 5 For students who have difficulty coming up with a number story to go with the problem, give them a particular scenario to work with, such as the chores in the sample answer.

Exercise 6 Ask what operation can be used to find how much money Omar's aunt gave him.
Subtraction; $18 - 13 = 5$

Lesson 11-4 Solve Equations with Unknowns

1. Use mental math to find each missing value.
   a. $? - 30 = 50$ $? = 80$
   b. $? + 5 = 25$ $? = 20$
   c. $18 - ? = 6$ $? = 12$

2. Use mental math or draw a picture to find each solution.
   a. $16 - e = 5$ $e = 11$
   b. $f + 8 = 13$ $f = 5$
   c. $10 + g = 19$ $g = 9$
   d. $j - 11 = 6$ $j = 17$

3. The park district plans to have 14 soccer teams this summer. They will also have some softball teams. All together, there will be 24 softball and soccer teams. Write an addition equation to represent this number story. Then find the number of softball teams.
   $14 + s = 24; s = 10$

4. Charlie had 35 music students. Some of them studied guitar. 26 studied piano. Write a subtraction equation to represent this number story. Then find the number of students who studied guitar.
   $35 - g = 26; g = 9$

5. Write a number story that could be represented by this equation. Then find the value for y.
   $y + 30 = 45$
   Sample: Kari cleaned the windows for some time. Then she mopped for 30 minutes. It took her 45 minutes to do all her work. How long did it take her to clean the windows? $y = 15$

6. Omar had $13. His aunt gave him some money for his birthday. Now Omar has $23. Write an addition number relationship using a ? to show how many dollars Omar's aunt gave him. Then solve.
   Sample: $13 + ? = 23$, $10

Student Booklet Page 46

Volume 4: Multiply and Divide

Topic 11: Expressions and Equations

Topic Summary

Objective: Review writing and solving expressions and equations in one variable.

Have students complete the summary page. You may want to have students work in groups of four with each student analyzing a different choice.

Ask students to share their ideas about each answer choice. Be sure they confirm the correct answer for each problem at the end of the discussion.

Answer Evaluation

1. **A** This equation models the number relationship correctly.

   B This picture models the number relationship correctly.

   C This choice is correct.

   D This picture models the number relationship correctly.

2. **A** Students used a variable without looking to see whether the variable has already been defined.

   B Students confused the phrase *7 more* with the phrase *7 times more*.

   C This choice is correct.

   D Students confused the phrase *7 more* with the phrase *7 less*, or they confused the meaning of *p*.

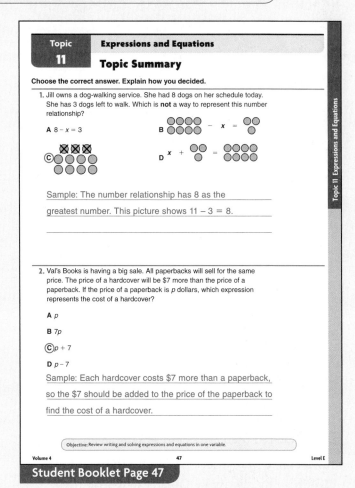

Student Booklet Page 47

 ## Error Analysis

Exercise 1 Point out that because students are looking for the way that does **not** show the number relationship, three answers will show ways to model the relationship.

Exercise 2 Students who answer incorrectly may need to list words and phrases that suggest each operation, such as *more than* and *increased by* for addition.

Progress Monitoring

When all assignments for this topic have been completed, assign the corresponding Progress Monitoring page for this topic (Assessment Resources Book, page 11). Be sure students complete the Progress Monitoring page before you administer the final assessment for this volume.

Volume 4, Topic 11: Multiply and Divide
Expressions and Equations
Mixed Review

Objective: Maintain concepts and skills.

Have students complete the Mixed Review page. Work with each student individually to review results. Identify strengths and weaknesses and correct any misunderstandings.

⭐ Error Analysis

Exercise 1 You may want to suggest that students align the two numbers in each problem vertically to more easily compare their values.

Exercise 2 Look over students' work to check for mastery of the multiplication algorithm.

Exercise 3 Have a student read the problem aloud. Ask students for the key word that tells them what to do. Have students circle *about* once it is identified. For students still having difficulty, review the rules for rounding.

Exercise 4 Allow students to write the number on a place-value chart before they look at the answer choices.

Exercise 6 Suggest students first write an empty set of parentheses for each variable, and then insert the appropriate value.

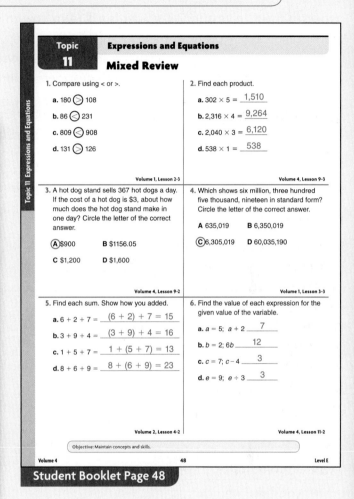

Student Booklet Page 48

Volume 5: Data, Geometry, and Measurement

Topic 12: Graphing

Topic Introduction

Lesson 12-1 gives students an introduction to several types of graphs and how to read and compare the data shown in them. In Lesson 12-2, students will learn how to record data in a table and in various graphs. The statistical measures of mean, median, and mode will be covered in Lesson 12-3. Lesson 12-4 returns to the various types of graphs the students studied early as they learn to show data in different ways. Lesson 12-5 provides students with guidance in graphing ordered pairs.

Lesson	Objective	Student Pages	Teacher Pages	Tutorials
Topic 12 Introduction	12.1 Represent and compare data by using pictures, bar graphs, tally charts, and picture graphs. 12.4 Represent the same data in more than one way. 12.6 Use two-dimensional coordinate graphs to represent points and graph lines and simple figures.	1	193	
12-1 Compare Data with Graphs	12.1 Represent and compare data by using pictures, bar graphs, tally charts, and picture graphs.	2–4	194–196	12a, 12b
12-2 Record Data	12.2 Record numerical data in systematic ways, keeping track of what has been counted.	5–7	197–199	12a, 12b
12-3 Mean, Median, and Mode	12.3 Find the mean, median, and mode of a set of data.	8–10	200–202	12c
12-4 Show Data in More Than One Way	12.4 Represent the same data in more than one way.	11–13	203–205	12a, 12b
12-5 Graph Ordered Pairs	12.5 Use two-dimensional coordinate graphs to represent points and graph lines and simple figures.	14–16	206–208	12d
Topic 12 Summary	Review graphing and interpreting graphs.	17	209	
Topic 12 Mixed Review	Maintain concepts and skills.	18	210	

Computer Tutorial

Some students may benefit from completing the computer tutorial before they attempt the Try It page of each lesson. If you are using the electronic components of *Pinpoint Math,* you will find a complete listing of Tutorial codes and titles when you access them either online or via CD-ROM.

Volume 5: Data, Geometry, and Measurement

Topic 12: Graphing

Topic Introduction

Objectives: 12.1 Represent and compare data by using pictures, bar graphs, tally charts, and picture graphs. **12.4** Represent the same data in more than one way. **12.5** Use two-dimensional coordinate graphs to represent points and graph lines and simple figures.

Materials ☐ Variety of graphs including bar graphs, line graphs, picture graphs, and circle graphs

Place the graphs on the board. Ask students what each graph shows and how the graphs are similar and different. Record responses on the board. Discuss what components a graph must have regardless of the type of graph. Components include: title, x- and y-axis labeled and titled, correct scale and key if necessary.

Informal Assessment

1. This graph shows the books read by 5 students during the last month. Who are the students? John, Elena, Vicky, Link, Tiva How is the vertical scale numbered? From 0 to 7 What do these numbers stand for? Number of books read in the past month

2. What kind of chart is shown? Tally chart What does a tally tell you? That a person chose that lunch How are the tallies organized? By groups of 5 How many students choose pizza? 8 hot dog? 7 How can you take this information and make a bar graph? Use the tallies to make the bars and title and label the graph.

3. A park ranger kept track of how many children used the playground at the park. What do the letters at the bottom of the graph stand for? Days of the week from Sunday through Saturday Which two days had the greatest number of children? Sunday and Saturday

4. Eleanor Jaycek has a shirt factory. They use 5 buttons for each shirt. The graph has no points on it because you will make those points. If 1 shirt has 5 buttons, how many buttons do 2 shirts have? 10 Make a dot on the graph to show this.

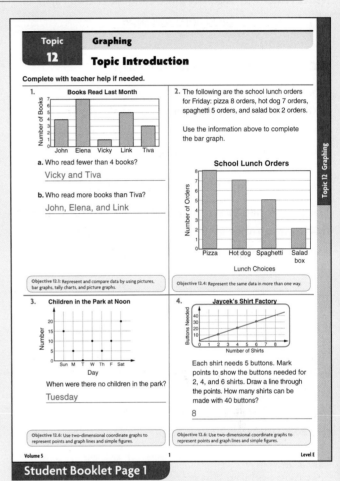

Another Way For Example 4, have students record the ordered pairs before they plot the points. If 5 buttons are needed for each shirt than the first ordered pair would be (1, 5). Move on to two shirts and the number of buttons until you reach the eighth shirt and number of buttons.

Lesson 12-1 Compare Data with Graphs

Model It

Objective 12.1: Represent and compare data by using pictures, bar graphs, tally charts, and picture graphs.

Teach the Lesson

Materials
- ☐ Teaching Aid 4 Centimeter Grid Paper
- ☐ Lined notebook paper

Activate Prior Knowledge

Ask students what kinds of graphs they have seen in textbooks, newspapers, and so on. Ask why people use graphs. Elicit the idea that it is an easy way to display and compare information.

Develop Academic Language

The difference between the greatest number and the least number in a set of data is called the *range*. Ask for students' ages. Write them on the board. Then show them how to find the range of their ages.

Model the Activities

Activity 1 Provide lined notebook paper. Have them use it to copy the graph given in the Activity. **What does each horizontal grid line represent? 4 students How many students chose cats? 18 How many students chose birds? 10 How do we compare the two numbers? Subtract.**

Activity 2 Remind students that the range is not how high and low the graph goes, but the difference between the greatest and least numbers in the data set. So we subtract 18 − 4 to find the range, not 20 − 0. **What does the "other" category represent? The number of students who preferred pets that were not named in the rest of the categories**

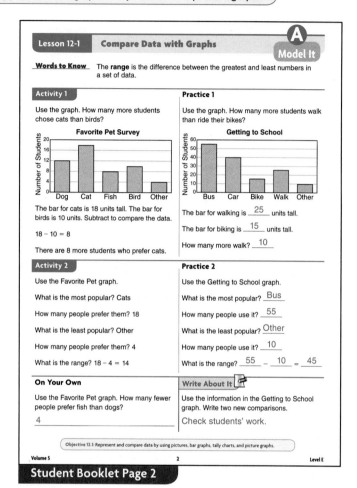
Student Booklet Page 2

⭐ Progress Monitoring

As a class, choose a survey topic such as favorite color, or sport. Make a chart or list on the board to show the data. Then have each student use lined notebook paper turned sideways to make a horizontal bar graph.

Error Analysis

Students will probably need help choosing the scales for their bar graphs.

Volume 5   194   Level E

Lesson 12-1 — Compare Data with Graphs

Objective 12.1: Represent and compare data by using pictures, bar graphs, tally charts, and picture graphs.

Facilitate Student Understanding

Develop Academic Language

Use the graphs on student pages 2 and 3 to show students examples of bar graphs and pictographs (also called picture graphs). Vocabulary terms include: *bar graph, axis, scale, pictograph, key.* Discuss, then list the terms on the board and refer to them as you discuss the examples.

Demonstrate the Examples

Example 1 Explain that one way in which a pictograph differs from a bar graph is the key. Have students find the key below the pictograph and explain what it means. **Each symbol stands for 10 people.** Ask how to find the number of customers in April. **Count the symbols on the pictograph—there are 6—and multiply by 10.** Explain that $\frac{1}{2}$ figure will be $\frac{1}{2}$ of 10 or 5 customers. Ask students to find how many customers there were in June. $(7 \times 10) + 5$ or 75.

Example 2 Instead of multiplying and then subtracting, we can find the difference in the number of symbols on the graph. Then we can multiply to find the number of customers. How

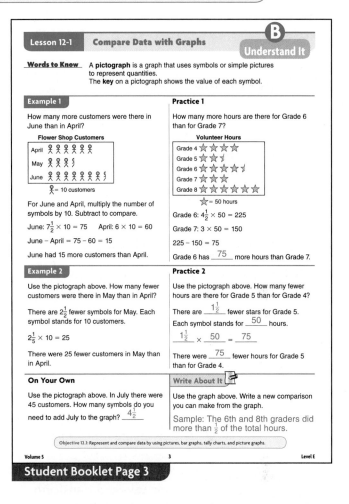

Student Booklet Page 3

many symbols are there for April? **6** How many for May? $3\frac{1}{2}$ What is the difference? $6 - 3\frac{1}{2} = 2\frac{1}{2}$

Progress Monitoring

Put this data chart on the board.

Distance from Home to School

Miles	0–1	2	3	4	More than 4
Number of Students	6	2	7	3	5

Have students make a pictograph. Discuss what the key will be; for example, 1 stick figure = 2 students.

Error Analysis

A simpler version of the distance-to-school data would use actual student names.

For example:

Distance from Home to School

Student	Ed	Kim	Juan	Sam	Alice
Miles	1	2	4	3	2

Have students make a pictograph using one symbol for each mile. Remind them to write their keys below the graph.

Lesson 12-1 Compare Data with Graphs

Objective 12.1: Represent and compare data by using pictures, bar graphs, tally charts, and picture graphs.

Observe Student Progress

Computer Tutorial

Some students may benefit from completing a computer tutorial before they attempt the Try It page. A list of the tutorials for each lesson can be found beginning on page x in the front of this book.

Develop Academic Language

Exercise 6 Remind students that the *range* of the data is the difference between the greatest and least values.

Error Analysis

Exercise 1 Have students color the bars for soccer and biking red; the bar for baseball blue. The colors will help them use the correct bars to answer the questions.

For extra practice, have students used lined notebook paper turned sideways and transform the data into a vertical bar graph.

Exercise 2 Before students answer the questions, have them count the trees planted each year by 20s and write the totals next to each row. Remind students that one-half of a tree stands for 10 real trees.

For extra practice, have students use lined notebook paper turned sideways and transform this data into a horizontal bar graph.

Exercise 5 Suggest that students apply what they have learned from solving simpler problems to help them solve this more complex problem. If they need a hint to get started, have them find the height of a plant that is twice as tall as Bart's plant. **30 cm**

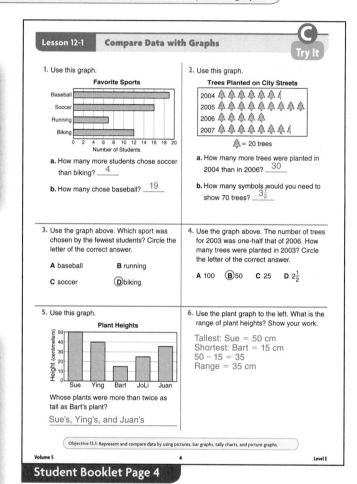

Student Booklet Page 4

For extra practice, have students use lined notebook paper turned sideways and transform this data into a horizontal bar graph.

Exercise 6 Ask students to explain how they can solve the problem by first breaking it into simpler parts. **Sample: Find the tallest plant, the shortest plant, and then subtract their heights.**

Volume 5   196   Level E

Lesson 12-2 Record Data

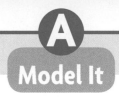

Objective 12.2: Record numerical data in systematic ways, keeping track of what has been counted.

Teach the Lesson

Activate Prior Knowledge

Four students are keeping track of the number of movies they watch in one week. Put this tally chart on the board.

Movies Watched Last Week

Student	Vince	Salvia	Kyle	Andrea
Number	||||	||||	|||| |||	||

What mark should Salvia make for her next movie? A fifth line that crosses the four lines already there. **Why do we use a line to cross the others?** To make the tallies easier to count

Develop Academic Language

Write the terms *tally mark, tally, tallies* on the board. Point out that the first term is a noun; the second and third terms are either nouns or verbs. Have students use the terms in sentences.

Model the Activities

Activity 1 Add the 5s. $5 + 5 + 5 + 5 + 5 + 5 = 30$ Now add the 1s. $1 + 3 + 2 = 6$ How many are there all together? $30 + 6 = 36$

Activity 2 Why do you think it is a good idea to write numbers in this chart rather than tallies? That would be a lot of tallies in a small space, and it might be hard to read.

ENGLISH LEARNERS For the Write About It, ask students what letters or characters are used most frequently in their native languages.

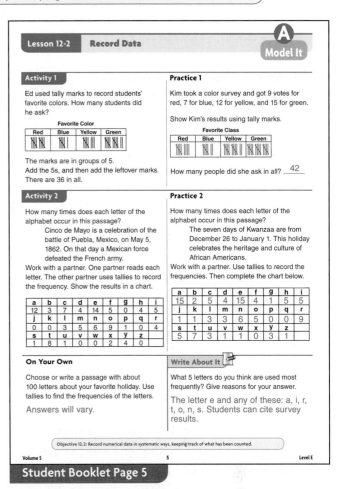

Student Booklet Page 5

Progress Monitoring

Take a pizza survey as a class activity. Have four students record the tallies for each type of pizza. For example:

Favorite Pizza

Kind	Cheese	Veggie	Sausage	Other
Number				

Error Analysis

Remind students that a good way to count tally marks is to count the 5s first, then count the rest. For example, if there are 3 groups of 5 and 2 more marks, students can count, "5, 10, 15, 16, 17."

Lesson 12-2 Record Data

Objective 12.2: Record numerical data in systematic ways, keeping track of what has been counted.

Facilitate Student Progress

Develop Academic Language

Use a drawing of a branch with 5 leaves to help students learn the term *stem-and-leaf plot*. Put this drawing on the board.

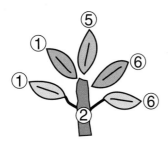

Point out that the drawing shows the numbers 21, 21, 25, 26, and 26. The tens digits 2 is on the *stem* of the branch. The ones digits are on the *leaves*.

Demonstrate the Example

Example Circle the least number in the box. What is it? 62 In stem-and-leaf plots, we organize the numbers by separating the tens and the ones. Then we order the numbers from least to greatest. How many tens are in 62? 6 How many ones? 2 Now we'll put the rest of the numbers in the plot. Suggest that students cross off the numbers in the box as they go so they don't use any number twice.

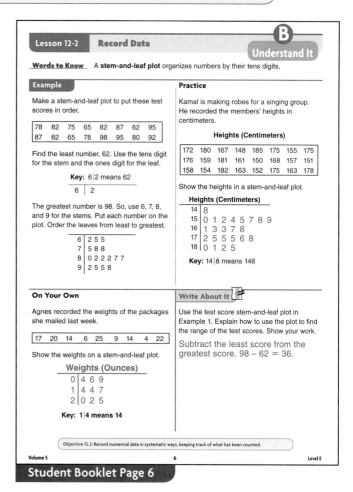

Student Booklet Page 6

✓ Progress Monitoring

Generate a list of two-digit numbers by having each student call out the last two digits of his or her phone number. Have students make stem-and-leaf plots to put the numbers in order from least to greatest.

Error Analysis

Check that students can identify the tens and ones digits in two-digit numbers. If they are unable to do so, provide several activities with base ten blocks or place-value charts.

Lesson 12-2 Record Data

Objective 12.2: Record numerical data in systematic ways, keeping track of what has been counted.

Observe Student Progress

Computer Tutorial

Some students may benefit from completing a computer tutorial before they attempt the Try It page. A list of the tutorials for each lesson can be found beginning on page x in the front of this book.

Error Analysis

Exercise 2 Have students check their stem-and-leaf plots from Exercise 1 before answering these comparison questions.

Exercise 3 Remind students to count the groups of 5 first.

Exercise 4 Ask how many students chose weather. **10** Point out that 10 is written in word form because it is at the beginning of a sentence.

Exercise 5b Students may need to actually make the graphs before they can answer this question.

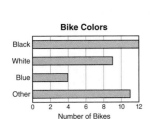

Point out that the stem-and-leaf plot does not represent the data very well.

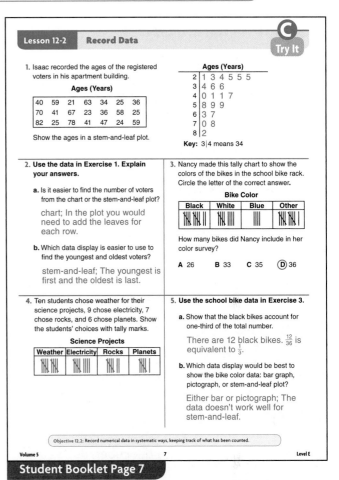

Lesson 12-3: Mean, Median, and Mode

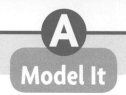

Objective 12.3: Find the mean, median, and mode of a data set.

Teach the Lesson

Materials ☐ LinkerCubes

Activate Prior Knowledge
Have students practice basic division facts and addition with more than 2 addends.

Develop Academic Language
When information is collected about people or things, the information is called *data*. *Mean, median,* and *mode* are calculations that help us make sense of data by evening out the highs and lows to reveal trends.

Model the Activities

Activity 1 Have students model the values with stacks of cubes. **How many cubes do you have? 15 How many stacks? 5** Without changing the total or the number of stacks, how can you arrange the cubes so each stack has an equal number? Allow students to experiment with the cubes. **Put 3 cubes in each row. 15 cubes can be divided evenly into 5 rows if you put 3 in each row. The mean of the data is the number of cubes in each row. How many cubes are in each row? 3 What is the mean? 3**

Activity 2 What is the sum of 3 + 5 + 5 + 7? **20** How many numbers are being added? **4** Explain the division equation you see. **20, the sum, is divided by the number of addends, 4.** What is 20 ÷ 4? **5 The mean of the numbers is 5.** If we put all the values together and then separate them into equal groups, one group per value, there will be 5 in each group.

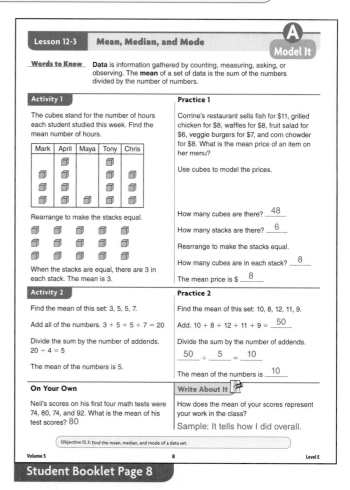

Student Booklet Page 8

Progress Monitoring
What is the mean of 8, 12, and 7? **9**

Error Analysis
Clarify that the total must not change, and that there must be one stack for each value in the data set.

ENGLISH LEARNERS Tell students that *mean* is a synonym for *average*. You may want to use *average* throughout the lesson to familiarize students with the term and to differentiate the noun *mean* from the adjective and verb.

Lesson 12-3: Mean, Median, and Mode

Objective 12.3: Find the mean, median, and mode of a data set.

Facilitate Student Understanding

Develop Academic Language

ENGLISH LEARNERS When you introduce *median*, ask 5 students to arrange themselves by height. Have the class find the student with the median height. Half of the students are taller and half are shorter than that student. Point out that *median* sounds like *medium*, the middle size.

Demonstrate the Examples

Example 1 To find the median of a set of numbers, you must first order the numbers from least to greatest. What is the order of the numbers? 2, 2, 3, 9, 11, 14, 16 Next, take turns crossing off one number from each end of the list. Which two numbers did you cross out first? 2 and 16 Keep crossing out numbers on each side until there is only one number left in the middle of the list. The number remaining is the median. You may want to add that if there are two numbers left, then their mean is the median.

Example 2 First, order the data. 57, 58, 59, 62, 62, 63, 65, 71, 74 The mode is the number that occurs the most in a data set. Which number occurs most often in the list? 62 What is the mode of the data? 62 The most common height in this set is 62 inches. Tell students that another set of data might have multiple modes or no mode. When there is no mode, it is incorrect to say "The mode is 0." The correct response would be "There is no mode."

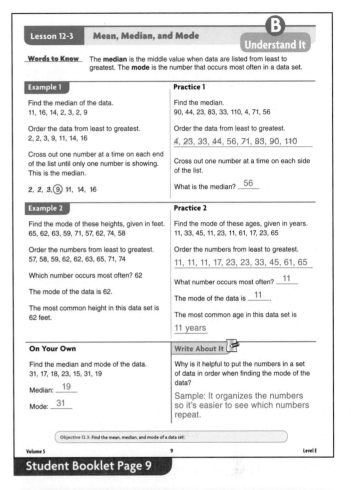

Student Booklet Page 9

 Progress Monitoring

Find the median and mode of the set 2, 8, 4, 9, 6, 8, 7, 4, 8. The median is 7; the mode is 8.

Error Analysis

When ordering a list, students might avoid listing the numbers that repeat. Tell students that each number needs to be listed, even if it is in the set more than once.

Lesson 12-3: Mean, Median, and Mode

Objective 12.3: Find the mean, median, and mode of a data set.

Observe Student Progress

Computer Tutorial

Some students may benefit from completing the computer tutorial before they attempt the Try It page of each lesson. A list of the tutorials for each lesson can be found beginning on page x in the front of this book.

Error Analysis

Exercise 1a Students who give 18 as an answer found the mean of the data and not the median. Have students review the meanings of *mean* and *median* and redo the problem.

Exercise 3 If students confuse the procedures for finding each measure, have them make and use flashcards showing the procedure on the front and the name of the measure of central tendency on the back.

Exercise 5 Help students see that the numbers in the "week" row do not count as part of the problem. They should be interpreted as names of the weeks rather than as values. Since the addends are greater than in previous problems, check students' calculations carefully.

Exercise 8 Students might think they are being asked to compare the two boys' mean scores. Show that the question asks students to consider each boy's scores separately and to determine for which boy the mean is a better measure of how well he is doing in math. You can ask students to use the individual scores first to determine how well each boy is doing. Then students can determine whether the mean supports this evaluation. Point out that the mean is most useful when the values in a data set are all very close together.

Student Booklet Page 10

Lesson 12-3: Mean, Median, and Mode — Try It

1. Find the median.
   a. 14, 12, 6, 4, 55   **12**
   b. 17, 34, 29, 16, 34, 3, 20, 5, 32   **20**
   c. 9, 90, 72, 42, 15, 38, 23, 102, 2, 7, 81   **38**

2. Find the mean.
   a. 12, 12, 7, 9, 10   **10**
   b. 3, 5, 5, 6, 3, 2,   **4**
   c. 31, 17, 18, 15, 19, 23, 31   **22**

3. Find the mean, median, and mode of the data set: 30, 16, 17, 15, 18, 21, 30.
   Mean: **21**
   Median: **18**
   Mode: **30**

4. What is the median amount that Sarah earned each week?

Sarah's Weekly Earnings				
Week 1	Week 2	Week 3	Week 4	Week 5
$20	$40	$25	$35	$50

   A $30   B $40   **C $35**   D $25

5. Find the mean and median.

Play Tickets Sold					
Week	1	2	3	4	5
Tickets	150	170	220	160	220

   Mean: **184**
   Median: **170**

6. Karl's scores on the last 4 math tests were 100, 92, 65, and 63. What is the mean of these scores?   **80**

7. Pedro's scores on the last 4 math tests were 88, 90, 84, and 82. What is the mean of these scores?   **86**

8. Use your answers to Exercises 6 and 7. Which student's mean score is a better representation of his performance?
   Sample: Pedro's, because the individual scores are closer to the mean. Karl's mean score doesn't reflect his very low performance on the last two tests.

Lesson 12-4 Show Data in More Than One Way

Objective 12.4: Represent the same data set in more than one way.

Teach the Lesson

Materials
☐ Lined notebook paper
☐ Teaching Aid 4 Centimeter Grid Paper

Activate Prior Knowledge

A student keeps track of the number of pages she reads each day. Put this tally chart on the board.

Pages Read Last Week

Mon	Tues	Wed	Thurs	Fri	Sat	Sun

Count the tally marks in the chart and make a new chart that has numbers instead of tally marks.

Pages Read Last Week

Mon	Tues	Wed	Thurs	Fri	Sat	Sun
3	7	4	6	6	2	3

Could you show the same data with a different kind of chart? Sample: A two-column chart with days in the left column and number of pages in the right column

Develop Academic Language

Differentiate between *vertical* and *horizontal* by making a poster with both terms illustrated that students can refer to.

Model the Activities

Activity 1 Provide lined notebook paper. Have students create a line plot based on the data. **What are the advantages of a line plot?** Sample: It is easy to make. **What are the disadvantages?** Sample: If your lines are crooked or your X's aren't the same size, they can be hard to read.

Activity 2 Have students make a vertical bar graph of the same data. **Draw the bars up from the days to the appropriate number of miles. Leave space between the bars so you can read them clearly.**

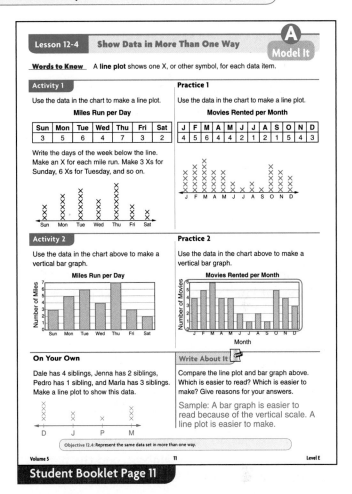

Student Booklet Page 11

What similarities do you notice between the line plot and the bar graph? Sample: Both display the data vertically. **What differences?** Sample: There is no scale on the line plot.

Progress Monitoring

Have students use tally charts to keep track of the number of glasses of water they drink per day for a week. Then have them show the data as bar graphs and as line plots.

Error Analysis

Point out that line plots get too large to be practical if there are more than about 10 Xs per category. For numbers greater than 10, bar graphs are a better choice.

Volume 5   203   Level E

Lesson 12-4: Show Data in More Than One Way

Objective 12.4: Represent the same data set in more than one way.

Facilitate Student Understanding

Develop Academic Language
Discuss the terms *pictograph* and *key*. Label an example to give students a visual model.

Demonstrate the Examples

Example 1 Look at the pictograph. What does one symbol represent? **10 students** Look at the row for Drama. What does the half symbol at the end stand for? **Half of 10, or 5, students** If we needed to represent 1 or 2 students, would this pictograph be a good choice? **Sample: No, because it would be hard to show $\frac{1}{10}$ or $\frac{2}{10}$ of a symbol.**

Example 2 Guide students to put the data in the bar graph. What similarities do you notice between the pictograph and the horizontal bar graph? **Sample: They both display the data horizontally.** What differences do you notice? **Sample: The pictograph uses a key and symbols, and the bar graph uses a scale.**

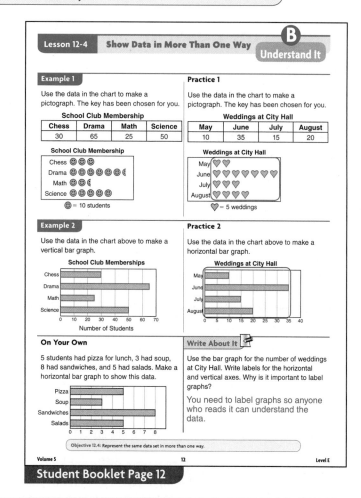

Student Booklet Page 12

 Progress Monitoring

Put this chart on the board.

Favorite Season

Season	Fall	Winter	Spring	Summer
Number				

Take a class survey by having each student call out his or her favorite season. Make tally marks as the seasons are chosen. Have students make pictographs and bar graphs for the data.

Error Analysis

If each symbol on the pictograph shows 2 people, have students number their bar graphs by 2s. Then one unit on the bar graph shows the same number as one symbol. If each symbol equals 5, have students number by 5s.

| Lesson 12-4 | **Show Data in More Than One Way** | |

Objective 12.4: Represent the same data set in more than one way.

Observe Student Progress

Computer Tutorial

Some students may benefit from completing a computer tutorial before they attempt the Try It page. A list of the tutorials for each lesson can be found beginning on page x in the front of this book.

 Error Analysis

Exercise 1 To help students get started, ask them how many people on the graph have no brothers or sisters. **5** Point out that this is the height of the first bar. Repeat the question for a few more bars until you are sure students understand the bar graph. For extra practice, have students show the data in a tally chart.

Exercise 2 Remind students to begin by finding the key. It shows that each ball stands for 2 games. Students can count by 2s, or, they can count the balls and multiply by 2.

Exercise 3 If students need a hint, tell them to begin by find the total number of people in the survey. To do this, they can count the X's on the line plot.

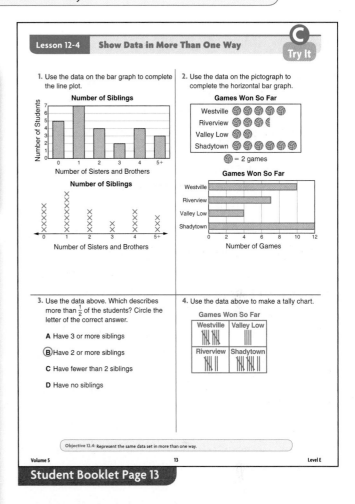

Student Booklet Page 13

ENGLISH LEARNERS Students may be unfamiliar with the word *sibling*. Tell them that a sibling is either a brother or a sister. Have students practice the word, by asking several students how many *siblings* they have and then how many brothers and how many sisters they have.

Lesson 12-5: Graph Ordered Pairs

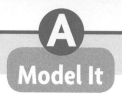

Objective 12.5: Use two-dimensional coordinate graphs to represent points and graph lines and simple figures.

Teach the Lesson

Materials ☐ Teaching Aid 15 First-Quadrant Coordinate Grid

Activate Prior Knowledge

Review how to identify points on the coordinate grid. **Identify the points located at (1, 5) and (4, 0).** Remember, the first number in the ordered pair is the *x*-coordinate and the second number in the ordered pair is the *y*-coordinate. Point *F* is located at (1, 5) and point *Z* is located at (4, 0).

Model the Activity

Activity 1 Draw the graph on the board. **What are the two variables for this problem?** The number of students and the grade level **How do we know which to use as the *x*-coordinate and which to use as the *y*-coordinate?** The graph is labeled "Grade" along the *x*-axis and "Number of Students" along the *y*-axis. Demonstrate how to graph the point (3, 25) on the grid.

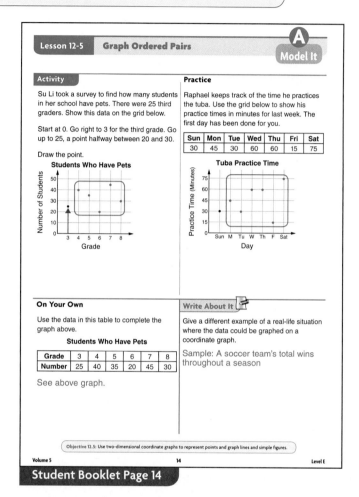

Student Booklet Page 14

Progress Monitoring

Have students use Teaching Aid 15 to make a grid for the previous week. **Number the vertical scale $\frac{1}{2}$, 1, 1$\frac{1}{2}$, 2, and so on.** Label the vertical axis "Number of Hours" and the horizontal axis "Day of the Week." Have students estimate the number of hours of homework they did each day. Graph the data.

Error Analysis

Students are likely to number all graph scales by 1s. Point out that this may make some graphs really huge. The numbering is usually chosen so the graph will be a reasonable size.

Lesson 12-5 Graph Ordered Pairs

Objective 12.5: Use two-dimensional coordinate graphs to represent points and graph lines and simple figures.

Facilitate Student Understanding

Materials ☐ Teaching Aid 15 First-Quadrant Coordinate Grid

Develop Academic Language

On the board, draw a line segment and a triangle with large dots at the vertices. Remind students to use letters to describe these figures. Ask for suggestions on letter names for the two items. Use the drawings to review this geometric vocabulary, writing each term on the board as you discuss it: *line segment, point, endpoint, triangle, vertex, vertices, side.*

Demonstrate the Examples

Example 1a Remind students that this is a coordinate grid. **Which number in the ordered pair is *x*-coordinate? 6 Which number is the *y*-coordinate? 5** Demonstrate how to graph the ordered pair.

Example 1b Write *B*(4, 1) on the board and remind students that this notation means "point *B* is at the location (4, 1)." **This notation is often used to give the vertices of triangles, rectangles, and other polygons.**

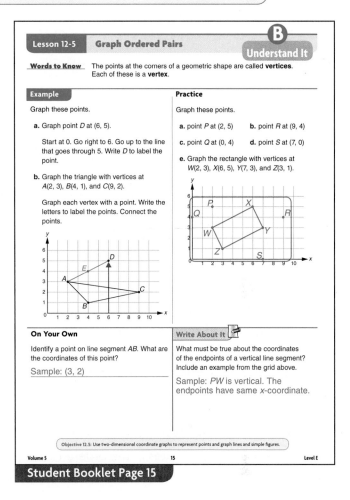

Student Booklet Page 15

ENGLISH LEARNERS Students may have difficulty with the terms *vertex* and *vertices*. Make sure students understand that these are two forms of the same noun. Also make sure students understand an *endpoint* is similar to, but different than, a *vertex*.

 ## Progress Monitoring

Have students use the grid in Practice 1. Tell them to graph point *K* at (6, 3) and describe figure *PXKW*. **It is a rectangle.** Then ask students to use points *Y* and *S* and add two more points to make a rectangle 3 units tall and 2 units wide. **There are two possibilities: (5, 0) and (5, 3); (9, 0) and (9, 3).**

Error Analysis

Students may need extra instruction for graphing points on the two axes. Make a grid and put a large dot at (0, 4). Explain that we get to this point by starting at 0 and not going to the right at all. We go 0 units right and then 4 units up. Then put a large dot at (5, 0). For this point we go 5 units to the right and then stop. We go over 5 and 0 units up.

| Lesson 12-5 | **Graph Ordered Pairs** |

Objective 12.5: Use two-dimensional coordinate graphs to represent points and graph lines and simple figures.

Observe Student Progress

Materials ☐ Teaching Aid 15 First-Quadrant Coordinate Grid

Computer Tutorial

Some students may benefit from completing a computer tutorial before they attempt the Try It page. A list of the tutorials for each lesson can be found beginning on page x in the front of this book.

⭐ Error Analysis

Exercise 1 Remind students that the first column in the chart is another way of showing the ordered pair (Sunday, 7) or (Sun, 7). Ask what this ordered pair means. **On Sunday there were 7 people working in the bike shop.**

Exercise 2 Watch to make sure that students are not reversing coordinates in the ordered pairs.

For extra practice have students graph the square with vertices at F(6, 6), Y(6, 2), X(10, 2), and Z(10, 6).

Exercise 4 Provide students with Teaching Aid 15. If students have difficulty identifying the shape as a square, have them double check their graphs. They may be placing the points incorrectly or have forgotten to connect the first and last points.

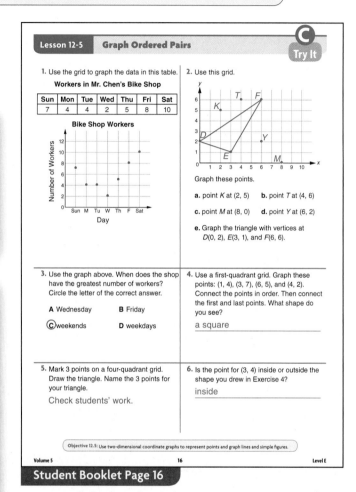

Student Booklet Page 16

ENGLISH LEARNERS Make sure students understand the *sides* of the figures they graphed are also called *line segments*. Remind students that a triangle is made of 3 *line segments*.

Volume 5 — Data, Geometry, and Measurement

Topic 12: Graphing

Topic Summary

Objective: Review creating and interpreting graphs.

Have students complete the summary page. You may want to have students work in groups of four with each student analyzing a different choice.

Ask students to share their ideas about each answer choice. Be sure they confirm the correct answer for each problem at the end of the discussion.

Answer Evaluation

1. **A** This choice is correct.

   B Students compared the quantities for 2006 and 2007 instead of for 2006 and 2004.

   C Students compared the quantities for 2005 and 2006 instead of for 2006 and 2004.

   D Students compared the quantities for 2005 and 2004 instead of for 2006 and 2004.

2. **A** Students misread the graph, thinking that the hour from 1 to 2 is the first hour rather than the second hour.

   B This choice is correct.

   C Students chose the hour with the second greatest distance.

   D Students chose the hour with the greatest number on the scale for the hours.

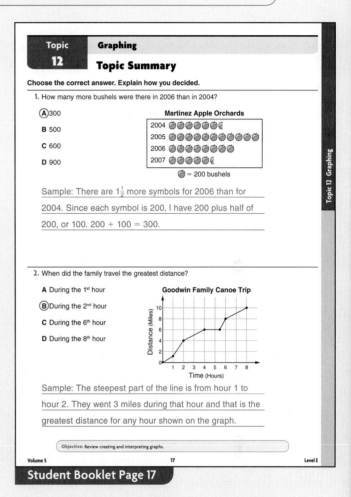

Student Booklet Page 17

Error Analysis

Exercise 1 Make sure that students can explain the meaning of the key. **Each apple stands for 200 bushels.** Ask what is meant by one-half an apple and which years include that symbol. **100 bushels; 2004 and 2007** Have students find the totals for each year and write them at the ends of the rows of symbols. **1,300; 2,200; 1,600; 1,100**

Exercise 2 Start at the left and ask students how far they had gone in 1 hour, 2 hours, and so on. **1 mile, 4 miles, and so on.** Ask when they stopped for lunch and how the graph shows this. **From the beginning of hour 5 until halfway through hour 6. The graph is flat during this time.**

Progress Monitoring

When all assignments for this topic have been completed, assign the corresponding Progress Monitoring page for this topic (Assessment Resources Book, page 12). Be sure students complete the Progress Monitoring page before you administer the final assessment for this volume.

Volume 5 — Data, Geometry, and Measurement

Topic 12: Graphing

Mixed Review

Objective: Maintain concepts and skills.

Have students complete the Mixed Review page. Work with each student individually to review results. Identify strengths and weaknesses and correct any misunderstandings.

Error Analysis

Exercise 1 Have students study the graph before reading the questions. Ask them to identify the title, the labels on the scales, and describe how the scales are numbered.

Exercise 2 Make sure students are lining up the place values of the addends correctly. They may be confused in Part b because the 2-digit number comes before the 3-digit number.

Exercise 5 Remind students that in expanded notation the value of each place is written out in standard form. If necessary, model an example using smaller numbers.

Exercise 6 Encourage students to draw a model or use manipulatives to act out this problem if they are having difficulty visualizing it.

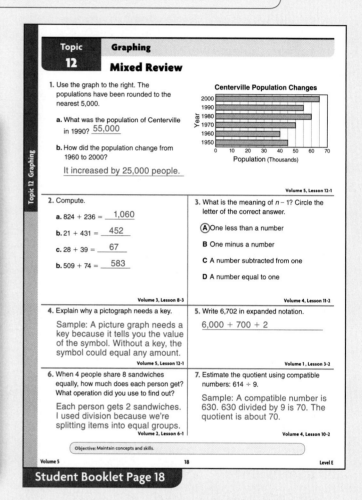

Student Booklet Page 18

Volume 5: Data, Geometry, and Measurement

Topic 13: Basic Geometric Figures

Topic Introduction

This topic introduces students to the concept of two-dimensional plane figures. Lesson 13-1 begins with ways to draw and classify lines, line segments, rays, and angles. Lesson 13-2 moves on to a discussion of different types of polygons. In Lessons 13-3 and 13-4, this discussion of polygons focuses on the different types of triangles and quadrilaterals and their properties. Finally, in Lesson 13-5, students define and explore the parts of a circle.

Lesson	Objective	Student Pages	Teacher Pages	Tutorials
Topic 13 Introduction	**13.1** Draw, measure, and classify different types of angles and lines. **13.2** Define *polygon* and classify the different types of polygons. **13.3** Explore, compare, and classify different types of triangles. **13.5** Explore circles and define their parts.	19	212	
13-1 Draw Angles and Lines	**13.1** Multiply decimals by whole numbers.	20–22	213–215	13a, 13b, 13c, 13d
13-2 Types of Polygons	**13.2** Define *polygon* and classify the different types of polygons.	23–25	216–218	13e
13-3 Triangles	**13.3** Explore, compare, and classify different types of triangles.	26–28	219–221	13f, 13g
13-4 Quadrilaterals	**13.4** Explore, compare, and classify different types of quadrilaterals.	29–31	222–224	13h
13-5 Circles	**13.5** Explore circles and define its parts.	32–34	225–227	13i
Topic 13 Summary	Review basic geometric figures.	35	228	
Topic 13 Mixed Review	Maintain concepts and skills.	36	229	

Computer Tutorial

Some students may benefit from completing the computer tutorial before they attempt the Try It page of each lesson. If you are using the electronic components of *Pinpoint Math*, you will find a complete listing of Tutorial codes and titles when you access them either online or via CD-ROM.

Volume 5 | **Data, Geometry, and Measurement**

Topic 13 | **Basic Geometric Figures**

Topic Introduction

Objectives: 13.1 Draw, measure, and classify different types of angles and lines. **13.2** Define *polygon* and classify the different types of polygons. **13.3** Explore, compare, and classify different types of triangles. **13.5** Explore circles and define their parts.

Materials ☐ Attributes blocks

Have students review different attribute blocks. Ask students to notice number of sides and vertices of the different shapes.

Informal Assessment

1. **An angle is made up of two rays or line segments that meet at a common endpoint. To find the angle, you need to use a protractor. Point to the vertex of the angle.** Check that students have identified the vertex. **Place the center point of the protractor on the vertex and align 0° with one of the rays forming the angle. Read where the other ray crosses the inside scale. What is the measure of the angle?** 60°

2. **Count the sides and vertices of the polygon. How many sides does it have?** 5 **How many vertices?** 5 **What is the name of the polygon?** Pentagon

3. **All the sides of the first triangle are congruent. What does this mean?** All of the sides are the same length. **What kind of triangle has congruent sides?** An equilateral triangle **The next triangle has an angle that measures 90°. What kind of angle is this?** Right **What is the name of this triangle?** Right triangle

4. **What are the parts of a circle?** Diameter, radius, chord **What is the diameter of this Circle C?** $\overline{AE}$ **How do you know?** The line passes through the center of the circle.

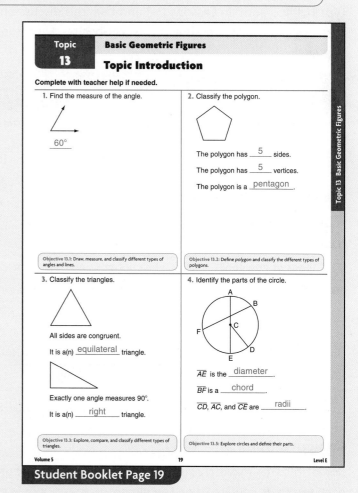

Student Booklet Page 19

Another Way Have pairs of students identify and sort shapes based on the number of sides and vertices.

ENGLISH LEARNERS Help students connect the vocabulary to the concept by providing illustrations. Explain the singular and plural *vertex/vertices* for polygons and *radius/radii* of circles.

Lesson 13-1 Angles and Lines

Objective 13.1: Draw, measure, and classify different types of angles and lines.

Teach the Lesson

Materials ☐ Protractors

Activate Prior Knowledge

Have students draw and name examples of lines, segments, and rays. Check that students label the figures correctly. A line's name is topped by a figure of a line (↔); a segment's, by a figure of a segment, and a ray's, by a figure of a ray.

Develop Academic Language

Make sure students use *line, line segment*, and *ray* appropriately.

Point out that protractors show two *scales*, like the scale on a ruler. Demonstrate using the correct scale, depending on the direction of the ray at 0°.

Model the Activities

Activity 1 Ask students to identify the vertex of the angle and use the instructions on the page to measure the angle with their protractors. Check that students have aligned their protractors correctly. **What is the measure of the angle?** 110°

Activity 2 Use the edge of the protractor to draw one side of the angle. Align the endpoint of the ray with the center point of the protractor. Align 0° with the other end of the ray. **Which scale do you use to draw the other side of the angle?** The answer depends on direction of the ray.

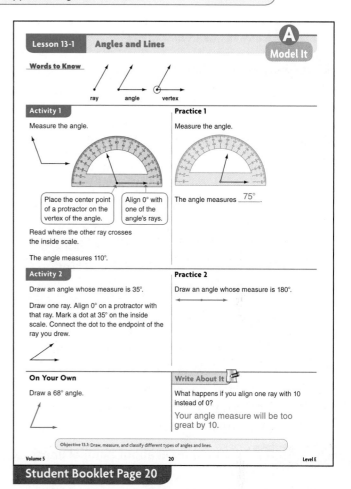

Student Booklet Page 20

Progress Monitoring

Have students draw a ray pointing to the left end of the page and use the outside scale of a protractor to make a 90° angle.

Error Analysis

Monitor students' use of protractors to ensure proper alignment and use of the appropriate scale.

ENGLISH LEARNERS Have students copy the Words to Know and diagrams in their word banks.

Volume 5   213   Level E

Lesson 13-1 — Angles and Lines

Objective 13.1: Draw, measure, and classify different types of angles and lines.

Facilitate Student Understanding

Materials ☐ Protractors

Develop Academic Language

Provide definitions of the Words to Know and ask students to find examples of each. **Find a right angle in the room. Find a straight angle in the room. Find a pair of parallel lines in the room. Find a pair of perpendicular lines in the room. Do you see any other kinds of angles in the room?** Call attention to lines that cross but are not perpendicular.

Demonstrate the Examples

Example 1 **Measure each angle. What is the measure of angle A?** 85° **What type of angle is angle A?** Acute **How do you know?** Its measure is less than 90°. Emphasize the relationship of each angle to 90° as students continue.

Example 2 **What is special about parallel lines?** They never meet. Point out that the lines must lie in the same plane. You may want to introduce the word *intersect*. **Which lines in the diagram appear to be the same distance from each other at all points?** Lines *AB* and *CD* **Why aren't lines AC and BD parallel?** They will cross if extended.

Practice 2 Point out that students can measure the angles at the intersections.

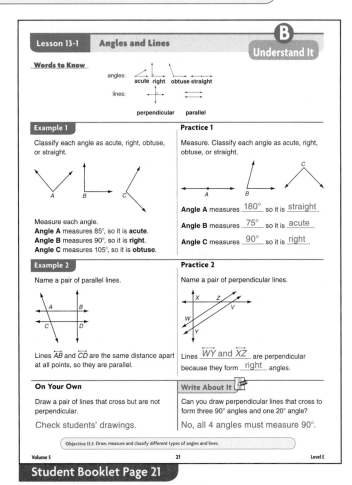

Student Booklet Page 21

⭐ Progress Monitoring

How many right angles are formed by the intersection of two perpendicular lines? 4

Error Analysis

In Example 2, students might assume that lines *AC* and *BD* are parallel because they do not intersect on the diagram. Remind them that lines continue infinitely in both directions, and students can extend the lines to see whether they will ever cross.

Volume 5 214 Level E

Lesson 13-1 Angles and Lines

Objective 13.1: Draw, measure, and classify different types of angles and lines.

Observe Student Progress

Computer Tutorial

Some students may benefit from completing the computer tutorial before they attempt the Try It page of each lesson. A list of the tutorials for each lesson can be found beginning on page x in the front of this book.

★ Error Analysis

Exercise 1 Watch for students who might orient the protractor incorrectly or use the wrong scale when reading the measure of the angle. If students notice that the angle is wider than a right angle, they can conclude that the angle measure must be greater than 90° and use this to check that they're using the right scale.

Exercise 3 Remind students to extend lines if necessary to determine whether they will ever meet. The lines in answer D might appear to be parallel, but if they are extended, these lines will meet.

Exercise 4 Point out that just as an acute angle must measure greater than 0° and less than 90°, an obtuse angle must measure greater than 90° and less than 180°. Ask students to explain why the straight angle in answer D is not correct.

Exercise 6 Remind students that a straight angle looks like a straight line, which measures 180 degrees.

Exercise 7 Help students identify and measure the angles of intersection.

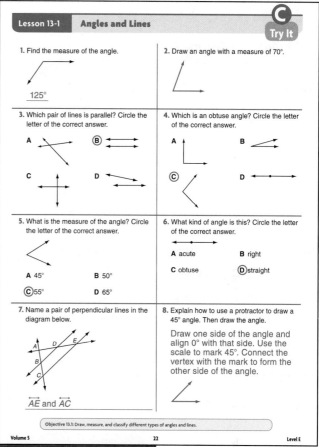

Student Booklet Page 22

Lesson 13-2 Types of Polygons

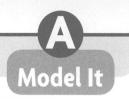

Objective 13.2: Define *polygon* and classify the different types of polygons.

Teach the Lesson

Materials ☐ Polygon Tiles

Activate Prior Knowledge

Think of words that start with the prefix *tri-*. Samples: triangle, tripod, tricycle, triple List students' suggestions on the board. Discuss the meaning of each word. **What does the prefix *tri-* mean?** Three Repeat with the prefixes *quad-*, *pent-*, and *oct-*.

Develop Academic Language

A *closed figure* can be traced by starting at one point on the figure and ending at the same point. **Draw a figure that is NOT closed. These figures are NOT polygons.** Point out that polygons are also 2-dimensional, meaning they have length and width but not thickness.

ENGLISH LEARNERS Remind students that the plural of *vertex* is *vertices*.

Model the Activities

Activity 1 **Find this quadrilateral polygon tile.** Have students trace the sides. **Is the figure closed?** Yes. **How many line segments make up the quadrilateral?** 4 **How many vertices are there?** 4 **How many sides and vertices does a quadrilateral have?** 4

Activity 2 **Use a polygon tile to model the pentagon. Is the figure closed?** Yes. **How many line segments make up the pentagon?** 5 **How many vertices are there?** 5 **How many sides and vertices does a pentagon have?** 5

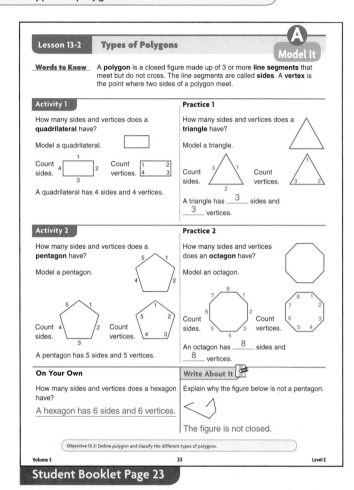

Student Booklet Page 23

Progress Monitoring

Write the following statements on the board:

1. All squares are quadrilaterals.
2. All rectangles are polygons.
3. All quadrilaterals are squares.
4. All polygons are quadrilaterals.

Which statements are true? 1 and 2

Error Analysis

Students might assume that all polygons are regular polygons. Use the quadrilateral in Activity 1 to show that this assumption is not true.

Volume 5   216   Level E

Lesson 13-2: Types of Polygons

Objective 13.2: Define *polygon* and classify the different types of polygons.

Facilitate Student Understanding

Develop Academic Language

What are the sides of a polygon? Line segments Draw some figures with curved sides on the board. These figures are not polygons, because not all of the sides are line segments.

ENGLISH LEARNERS Point out that *classify* means to put something in a named group. On this page, students are being asked to name each shape.

Demonstrate the Examples

Example 1 How can you determine what kind of polygon this figure is? **Count the sides or the vertices.** How many sides does it have? **7** What prefix means seven? **Hept-** What is the name of the polygon? **Heptagon** You may want to point out that this is an irregular heptagon; not all the sides are the same length.

Example 2 How many sides does this figure have? **4** What prefix means four? **Quad-** What is the name of the polygon? **Quadrilateral** Ask students to explain why *rectangle* and *square* are not good names for this shape; it has no right angles. Explain that *quadrilateral* is a general term that includes rhombuses, rectangles, trapezoids, and so on.

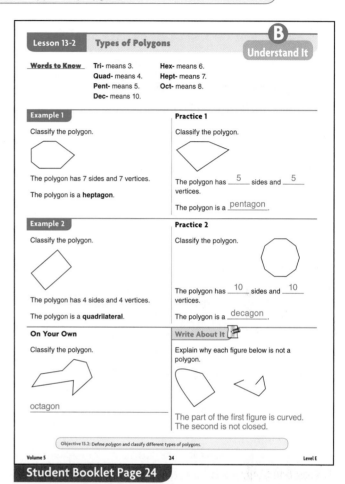

Student Booklet Page 24

✓ Progress Monitoring

Draw a quadrilateral with four equal sides and four right angles. What is its name? **Square**

Error Analysis

Encourage students to make flash cards with a picture of the figure on the front and characteristics listed on the back.

ENGLISH LEARNERS If students have trouble pronouncing the names of certain polygons, allow them to hold up the correct flash card when answering a question.

Lesson 13-2: Types of Polygons

Objective 13.2: Define *polygon* and classify the different types of polygons.

Observe Student Progress

Computer Tutorial

Some students may benefit from completing the computer tutorial before they attempt the Try It page of each lesson. A list of the tutorials for each lesson can be found beginning on page x in the front of this book.

 Error Analysis

Exercise 1 Help students relate *oct-* to *octopus* to remember that it means "eight," and note that *six* and *hex-* both have the letter *x*.

Exercise 2 Remind students that any 4-sided polygon is a quadrilateral. Have students name as many 4-sided figures as they can. Remind students that a polygon has as many vertices as sides.

Exercises 3, 5, and 6 Ask students to explain why each incorrect answer choice does not meet the requirements of the question.

Exercise 7 Ask students to think of other street signs and the polygon used for each. **Samples:** one-way—quadrilateral; school crossing—pentagon

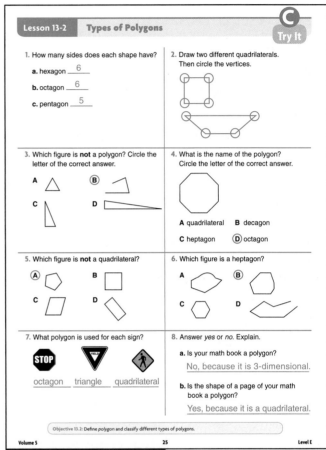

Student Booklet Page 25

Lesson 13-3 Triangles

Objective 13.3: Explore, compare, and classify different types of triangles.

Teach the Lesson

Materials ☐ Protractor and ruler
☐ Teaching Aid 18 Angles and Triangles

Activate Prior Knowledge

Have students use Teaching Aid 18 to practice measuring length in centimeters and angles at different orientations.

Develop Academic Language

Triangles can be classified according to the lengths of their sides, the measures of their angles, or both. *Congruent* means having the same measure. What is a triangle with 3 congruent sides called? **Equilateral** A triangle with exactly 2 congruent sides? **Isosceles**

Model the Activities

Activity 1 How can we use a ruler to determine whether this triangle is equilateral? **Measure each side. If the lengths are equal, the triangle is equilateral.** Have students measure. **The marks on the sides show that those sides are equal. The triangle is equilateral.**

Practice 1 Show students how to mark equal angles.

ENGLISH LEARNERS Relate *equi-* to *equivalent*.

Activity 2 How can we use a ruler to determine whether this triangle is isosceles? **Measure each side. Which sides have to be equal? Any two** Have students measure.

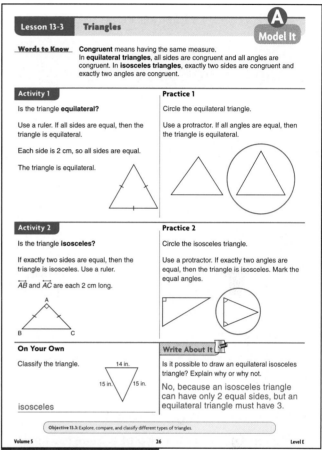

Student Booklet Page 26

Progress Monitoring

How many different equilateral triangles can you draw if one of the sides must be 5 cm long? **1** If the sides can be any length? **An infinite number, as long as the sides are equal**

Error Analysis

Have students label the angles in Practice 1 and Practice 2 to ensure that they do not try to answer the questions without measuring.

Volume 5   219   Level E

Lesson 13-3 Triangles

Objective 13.3: Explore, compare, and classify different types of triangles.

Facilitate Student Understanding

Materials ☐ Teaching Aid 18 Angles and Triangles

Activate Prior Knowledge

Have students find the sum of the angle measures in triangles on Teaching Aid 18.

Develop Academic Language

A *right triangle* has exactly 1 right angle. Show that 180° − 90° leaves 90° for the combined measures of the other angles. Draw a right triangle on the board for students and point out to them the symbol used to mark a right triangle in a model.

ENGLISH LEARNERS Clarify that *measure* is used as a noun. On this page, students should not have to measure.

Demonstrate the Examples

Example 1 If you know two of a triangle's angle measures, you don't need a protractor to find the third measure. What does this example tell us about the angle measures of any triangle? They sum to 180°. We have to take each of the known angle measures away from 180°. What are some ways to do this? We can subtract one at a time or add and then subtract. Model each method. Mentally, it may be easier to add and then subtract.

Example 2 What does the square mark mean about that angle? Allow students to measure. It's a right angle. Ask students to describe the 3 types of triangles they have studied. If you had to choose to measure sides OR angles to classify a triangle, which would allow you to classify all those types of triangles? Angle measure

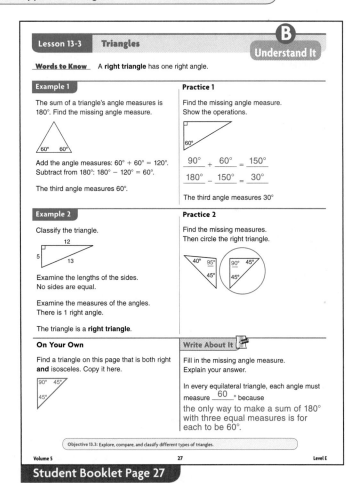

Student Booklet Page 27

★ Progress Monitoring

A triangle has two angles measuring 30° and 60°. What do you know about the triangle? It's right, since the third angle must measure 90°.

Error Analysis

Allow students to check answers with a protractor. Memorizing the fact that equilateral triangles have total angle measures of 60° × 3 may provide a tool for remembering 180° as the sum of any triangle's angles.

Lesson 13-3 Triangles

Objective 13.3: Explore, compare, and classify different types of triangles.

Observe Student Progress

Computer Tutorial

Some students may benefit from completing the computer tutorial before they attempt the Try It page of each lesson. A list of the tutorials for each lesson can be found beginning on page x in the front of this book.

⭐ Error Analysis

ENGLISH LEARNERS Have students make flash cards showing examples of the different types of triangles.

Exercises 1 and 2 Point out that students are given enough information to solve without measuring. Allow students to find angle measures to check their answers.

Exercise 3 Help students recall the meaning of the right-angle mark and the marks on equal sides. Watch for students who notice only the right angle and choose answer C. Have students list every attribute of the triangle before choosing an answer.

Exercise 4 Remind students that if two angle measures are known, there is a way to find the third angle measure without using a protractor. Help students recall the operations.

Exercises 5 and 6 Point out that students are given enough information to solve without measuring. Allow students to find side lengths to check their answers.

Exercises 6 and 7 Have students predict the measure of the third angle before measuring it.

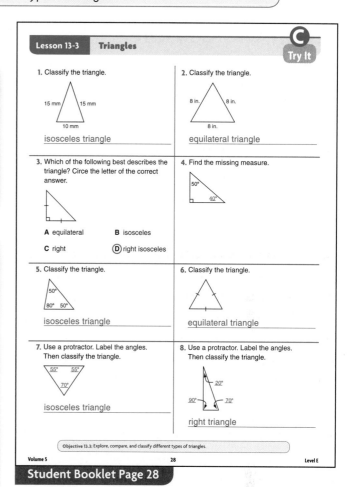

Student Booklet Page 28

Lesson 13-4 Quadrilaterals

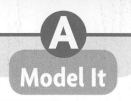

Objective 13.4: Explore, compare, and classify different types of quadrilaterals.

Teach the Lesson

Materials
- ☐ Protractors
- ☐ Rulers
- ☐ Polygon Tiles

Activate Prior Knowledge
Have students draw and define *parallel lines* and *right angles*. Lines that lie in the same plane and do not intersect; 90° angles

Develop Academic Language
A *quadrilateral* is a polygon with 4 sides. Help students write definitions to their word banks. A *parallelogram* is a quadrilateral with 2 pairs of parallel sides. A *rectangle* is a parallelogram with 4 right angles. A *square* is a parallelogram with 4 congruent sides and 4 right angles. A *rhombus* is a parallelogram with 4 congruent sides.

ENGLISH LEARNERS The prefix *quad-* means 4. Ask students to find another *quad-* word. Samples: quadruplets, quadruple

Model the Activities

Activity 1 Use a protractor and a ruler to measure the angles and sides. What is true of the sides and angles of a rectangle? Opposite sides are congruent and each angle measures 90°. In Practice 1, you'll find one shape that has a different name but is also a rectangle.

Activity 2 How can you use a ruler to test whether opposite sides of a shape are parallel? Measure at different points to see whether they're always the same distance apart. Why isn't the shape also a rectangle? There are no 90° angles. A square? All sides aren't congruent.

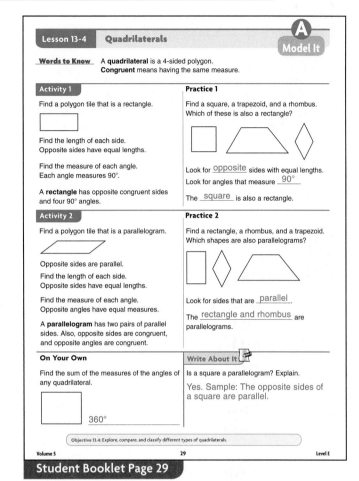

Student Booklet Page 29

 Progress Monitoring

Which quadrilaterals have AT LEAST one pair of opposite sides that are parallel? Rectangles, rhombuses, squares

Error Analysis

Show this chart to help students classify the quadrilaterals shown on this page:

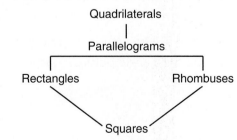

Lesson 13-4 Quadrilaterals

Objective 13.4: Explore, compare, and classify different types of quadrilaterals.

Facilitate Student Understanding

Develop Academic Language

Have students add the new definition to their word banks: A *trapezoid* is a quadrilateral with exactly 1 pair of parallel sides. You can add to the tree diagram shown on the preceding page a branch labeled *Not parallelograms* leading to *Trapezoids* and *Other*. As students classify quadrilaterals, you may want to have them work with definitions from the top of the diagram down.

Demonstrate the Examples

Example 1 Guide students to examine each trapezoid. **What do these shapes have in common?** Each has 1 pair of parallel sides. **What is different about the shapes?** The angle measures can be different in each shape; the sides' lengths can be different or 1 pair can be congruent.

Practice 1 What do the marks on the sides tell you? Opposite sides are congruent. What does the mark on each angle tell you? The angle measures 90°.

Example 2 What is the sum of the angle measures of any quadrilateral? 360° You may want to point out that this is twice the sum of a triangle's angle measures. If you know the measures of 3 angles of a quadrilateral, what are some ways to find the measure of the fourth angle without measuring? Add the measures of the 3 angles and subtract the sum from 360°; subtract each angle measure from 360°. Help students interpret the equation and solve for *x*. 70°

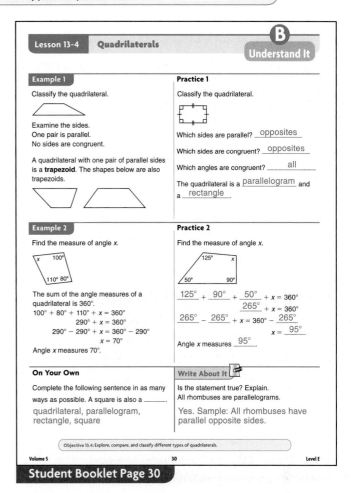

Student Booklet Page 30

Progress Monitoring

Complete each sentence with *All*, *No*, or *Some*.
1. All squares are rhombuses. 2. No trapezoids are parallelograms. 3. All squares are rectangles.

Error Analysis

Remind students not to assume that a figure is a special quadrilateral simply because it looks like one. For example, unless the problem states a figure is a square, the sides of a square should be marked congruent and the angles should be marked as right angles.

Volume 5   223   Level E

Lesson 13-4: Quadrilaterals

Objective 13.4: Explore, compare, and classify different types of quadrilaterals.

Observe Student Progress

Computer Tutorial

Some students may benefit from completing the computer tutorial before they attempt the Try It page of each lesson. A list of the tutorials for each lesson can be found beginning on page x in the front of this book.

Error Analysis

Exercise 1 Display the Quadrilaterals tree diagram described on the previous pages to help students test the options systematically.

Exercise 2 If students label the figure a square because it has 4 congruent sides, remind them to check angles as well as sides. Review the definitions of *square* and *rhombus*.

Exercise 3 Students who choose B have not read the question correctly. Remind students to read carefully and underline important words such as *cannot*.

Exercise 5 Suggest that students draw the quadrilateral described in the problem.

Exercise 6 Show a diagram of a rectangle divided diagonally in half. Show that the sum of the angle measures in a triangle, 180°, is half of 360°, the sum of the angle measures in a quadrilateral.

Exercise 8 Suggest that students try to draw a quadrilateral that is not a square to determine whether the statement is true.

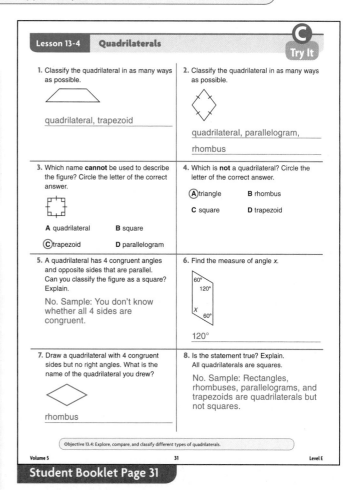

Student Booklet Page 31

ENGLISH LEARNERS Help students devise mnemonics for the shapes' names. The double *l* in *parallelogram* may help students think of *parallel*, students can draw a trap in the shape of a trapezoid to help them recall the meaning of *trapezoid*, and so on.

Lesson 13-5 Circles

Objective 13.5: Explore circles and define their parts.

Teach the Lesson

Materials ☐ Compass
☐ Ruler

Activate Prior Knowledge

What is a polygon? A closed figure made up of line segments that don't cross **Is a circle a polygon?** No. **Why?** It's a closed figure, but it's not made of line segments.

Develop Academic Language

Draw a circle on the board. Ask for volunteers to draw a radius and diameter as other students read the definitions.

ENGLISH LEARNERS Point out that the compass used in this lesson is different from the compass used for determining direction. Ask students how the directional compass is also related to a circle.

Model the Activities

Activity 1 To draw a circle, begin with a point for the center. How long does the radius have to be? $\frac{3}{4}$ inch Have students follow the directions in Activity 1. **Does it matter where you put your starting point?** No. **How can you check your answer?** Measure the radius. Align the 0 mark on a ruler with the center and measure to any point on the circle.

Activity 2 To use the compass, we have to know the length of the radius. Allow students to test what happens when the compass is set to the length of the diameter. **How are the diameter and radius of a circle related?** The diameter is twice the radius. Draw a diameter on a circle, and darken half of the diameter to show a radius as you ask: **If you know the length of the diameter, how do you find the length of the radius?** Divide the diameter by 2. Have students draw the circle.

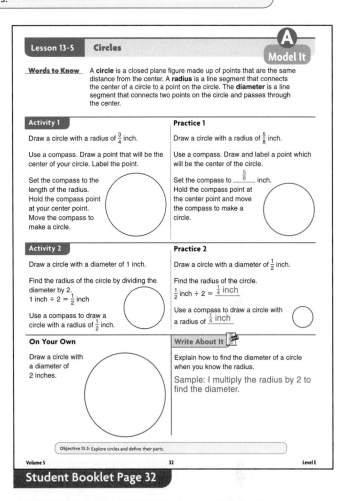

Student Booklet Page 32

Progress Monitoring

What happens if you set your compass to 1 inch to draw a circle with a 1-inch diameter? The diameter will be 2 inches.

Error Analysis

Check that students correctly use the pivot point of the compass as the center of the circle.

Volume 5   225   Level E

Lesson 13-5 Circles

Objective 13.5: Explore circles and define their parts.

Facilitate Student Understanding

Develop Academic Language

Clarify that *on the circle* means points on the curve rather than points within the curve. Remind students that a circle is named by its center, which is not a point on the circle.

The plural of *radius* is *radii*. How many radii can a circle have? An infinite number **Diameters?** An infinite number **Chords?** An infinite number

Demonstrate the Examples

Example 1 **What does *radius* mean?** A line segment that connects the center to one point on the circle **What is the center of this circle?** C **Trace each segment and decide whether it meets the definition.** Students who start at the center will see that $\overline{AC}$ and $\overline{BC}$ can be traced separately. Point out to other students that the diameter $\overline{AD}$ can be separated into 2 radii. **How many radii are shown?** 3 **What are they?** $\overline{BC}, \overline{AC},$ and $\overline{DC}$ **Which segment is a diameter?** $\overline{AD}$ **Which segments do NOT include the center?** $\overline{AD}, \overline{AE},$ and $\overline{DF}$ Any segments that connect 2 points on the circle without passing through the center are chords.

Example 2 **How can you use the radius to draw a diameter on this circle?** Extend the segment through the center to touch another point on the circle. **How can we find the length of the diameter?** Double the radius' length. **How can you write this relationship as an equation?** $d = 2 \times r$ Have students complete the calculations.

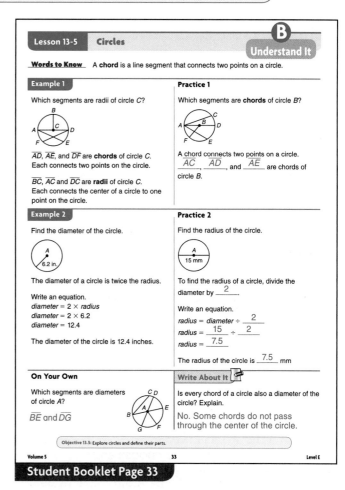

Student Booklet Page 33

Progress Monitoring

Can you use a compass to draw a circle if you know the length of a chord? Explain. No; a chord can connect any two points on the circle, so it can be many different lengths. Help students draw examples.

Error Analysis

Remind students that a diameter is also a chord and that 2 radii form each diameter.

ENGLISH LEARNERS Carefully model the pronunciation of *radii* (RAY-dee-eye) and *chord* (kord).

Lesson 13-5 Circles

Objective 13.5: Explore circles and define their parts.

Observe Student Progress

Computer Tutorial
Some students may benefit from completing the computer tutorial before they attempt the Try It page of each lesson. A list of the tutorials for each lesson can be found beginning on page x in the front of this book.

★ Error Analysis

Exercises 1 and 2 Will you set your compass to the length of the radius or the length of the diameter? The length of the radius Have students draw the circles and use a ruler to check their measurements.

Exercise 3 Students who choose answers C or D are confusing *radius* and *diameter*. Review the definitions.

Exercise 4 Students who choose answer A have misread the question. Remind students to read carefully and underline any important words, such as *not*.

Exercise 5 Some students might need visual cues to help them solve this problem. Encourage students to draw each figure and label the radius and diameter.

Exercise 6 Have students draw a diagram of the pool. What are you asked to find: the length of the radius, the diameter, or a chord? The length of a radius, from the center to a point on the circle

Exercise 8 Have students draw or construct two sample circles to check. You may want to provide a radius to start with.

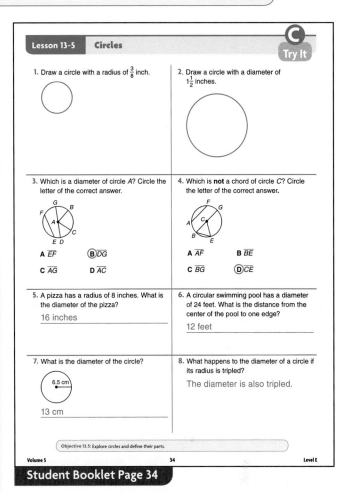

Student Booklet Page 34

Volume 5 — Data, Geometry, and Measurement

Topic 13: Basic Geometric Figures

Topic Summary

Objective: Review basic geometric figures.

Have students complete the student summary page. You may want to have students work in groups of four with each student analyzing a different choice.

Ask students to share their ideas about each answer choice. Be sure they confirm the correct answer for each problem at the end of the discussion.

Answer Evaluation

1. **A** Students thought 90° angle was acute (right angle).
   B This choice is correct.
   C Students thought 180° angle was acute (straight angle).
   D Students thought an angle over 90° and less than 180° was acute (obtuse angle).

2. **A** Students were unaware of quadrilateral properties.
   B Students were unaware of quadrilateral properties.
   C This choice is correct.
   D Students were unaware of quadrilateral properties.

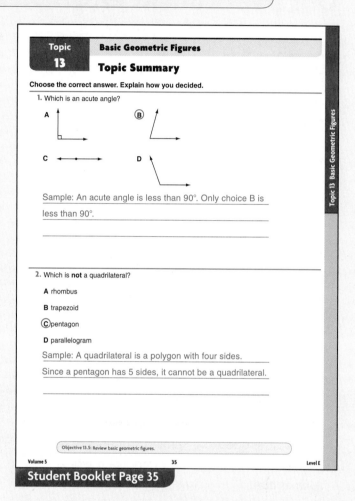
Student Booklet Page 35

 ## Error Analysis

Exercise 1 Review the Words to Know so students know the difference in measurements of the four types of angles.

Exercise 2 Review the names of polygons and properties of quadrilaterals.

Progress Monitoring

When all assignments for this topic have been completed, assign the corresponding Progress Monitoring page for this topic (Assessment Resources Book, page 13). Be sure students complete the Progress Monitoring page before you administer the final assessment for this volume.

Volume 5: Data, Geometry, and Measurement
Topic 13: Basic Geometric Figures
Mixed Review

Objective: Maintain concepts and skills.

Have students complete the Mixed Review page. Work with students individually to review results. Identify strengths and weaknesses and correct any misunderstandings.

 Error Analysis

Exercises 1 and 5 If students have difficulty determining which mathematical operation is necessary, review key terms in exercises.

Exercise 2 Review key terms to determine placement of variable.

Exercise 4 Review the terms *mode* and *median* with any students who use the incorrect term for the data. Suggest a strategy to find the median by first placing the data in numerical order and then crossing out the data alternately at each end.

Exercise 6 Remind students to group addends in a way that makes the problem easier to solve. **Which two addends should you add first?** Sample: 6 and 4, because they make 10, and it is easy to add another number to 10

Topic 13 Basic Geometric Figures — Mixed Review

1. Élan ships gift packages to friends. He ships 9 packages. Each package costs $15 to ship. How much does Élan pay to ship his packages?
   $135

2. Which equation represents the following statement? Circle the letter of the correct answer.

   Eight more than some number is 13.

   A $n \div 8 = 13$ B $8 \div n = 13$
   C $13 + n = 8$ (D) $n + 8 = 13$

3. What is $78 \div 6$? Circle the letter of the correct answer.
   (A) 13   B 12
   C 14   D 15

4. Find the median and mode of the data.
   26, 29, 18, 31, 20, 29, 22
   Median: 26
   Mode: 29

5. Glenn worked 40 hours in 5 days. He worked the same schedule each day. How many hours did he work each day?
   8

6. Brianna walked 6 miles on Saturday, 4 miles on Sunday, and 5 miles on Monday. How many miles did she walk all together? 15

Student Booklet Page 36

ENGLISH LEARNERS The word problems in Exercises 1, 5, and 6 may pose difficulties for students. Pair them with fluent English readers to translate the word problems into number sentences.

Volume 5: Data, Geometry, and Measurement

Topic 14: Measurement Conversion

Topic Introduction

Measurement is the focus of this lesson. In Lesson 14-1 students are introduced to the basic units of measurement used in the U.S. In Lesson 14-2, students define and compare metric prefixes like *centi-*, *milli-*, *deci-*, and *kilo-* to prepare them for Lesson 14-3. Lesson 14-3 gives students an opportunity to use the metric system to explore the basic metric units and their relationships in terms of multiples of 10. Lesson 14-4 teaches students the conversion factors they will need to move on to Lesson 14-5, where students will perform basic unit conversions within a system of measurement.

Lesson	Objective	Student Pages	Teacher Pages	Tutorials
Topic 14 Introduction	14.5 Carry out simple unit conversions within a system of measurement.	37	231	
14-1 U.S. Customary Units	14.1 Explore the basic units of measure in the United States.	38–40	232–234	14a
14-2 Basic Metric Prefixes	14.2 Explore the basic metric prefixes and what they mean.	41–43	235–237	14b, 14c
14-3 Use the Metric System	14.3 Explore the basic metric units and their relationships.	44–46	238–240	14b, 14c
14-4 Factors in Unit Conversions	14.4 Express simple unit conversions in symbolic form.	47–49	241–243	14b, 14d, 14e, 14a
14-5 Convert Units within a System	14.5 Carry out simple unit conversions within a system of measurement.	50–52	244–246	14b, 14d, 14e
Topic 14 Summary	Review measurement conversions.	53	247	
Topic 14 Mixed Review	Maintain concepts and skills.	54	248	

Computer Tutorial

Some students may benefit from completing the computer tutorial before they attempt the Try It page of each lesson. If you are using the electronic components of *Pinpoint Math*, you will find a complete listing of Tutorial codes and titles when you access them either online or via CD-ROM.

Volume 5: Data, Geometry, and Measurement

Topic 14: Measurement Conversion

Topic Introduction

> **Objective: 14.5** Carry out simple unit conversions within a system of measurement.

Materials ☐ Foot ruler, yardstick, and meterstick

Have students refresh the skills necessary for this lesson by solving various multidigit number multiplication problems.

Informal Assessment

1. **How many inches are in 1 foot?** 12 **If we put two feet end-to-end, how many total inches is that length?** 24 **How could you find the number of inches in 7 feet?** Multiply 7 and 12.

   How many centimeters are in a meter? 100 **How many centimeters are in 2 meters?** 200 **If a length is 600 centimeters, how many meters is that?** 6

2. Display a foot ruler, a yardstick, and a meterstick. **Which of these is the shortest?** Foot ruler **Which is the longest?** Meterstick **Which are used for U.S. measurement?** Foot ruler and yardstick **What measurement system is the meter used for?** Metric

3. **How many cups are in a pint?** 2 **How many ounces are in a cup?** 8 **How would I find out how many ounces are in a pint?** Multiply 2 × 8 to get 16 ounces.

4. **How many ounces are in 8 tablespoons?** 4 **How else can you measure 4 ounces?** $\frac{1}{2}$ cup **If you were a baker, would it be easier to measure 8 tablespoons, or $\frac{1}{2}$ cup?** $\frac{1}{2}$ cup

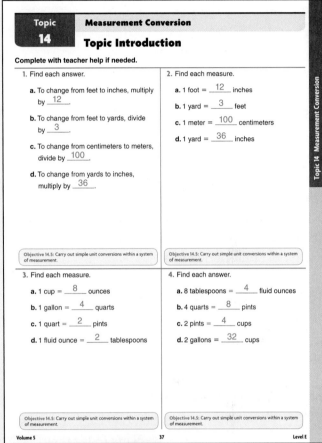

Student Booklet Page 37

Another Way To help students understand the uses of measurement conversion, have them give examples of when they would use inches versus feet and tablespoons versus gallons.

Lesson 14-1 U.S. Customary Units

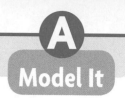

Objective 14.1: Explore the basic units of measure in the United States.

Teach the Lesson

Materials
- ☐ Paper clips
- ☐ Paper
- ☐ Inch ruler
- ☐ 5 quarters

Activate Prior Knowledge

Distribute paper clips, paper, and rulers. **About how long is the paper clip?** 1 inch **The paper?** 11 inches Have students find things in the classroom that measure about an inch or a foot. Next, have students hold 5 quarters to feel the weight of an ounce. Ask students to name objects that weigh about 1 pound. Mention 2 rolls of quarters.

Develop Academic Language

Length is usually used to refer to the longer dimension of something, but *length* and *width* are also often used interchangeably.

Model the Activities

Activity 1 Have students find one inch, 12 inches, and 1 foot on a ruler. Ask 3 students to lay their rulers end to end. **What is the total length?** 3 feet or 36 inches **What is another name for 3 feet?** 1 yard **Why isn't a mile a good measure for the height of a house?** 1 mile is too big. **Why isn't a pound a good measure for the cell phone?** A cell phone is probably lighter than 1 baseball, and there are about 3 baseballs in a pound.

Activity 2 Have students refer to the chart at the top of the student page. **There are 16 ounces in 1 pound. How many ounces are in 2 pounds?** 32 **Which is bigger, 2 pounds or 16 ounces?** 2 pounds

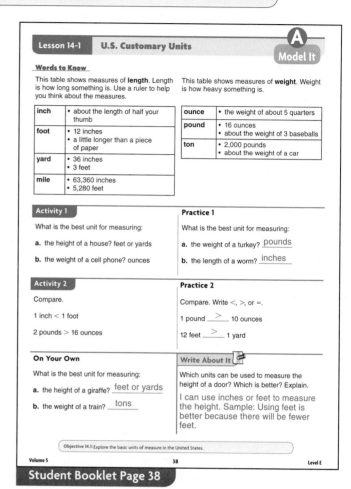

Student Booklet Page 38

Progress Monitoring

What is the best measure for weighing a spoon? Ounces **The height of a bottle?** Inches

Error Analysis

Often several different units can be used to measure length. Have students choose the largest possible unit, but clarify that the smaller units might produce numbers too great to work with, while the larger units might be too imprecise.

Volume 5   232   Level E

Lesson 14-1 U.S. Customary Units

Objective 14.1: Explore the basic units of measure in the United States.

Facilitate Student Understanding

Materials ☐ Measuring cups and spoons

Develop Academic Language

Have students use water with the measuring cups and spoons to model each unit in the table. Help students think of and record the names of familiar containers that hold each amount.

ENGLISH LEARNERS Students may have seen *capacity* printed in buses—CAP 71 PASS, for example—in reference to the number of passengers the bus can hold.

Distinguish between fluid ounces (8 per cup) and ounces measuring weight (16 per pound). Students have probably seen the abbreviation *fl oz* on liquid products. Point out that *fluid* means "liquid."

Demonstrate the Examples

Example 1 Remind students to think of standard-size items when determining the best unit. Have volunteers use the measuring tools to model each amount given as an answer.

Example 2 Have students refer to the chart at the top of the page. **Which is more, 1 quart or 1 pint?** 1 quart There are 8 pints in a gallon. **How many pints are in $\frac{1}{2}$ gallon?** 4 pints **Which is more, $\frac{1}{2}$ gallon or 1 pint?** $\frac{1}{2}$ gallon

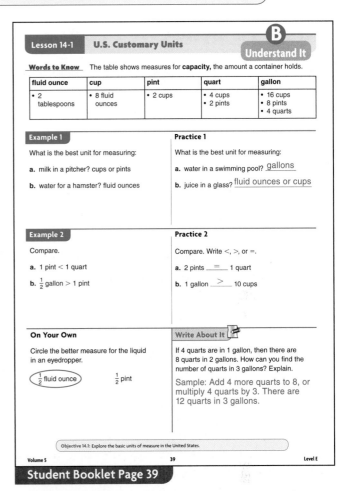

Student Booklet Page 39

✓ Progress Monitoring

State whether each question asks you to measure length, weight, or capacity. Then state the best unit to measure with. **How heavy is a bowling ball?** Weight; pounds **How long is the street?** Length; feet, miles, or yards **How much water fits in a sink?** Capacity; quarts or gallons

Error Analysis

Help students illustrate the meanings of the units. They can label pictures of objects reminding them of the measures: a gallon of milk, a quart of juice, and so on, or they can draw schematics such as a large *G* for *gallon* holding 4 *Q*'s for *quart*.

Lesson 14-1: U.S. Customary Units

Objective 14.1: Explore the basic units of measure in the United States.

Observe Student Progress

Computer Tutorial

Some students may benefit from completing the computer tutorial before they attempt the Try It page of each lesson. A list of the tutorials for each lesson can be found beginning on page x in the front of this book.

Error Analysis

Exercise 2 Have students list the units from smallest to largest.

Exercise 3 Students using an equal symbol for the first comparison have mistaken quarts for pints. Help students think of reference objects, such as a half-pint of milk or a quart of juice, to help them visualize the relationships.

Exercise 4 Students answering incorrectly have not memorized the capacity measures by order of size. Students can try memorizing initials—CPQG—or making a sentence with the initials to memorize their order.

Exercise 5 Students selecting B, C, or D do not understand the definition of *capacity*. Remind students that capacity is how much something holds; a ton is a measure of how much something weighs.

Exercise 7 Help students interpret the first part of the picture. **What unit is equal to 2 pints?** 1 quart Then the box represents 1 quart. **How many quarts do you need to show to make a gallon?** 4 Draw the rest of the quarts, and make your picture show 2 pints in each.

Student Booklet Page 40

Lesson 14-1 U.S. Customary Units — Try It

1. Which estimate of the weight of a pair of boots is best? Circle the letter of the correct answer.
   - A about 2 ounces
   - B about 2 gallons
   - **(C) about 2 pounds**
   - D about 2 tons

2. Which of the following is **not** true? Circle the letter of the correct answer.
   - A 1 pint > 1 cup
   - B 1 inch < 1 yard
   - C 1 quart > 1 cup
   - **(D) 1 gallon < 1 fluid ounce**

3. Compare. Write <, >, or =.
   - 4 pints __<__ 1 gallon
   - 1 yard __>__ 8 feet

4. Put in order from smallest to largest: pints, quarts, cups, gallons.
   cups, pints, quarts, gallons

5. Which is **not** a unit for measuring capacity? Circle the letter of the correct answer.
   - **(A) ton**
   - B quart
   - C cup
   - D gallon

6. What is the best unit for measuring:
   a. the distance across an ocean? __miles__
   b. water in a wet kitchen sponge? __fluid ounces__

7. There are 2 pints in a quart. There are 4 quarts in a gallon. Finish the picture to show how many pints are in a gallon. Explain.

   | Pint | Pint | Pint | Pint |
   | Pint | Pint | Pint | Pint |

   Sample: Each box is a quart. Each quart holds 2 pints, so there are 4 × 2 pints = 8 pints in 1 gallon.

8. Write the most reasonable unit of length.
   a. The distance from New York to Los Angeles is 2,794 __miles__.
   b. The door of your classroom is about 2 __feet__ wide.
   c. The amount of weight a truck can pull is measured in __pounds or tons__.

ENGLISH LEARNERS If students are more familiar with the metric system, provide or have them use the Internet to find metric equivalents of each unit.

Lesson 14-2: Basic Metric Prefixes

Objective 14.2: Explore the basic metric prefixes and what they mean.

Teach the Lesson

Activate Prior Knowledge

Have students count by tens up to 100. **How many tens are in 100?** 10 Next have students count up to 1,000 by hundreds. **How many hundreds are in 1,000?** 10 Point out that the numbers 10, 100, and 1,000 are related by tens or multiples of tens.

Develop Academic Language

The base units in metric measurement are gram, meter, and liter. Grams measure weight, meters measure length, and liters measure capacity.

Model the Activities

Activity 1 **What two units are used in Part a?** Kilograms and grams Point out the Grams Table at the top of the page. **How many grams are in 1 kilogram?** 1,000 **Which amount is smaller, 1 kilogram or 1 gram?** 1 gram **What two units are used in Part b?** Centimeters and meters Point out the Meters Table at the top of the page. **How many centimeters are in 1 meter?** 100 **Which is smaller, 1,000 centimeters or 1 meter?** 1 meter

Activity 2 **How many millimeters are in 1 meter?** 1,000 **How many meters are in 1 kilometer?** 1,000 **How many meters are in $\frac{1}{2}$ kilometer?** 1,000 ÷ 2 = 500 **Which is smaller, 1 meter or 500 meters?** 1 meter **Which measure is smaller, 1,000 millimeters or $\frac{1}{2}$ kilometer?** 1,000 millimeters **How many centimeters are in 1 meter?** 100 **How many centimeters are in $\frac{1}{2}$ meter? Explain.** 50, because half of 100 is 50. **Which is smaller, 100 centimeters or $\frac{1}{2}$ meter?** $\frac{1}{2}$ meter

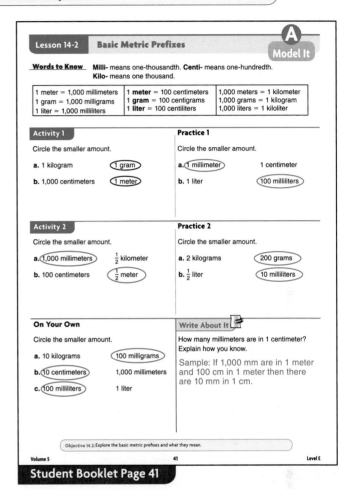

Student Booklet Page 41

 Progress Monitoring

Put the following amounts in order from smallest to largest: meter, kilometer, centimeter, millimeter. Millimeter, centimeter, meter, kilometer

Error Analysis

When deciding the size of units, students might look at the word measure only. Remind them to look at the number in front to make the comparison.

Lesson 14-2: Basic Metric Prefixes

Objective 14.2: Explore the basic metric prefixes and what they mean.

Facilitate Student Understanding

Activate Prior Knowledge

Review with students multiplication and division patterns with 10, 100, and 1,000. Review decimal point placement when solving these types of problems.

Develop Academic Language

Discuss the metric prefixes and base units of measure used in the metric system: *milli-* means one-thousandth; *centi-* means one-hundredth; *kilo-* means one thousand. Discuss conversion among units with these prefixes.

Demonstrate the Examples

Example 1 What measures are in the table? **Grams and centigrams** How are grams changed to centigrams? **Multiply the number of grams by 100.** How many centigrams are in 1 gram? **100** 2 grams? **200** 3 grams? **300** 4 grams? **400**

Example 2 What measures are in the table? **Meters and kilometers** How are meters changed to kilometers? **Divide the number of meters by 1,000.** How many kilometers in 1,000 meters? **1** 4,000 meters? **4** 7,000 meters? **7** 10,000 meters? **10**

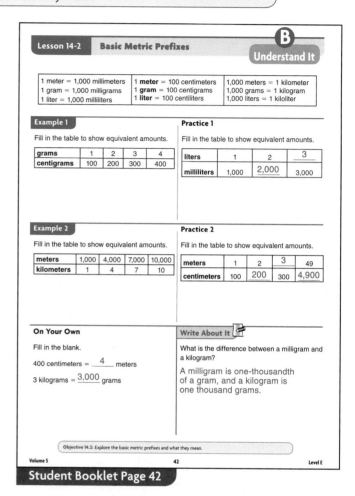

Student Booklet Page 42

Progress Monitoring

How do you change milliliters to liters? **Divide the number of milliliters by 1,000.**

Error Analysis

Students might place the decimal point incorrectly when multiplying or dividing by 10, 100, or 1,000. Point out that multiplying moves the decimal point to the right and dividing moves it to the left.

ENGLISH LEARNERS Students might be confused by changing grams and liters because the table mentions meters only. Tell students to create new tables for grams and liters.

Lesson 14-2 — Basic Metric Prefixes

Objective 14.2: Explore the basic metric prefixes and what they mean.

Observe Student Progress

Computer Tutorial

Some students may benefit from completing the computer tutorial before they attempt the Try It page of each lesson. A list of the tutorials for each lesson can be found beginning on page x on in the front of this book.

✓ Error Analysis

Exercise 1 Students might multiply or divide by an incorrect multiple of ten. Tell students to review the chart at the top of page B to determine how to complete the exercise.

Exercise 2 Students who chose answer C confused *thousands* and *thousandths*. Remind students that milligrams are smaller than 1 gram and kilograms are larger than 1 gram.

Exercise 3 Students answering incorrectly forgot that *kilo-* is a prefix indicating size and that kilometer is not a base unit.

Exercise 7 Students might not realize that 10 millimeters is equal to 1 centimeter. Refer students to the picture on page A and show them a ruler with millimeters and centimeters to show them that this is true.

Exercise 8 Remind students about the meanings of the prefixes, and have them refer to the table on page 41 if they have difficulty with this exercise.

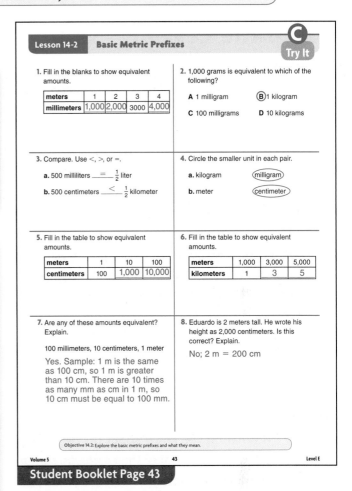

Student Booklet Page 43

ENGLISH LEARNERS Write the words *milligram, milliliter,* and *millimeter* on the board. Show students the common prefix and remind them it has the same meaning with different base units.

Lesson 14-3 Use the Metric System

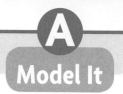

Objective 14.3: Explore the basic metric units and their relationships.

Teach the Lesson

Materials
- ☐ Metric cylinders
- ☐ Tape measure
- ☐ Balance

Activate Prior Knowledge

Have students use tape measure to measure a paper clip, height of a desk, and the height of a door. Be sure students use millimeters, centimeters, and meters when measuring. Point out that metric measures are always related in terms of 10. A window measuring 3 meters also measures 300 centimeters and 3,000 millimeters.

Develop Academic Language

Review the order of the metric units from smallest to largest as millimeter, centimeter, and meter.

Model the Activities

Activity 1 What do you know about the depth of a diving pool? It should be taller than a person so it is safe to dive into. Remind students of the objects they measured earlier. How big is a centimeter? The width of a finger Name something that measures 3 cm. The length of a small paper clip How big is a meter? A little more than 3 feet How big is 3 meters? Between 9 and 10 feet What is the best unit for measuring the depth of a pool? 3 meters

Activity 2 How are millimeters related to centimeters? 10 mm equals 1 cm Which is larger, 5 mm or 1 cm? 1 cm How are centimeters related to meters? 100 cm equals 1 m. Which is bigger, 1,000 cm or 1 m? 1,000 cm

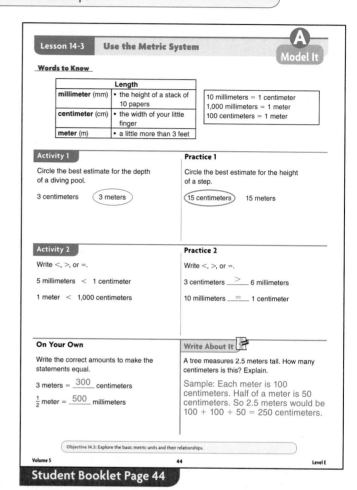

Student Booklet Page 44

Progress Monitoring

Put the measures below in order from smallest to largest: 100 cm, 20 mm, $\frac{1}{2}$ m, 0.1 cm 0.1 cm, 20 mm, $\frac{1}{2}$ m, 100 cm

Error Analysis

When choosing best estimates, tell students to think of standard objects to help them decide the appropriate measure. For example, a small paper clip measures 30 millimeters or 3 centimeters.

Lesson 14-3 | **Use the Metric System**

Objective 14.3: Explore the basic metric units and their relationships.

Facilitate Student Understanding

Activate Prior Knowledge

Have students find 1 mL, 250 mL, 500 mL and 1,000 mL on a metric cylinder. Remind them that 1,000 mL equals 1 liter. Tell students to think of things they might measure with these amounts. Next, have students use the balance to find the mass of different objects in metric units.

Develop Academic Language

Help students understand that *kilo-* means one thousand, and that a kilogram is 1,000 times larger than a gram.

Demonstrate the Examples

Example 1 What do you think is the mass of a textbook in U.S. units? About a few pounds Remind students of the objects they measured earlier. What has the mass of 1 gram? A paper clip What might have a mass of 2 grams? 2 paper clips What has the mass of 1 kilogram? A liter of water Then what might have the mass of 2 kilograms? 2 liters of water What is the best unit for measuring the mass of a textbook? 2 kilograms

Example 2 Have students refer to the chart at the top of the page. There are 1,000 grams in 1 kilogram. What is bigger, 100 grams or 10 kilograms? 10 kg There are 1,000 milliliters in 1 liter. What is bigger, 1 liter or 100 milliliters? 1 liter

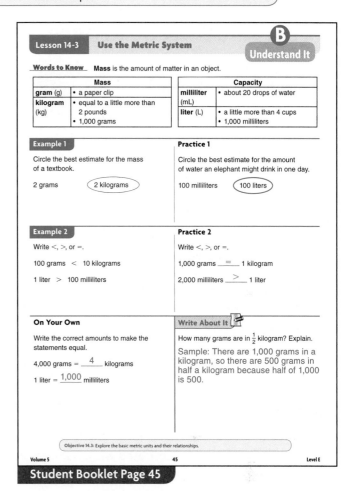

Student Booklet Page 45

Progress Monitoring

State whether each measure is a measure of length, mass, or capacity. Kilogram? Mass Millimeter? Length Liter? Capacity

Error Analysis

If students are having difficulty converting from one unit to another, have them practice multiplying and dividing by multiples of 10 for numbers without units first.

ENGLISH LEARNERS Students may not have seen the abbreviations used in this lesson before. Have students write each measure and its abbreviation and discuss how an abbreviation is just the shortened version of a word.

Lesson 14-3 Use the Metric System

Objective 14.3: Explore the basic metric units and their relationships.

Observe Student Progress

Computer Tutorial

Some students may benefit from completing the computer tutorial before they attempt the Try It page of each lesson. A list of the tutorials can be found beginning on page x in the front of this book.

 Error Analysis

Exercise 1 Remind students that 1 liter is a little more than 4 cups. So a glass of juice would be much less than 1 liter.

Exercise 4 Students who think that 100 milliliters is greater than or equal to 1 liter have mistaken the prefix *centi-* for *milli-*. Remind students that 1,000 milliliters is equal to 1 liter.

Exercise 5 Students answering this exercise incorrectly have become confused with multiples of ten. Remind students that *centi-* means "one hundredth" so 100 cm is equal to 1 meter.

Exercise 7 Students selecting choices A, C, or D did not look at the word ending. Point out that all three units of length end in "meter."

Exercise 8 Students answering incorrectly have not remembered the equivalency 1 kilogram = 1,000 grams. Encourage students to use the chart at the top of page 45 or to draw a picture to help them remember.

Student Booklet Page 46

ENGLISH LEARNERS Encourage students to look at word endings to help them determine the appropriate units for measure. Tell students that length will end in meters, mass will end in grams, and capacity will end in liters.

Lesson 14-4 Factors in Unit Conversions

Objective 14.4: Express simple unit conversions in symbolic form.

Teach the Lesson

Materials ☐ Rulers, yardsticks, and metersticks

Activate Prior Knowledge

Remind students of the equivalences in length measurement. In the U.S. system, 12 inches = 1 foot, 3 feet = 1 yard, and 36 inches = 1 yard. In the metric system, 100 centimeters = 1 meter.

Develop Academic Language

Remind students that the U.S. system of measurement uses length units such as inches, feet, and yards. The metric system uses length units such as centimeters and meters. Ask students whether they have had experience with both systems or only one system. Have students talk about the differences in the systems and the advantages and disadvantages of using each system.

Model the Activities

Activity 1 Display a 1-foot ruler. **What does each of the numbers 1–12 represent on the ruler? Inches Is a foot a smaller or a larger unit than an inch? Larger Did you multiply or divide to go from the larger unit, the foot, to the smaller unit, the inch? Multiply**

Activity 2 Display a meterstick. **What does each of the numbers 1–100 represent on the meterstick? Centimeters How many centimeters are equal to 1 meter?** 100

Write About It

ENGLISH LEARNERS Allow students of beginning proficiency levels to draw a diagram or symbols to explain, or to write in their primary languages.

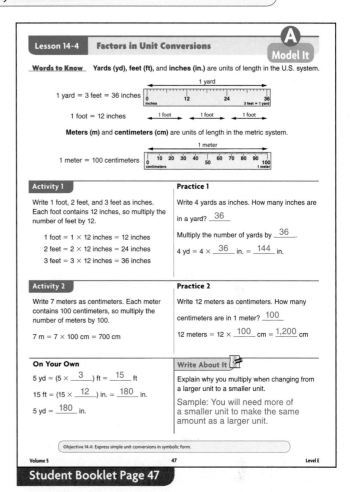

Student Booklet Page 47

Progress Monitoring

How would you change 10 yards to feet? Multiply by 3. **How many feet are in 10 yards?** 30

Error Analysis

Check to ensure that students understand that when changing to a smaller unit, they multiply because they will need more of the smaller units to represent the same length.

Volume 5   241   Level E

Lesson 14-4: Factors in Unit Conversions

Objective 14.4: Express simple unit conversions in symbolic form.

Facilitate Student Understanding

Develop Academic Language

Explain that another term for changing units is *converting*.

Demonstrate the Examples

Example 1 Ask students which is bigger, a foot or an inch. **Foot** **If the length of the classroom is measured in feet and then in inches, which value will be greater? Inches Why? Because inches are smaller than feet, it will take more inches to cover the same distance.**

If you know the number of inches, what will you do to find the number of feet? Divide by 12. An inch is a smaller unit than a foot. To change a smaller unit to a larger unit, you divide. To change a larger unit to a smaller unit, you multiply. Write these two rules on the board.

Example 2 For students unfamiliar with the metric system, provide references of lengths that would be appropriate to measure in meters, such as the height of a doorway or the length of the classroom, and lengths that would be appropriate to measure in centimeters, such as the width of their math books or the length of their pencils.

Computer Tutorial

Some students may benefit from completing a computer tutorial before they attempt the Try It page. A list of the tutorials for each lesson can be found beginning on page x in the front of this book.

Student Booklet Page 48

Lesson 14-4 Factors in Unit Conversions — Understand It

Words to Know
To change from feet to yards, divide the number of feet by 3 since 3 feet = 1 yard.
To change from inches to yards, divide the number of inches by 36 since 36 inches = 1 yard.
To change from centimeters to meters, divide the number of centimeters by 100 since 100 centimeters = 1 meter.

Example 1
How many feet are in 48 inches?
Each foot contains 12 inches.
How many 12s are in 48?
Divide. 48 ÷ 12 = 4
48 inches = 4 feet

Practice 1
How many yards are in 18 feet?
1 yard = __3__ feet
To find the number of yards in 18 feet, divide __18__ by __3__.
18 feet = __6__ yards

Example 2
How many meters are in 800 centimeters?
Each meter contains 100 centimeters.
How many 100s are in 800?
Divide. 800 ÷ 100 = 8
800 centimeters = 8 meters

Practice 2
How many meters are in 1,400 centimeters?
1 meter = __100__ centimeters
To find the number of meters, divide __1,400__ by __100__.
1,400 centimeters = __14__ meters

On Your Own
288 in. = (288 ÷ __12__) ft = __24__ ft
24 ft = (24 ÷ __3__) yd = __8__ yd
288 in. = __8__ yd

Write About It
Explain why you divide when changing from a smaller unit to a larger unit.
Sample: You need a lesser number of a larger unit to measure the same amount.

Progress Monitoring

How would you change 400 centimeters to meters? Divide by 100. How many meters are in 400 centimeters? 4

Error Analysis

Check to ensure that students understand that when changing to a larger unit, they divide because they will need fewer of the larger units to represent the same length.

Lesson 14-4: Factors in Unit Conversions

Objective 14.4: Express simple unit conversions in symbolic form.

Observe Student Progress

 Error Analysis

Exercise 1 Point out to students that some exercises require multiplication and others require division. Make sure students correctly identify the larger unit in each equation.

Exercise 2 Remind students that they multiply when changing to a smaller unit and divide when changing to a larger unit.

Exercise 4 Have students make a conversion using each of the answer options to find the one that is wrong.

Exercise 5 Ask students to explain how they can solve the problem by first breaking it into simpler parts. Emphasize that the number of units in the answer needs to be a whole number.

Exercise 6 Make sure students understand that we divide not just because centimeters are smaller than meters, but because we are making equal groups. You can model this with metersticks for students.

Student Booklet Page 49

Lesson 14-4 Factors in Unit Conversions Try It

1. Fill in each blank.
   a. feet = yards × __3__
   b. inches ÷ __36__ = yards
   c. meters = centimeters ÷ __100__
   d. feet × __12__ = inches
   e. yards = feet ÷ __3__
   f. inches ÷ __12__ = feet
   g. centimeters = meters × __100__
   h. inches = yards × __36__

2. a. Find the number of inches in 5 feet.
      Multiply 5 × 12.
      5 feet = __60__ inches
   b. Find the number of meters in 700 centimeters.
      Divide 700 by 100.
      700 centimeters = __7__ meters

3. Fill in each blank.
   a. To write yards as feet, multiply the number of yards by __3__.
   b. To write feet as inches, multiply the number of feet by __12__.
   c. To write inches as feet, divide the number of inches by __12__.

4. Which statement is **not** correct? Circle the letter of the correct answer.
   A To change feet to inches, multiply the number of feet by 12.
   B To change centimeters to meters, divide the number of centimeters by 100.
   C To change feet to yards, multiply the number of feet by 3.

5. Jolene needs to add 3 yards, 7 feet, and 144 inches. She wants her answer to be a whole number using the largest unit possible. What unit should she use? Explain. What is the total length?
   Feet; If she uses yards, 7 feet does not convert to a whole number of yards; 28 feet.

6. When you convert centimeters to meters, do you multiply or divide? Why?
   You divide. There are 100 centimeters in a meter, so you are making equal groups of 100 centimeters.

Lesson 14-5: Convert Units within a System

Objective 14.5: Carry out simple unit conversions within a system of measurement.

Teach the Lesson

Materials ☐ Rulers, metersticks, and yardsticks

Activate Prior Knowledge

Have students review basic length equivalences: 1 foot = 12 inches, 3 feet = 1 yard, 36 inches = 1 yard, and 100 centimeters = 1 meter.

Develop Academic Language

Point out that there are two major measurement systems in the world, the U.S. system and the metric system. Ask students whether they have had experience with the metric system. If so, have them describe some of the similarities and differences between the two systems.

Model the Activities

Activity 1 Lay out 4 yardsticks and 3 foot-long rulers end-to-end on the floor. **How many feet are there in each yard?** 3 **How many feet are there in 4 yards?** 12 **How many additional feet are there?** 3 **How many feet are there for the entire length?** 15

Activity 2 How many centimeters are in a meter? 100 Lay out 8 metersticks end-to-end on the floor. Place a mark or piece of tape at 85 cm on the last meterstick. **How many centimeters are in the 7 full metersticks?** 700 **How many additional centimeters are in the last meterstick?** 85 **What is the total number of centimeters shown?** 785

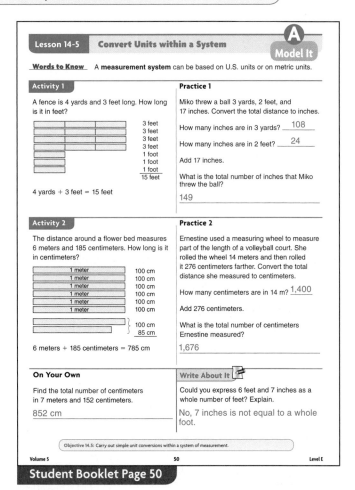

Student Booklet Page 50

⭐ Progress Monitoring

How many inches are there in 4 yards, 5 feet, 17 inches? 221 inches

Error Analysis

Check to be sure students are using the correct conversions and operations when changing from one unit to another.

Volume 5   Level E

Lesson 14-5: Convert Units within a System

Objective 14.5: Carry out simple unit conversions within a system of measurement.

Facilitate Student Understanding

Develop Academic Language

Remind students of the relationships between various units of time measurement: 1 day = 24 hours, 1 hour = 60 minutes, 1 minute = 60 seconds.

Demonstrate the Examples

Example 1 When you convert hours to seconds, why do you multiply by 60 × 60? There are 60 minutes in each hour and 60 seconds in each minute.

Remind students to add the additional number of seconds at the end of their calculations.

Example 2 Explain that the length of the trip includes travel time, meals, and rest time, thus 24 hours is counted for each day of travel.

Remind students that $\frac{1}{2}$ day is 12 hours.

Computer Tutorial

Some students may benefit from completing a computer tutorial before they attempt the Try It page. A list of the tutorials for each lesson can be found beginning on page x in the front of this book.

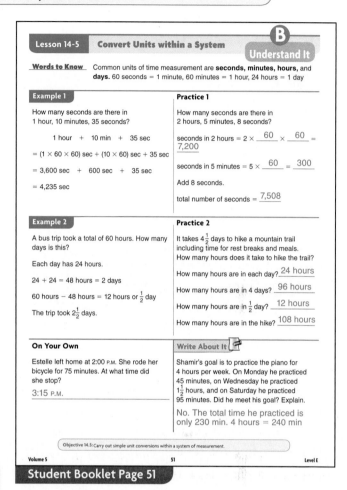

Student Booklet Page 51

Progress Monitoring

How many minutes are in 3 days? **4,320 minutes**

Error Analysis

Check to be sure students are using the correct units with their answers.

Lesson 14-5: Convert Units within a System

Objective 14.5: Carry out simple unit conversions within a system of measurement.

Observe Student Progress

 Error Analysis

Exercises 1 and 2 Check to see that students are expressing the sum of the measurements in terms of the larger unit in the first problem in each pair, and then in terms of the smaller unit.

Exercise 3 Remind students to verify what unit should be used for expressing their answers.

Exercise 4 If students answer incorrectly, have them first write each of the three amounts of time in terms of seconds and in terms of minutes. Have them find the total amount of seconds and the total amount of minutes, then compare those amounts to the answer choices.

Exercise 5 Have students first determine the number of minutes from 5:30 A.M. to 6:30 A.M. **60 min** Next, have them find the number of minutes from 6:30 A.M. to 6:45 A.M. **15 min** Then have them find the total time. **75 min**

Exercise 6 Remind students that the answer is to be given in terms of minutes, therefore 1 hour 30 minutes will have to be converted to minutes.

Exercise 8 Remind students that the answer is to be given in feet, therefore, 4 yards and 72 inches will each have to be converted to feet.

Student Booklet Page 52

Lesson 14-5 Convert Units within a System Try It

1. Complete each of the following.
   a. 9 ft + 192 in. = __25__ ft
   b. 9 ft + 192 in. = __300__ in.
   c. 248 cm + 11 m + 52 cm = __14__ m
   d. 248 cm + 11 m + 52 cm = __1,400__ cm

2. Complete each of the following.
   a. 150 min + 7 hr + 120 min = __$11\frac{1}{2}$__ hr
   b. 150 min + 7 hr + 120 min = __690__ min
   c. 18 hr + 5 days + 30 hr = __7__ days
   d. 18 hr + 5 days + 30 hr = __168__ hr

3. Complete each of the following.
   a. 5 hr + __90__ min = $6\frac{1}{2}$ hr
   b. 13 ft + __60__ in. = 18 ft
   c. 800 cm + __4__ m = 12 m
   d. 2 days + __12__ hr = $2\frac{1}{2}$ days

4. Which of the following is **more** than 120 sec + 3 min + 75 sec? Circle the letter of the correct answer
   A 350 sec   B 6 min
   Ⓒ 7 min   D 200 sec

5. Eugene began his paper route at 5:30 A.M. If he must be done no later than 6:45 A.M. to catch the bus to school, what is the longest amount of time, in minutes, that his paper route can take?
   75 min

6. John played soccer for 1 hour 30 minutes and jogged for 45 minutes. How long, in minutes, was his exercise time?
   135 min

7. How many inches is 1 yard 1 foot 5 inches? Show your work.
   53 inches; One yard is 36 inches. 1 foot is 12 inches. 36 + 12 + 5 = 53

8. Carmelita measured out three lengths of baseboard. The lengths of the baseboard pieces were 4 yards, 5 feet, and 72 inches. How many feet of baseboard did she have?
   23 feet

Objective 14.5: Carry out simple unit conversions within a system of measurement.

Volume 5 — Data, Geometry, and Measurement

Topic 14: Measurement Conversion

Topic Summary

Objective: Review measurement conversions.

Have students complete the student summary page. You may want to have students work in groups of four with each student analyzing a different choice.

Ask students to share their ideas about each answer choice. Be sure they confirm the correct answer for each problem at the end of the discussion.

Answer Evaluation

1. **A** This choice is correct.
   B Students divided by 10 instead of by 100.
   C Students multiplied by 10 instead of dividing by 100.
   D Students multiplied by 100 instead of dividing.

2. **A** 804 is the number of inches, not the number of feet.
   B Students added 16 + 12 + 84 without regard to the different units.
   C This choice is correct.
   D Students only used the number of feet in the original information and did not convert the yards and inches to feet.

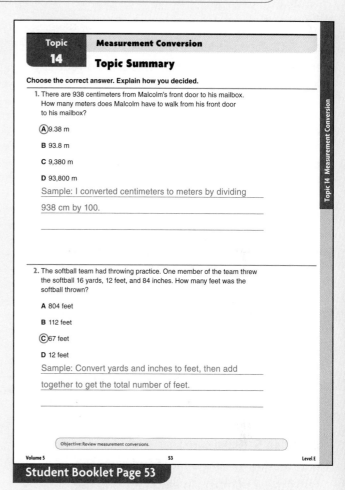

Student Booklet Page 53

 ## Error Analysis

Exercise 1 If students cannot remember what to divide by, encourage them to review the definition of a centimeter. Provide them with a meterstick if they still have difficulty.

Exercise 2 Remind students to convert the yards and inches to feet before adding the measurements together.

Progress Monitoring

When all assignments for this topic have been completed, assign the corresponding Progress Monitoring page for this topic (Assessment Resources Book, page 14). Be sure students complete the Progress Monitoring page before you administer the final assessment for this volume.

Volume 5 Data, Geometry, and Measurement

Topic 14 Measurement Conversion

Mixed Review

Objective: Maintain concepts and skills.

Have students complete the Mixed Review page. Work with each student individually to review results. Identify strengths and weaknesses and correct any misunderstandings.

Error Analysis

Exercise 1 Remind students that they are to find the sums and differences mentally. For students having difficulty, have them practice with addition and subtraction flash cards.

Exercise 2 Check for students who incorrectly express the places with a zero.

Exercise 3 Remind students that they are to find the products mentally. For students having difficulty, have them practice with multiplication flash cards.

Exercise 4 Ask students to explain how they can solve the problem by first breaking it into two parts.

Exercise 5 Make sure that students remember the rules for finding a fact family: use the same three numbers in each fact, and use inverse operations (in this case, division and multiplication).

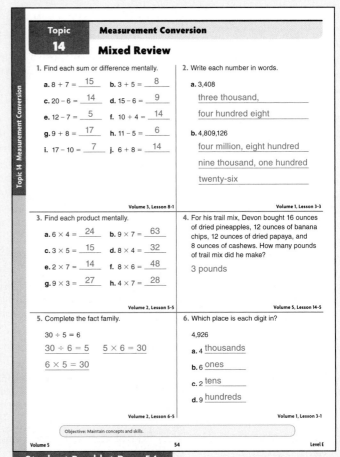

Student Booklet Page 54

Volume 5 — Topic 15: Data, Geometry, and Measurement

Measure Geometric Figures

Topic Introduction

In Lesson 15-1, students will measure objects using an inch or centimeter ruler. Lesson 15-2 has the students apply some of the concepts learned in Lesson 15-1 to find the perimeter of a polygon with integer sides. Students will learn how to determine the area of figures in Lesson 15-3, which is followed by the introduction of the formula for area of rectangles in Lesson 15-4. Finally, students learn to estimate volume using the cube-counting method in Lesson 15-5.

Lesson	Objective	Student Pages	Teacher Pages	Tutorials
Topic 15 Introduction	**15.1** Measure the length of an object to the nearest inch or centimeter.	55	250	
	15.2 Find the perimeter of a polygon with integer sides.			
	15.3 Estimate or determine the area of figures by covering them with squares.			
	15.4 Estimate or determine the volume of solid figures by counting the number of cubes that would fill them.			
15-1 Length	**15.1** Measure the length of an object to the nearest inch or centimeter.	56–58	251–253	15a
15-2 Perimeter	**15.2** Find the perimeter of a polygon with integer sides.	59–61	254–256	15b
15-3 Area	**15.3** Estimate or determine the area of figures by covering them with squares.	62–64	257–259	15c
15-4 Area of Rectangles	**15.4** Measure the area of rectangular shapes by using appropriate units.	65–67	260–262	15c
15-5 Volume	**15.5** Estimate or determine the volume of solid figures by counting the number of cubes that would fill them.	68–70	263–265	15d
Topic 15 Summary	Review measuring geometric figures.	71	266	
Topic 15 Mixed Review	Maintain concepts and skills.	72	267	

Computer Tutorial

Some students may benefit from completing the computer tutorial before they attempt the Try It page of each lesson. If you are using the electronic components of *Pinpoint Math*, you will find a complete listing of Tutorial codes and titles when you access them either online or via CD-ROM.

Volume 5: Data, Geometry, and Measurement

Topic 15: Measure Geometric Figures

Topic Introduction

Objectives: 15.1 Measure the length of an object to the nearest inch or centimeter. **15.3** Estimate or determine the area of figures by covering them with squares. **15.4** Estimate or determine the volume of solid figures by counting the number of cubes that would fill them. **15.2** Find the perimeter of a polygon with integer sides.

Materials ☐ Various small objects to use for measuring

Have students measure the length different objects around the classroom, such as desks, tables, books, and other students. Students should use different items to measure with, such as paper clips, pencils, shoes, hands (no rulers). Select a student to walk around the room heel-to-toe and count the number of steps it talks to get back to the starting point. **Would it be quicker to measure a house using your feet or pencils?** Your feet **If you do not have a ruler, what else could you use to measure the length of an object?** Samples: a block, a piece of string, or a small box

Informal Assessment

1. **Where does the point of the pencil fall on the ruler?** Between 3 inches and 4 inches **What measurement does the mark halfway between 3 and 4 represent?** $3\frac{1}{2}$ inches **To which inch marking is the point of the pencil closest?** 4 inches

2. **What shape is shown?** Rectangle **How can you tell how long the rectangle is?** Count the squares going across the top or bottom. **How can you tell how wide the rectangle is?** Count the squares along the left or the right side.

3. **How many dimensions does the box have?** 3 **What are they?** Length, width, and height **How long is the box?** 4 cm **How wide is the box?** 3 cm **How high is the box?** 2 cm

4. **What shape is shown?** Triangle **How many sides does it have?** 3 **Is each side a different length?** Yes. **How would you find the distance around the triangle?** Add the lengths of all the sides.

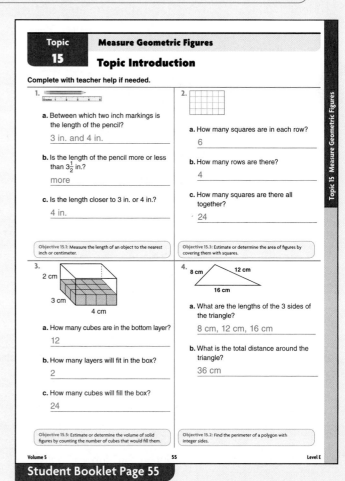

Student Booklet Page 55

Another Way Have students use attribute blocks and estimate how many squares or triangles it would take to cover the cover of a book. Continue activity using small pictures or small objects, estimating the number of attribute blocks needed to determine the length, area, or perimeter.

Lesson 15-1 | Length

Objective 15.1: Measure the length of an object to the nearest inch or centimeter.

Teach the Lesson

Materials
- ☐ Two-sided rulers
- ☐ Several small objects close to 1 inch long
- ☐ Teaching Aid 14 Centimeter and Inch Rulers

Activate Prior Knowledge

Give students a sheet of paper with several lines on it, some which measure an exact number of inches and others that measure an exact number of centimeters. Have them use their rulers to find those measurements.

Develop Academic Language

In the U.S. measurement system, the smallest unit commonly used for measuring length is the inch. An inch is about the distance from the knuckle on your thumb to the end of your thumb. In the metric system, the centimeter is used to measure things that could also be measured in inches. Some examples are a desk, a pencil, or a notebook.

Model the Activities

Activity 1 Give students objects that are slightly more or less than 1 inch long, like small paper clips. **Is your object at least 1 inch long?** Answers will vary. **Is it more than 1 inch?** Answers will vary. **Is it less than 1½ inches?** Yes. **Is the length closer to 1 inch or 2 inches?** 1 inch

Activity 2 Have students measure the same objects using the centimeter side of the ruler. **Is your object at least 2 centimeters long?** Answers will vary. **Is it more than 3 centimeters?** Answers will vary. **Is its length closer to 2 centimeters or to 3 centimeters?** Answers will vary.

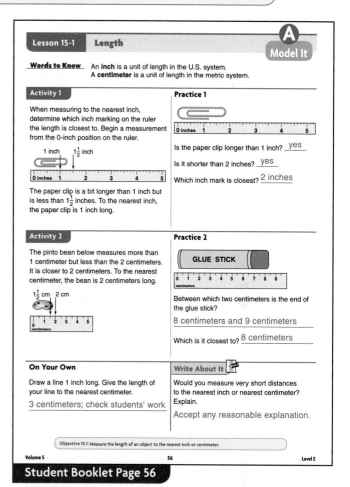

Student Booklet Page 56

Progress Monitoring

Give students an object and have them measure it to the nearest inch and to the nearest centimeter.

Error Analysis

Be sure students are not rounding all measurements up to the next unit but only those that are halfway or more. Remind them of using a number line to round. Suggest that they think of a ruler as a number line.

Lesson 15-1 Length

Objective 15.1: Measure the length of an object to the nearest inch or centimeter.

Facilitate Student Understanding

Demonstrate the Examples

Example 1 On the board, draw a line. Mark the left end 3 in. and the right end 4 in. and show the markings in between for $3\frac{1}{2}$ in. Explain that when you measure to the nearest inch, you find the inch that the measurement is closest to. If it comes to the left of the halfway mark, it is closer to the lesser number. If it comes to the right of the halfway mark, it is closer to the greater number.

Example 2 Be sure that students are careful about which units they are using when they measure, especially when using 2-sided rulers. Also make sure that they line up the edge of the ruler with the end of the object correctly, and that they are not measuring from the wrong end of the ruler.

Computer Tutorial

Some students may benefit from completing a computer tutorial before they attempt the Try It page. A list of the tutorials for each lesson can be found beginning on page x in the front of this book.

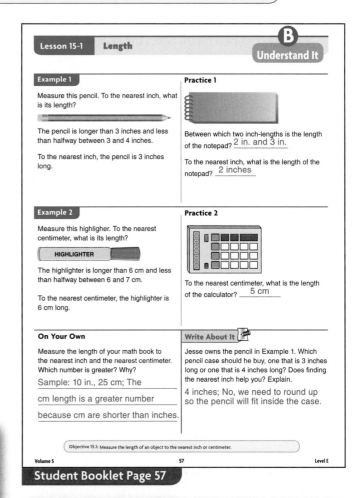

Student Booklet Page 57

Progress Monitoring

If a pen measures $5\frac{7}{8}$ in., what is its measure to the nearest inch? 6 in.

Error Analysis

Check to ensure that students give each answer with the appropriate unit, either inches or centimeters.

Lesson 15-1   Length

Objective 15.1: Measure the length of an object to the nearest inch or centimeter.

Observe Student Progress

 Error Analysis

Exercises 1 and 2 If students answer incorrectly, have them extend the lines for the markings up to each arrow. Point out that if the arrow does not extend that far, its measure to the nearest inch or centimeter is to the smaller unit. If the arrow reaches beyond that point, then the measure should be to the next larger unit.

Exercise 3 Students' answers for these will vary. Check to make sure that their arrows are clearly closest to the given measurement. Note that some may be shorter and others longer than the given unit.

Exercise 5 If this measurement of 12 inches does not seem appropriate for your classroom, you could give students another measurement to look for.

Exercise 6 Have students compare their answers to Exercise 4 and Exercise 6. **Which gave you the greater number?** Centimeters **Why is that?** Centimeters are smaller than inches, so it takes more of them to measure the same length.

Student Booklet Page 58

Lesson 15-2 Perimeter

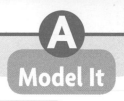

Objective 15.2: Find the perimeter of a polygon with integer sides.

Teach the Lesson

Materials
- ☐ Two-sided rulers
- ☐ Pattern blocks
- ☐ Teaching Aid 4 Centimeter Grid Paper
- ☐ Masking tape

Activate Prior Knowledge

Use pattern blocks to provide examples of polygons. Ask students to name the polygons. Then give them the names and have them answer with the number of sides.

Develop Academic Language

Use pattern blocks to discuss what polygons look like. Explain that all sides of a polygon are line segments. Have a volunteer walk around the edge of the room, and then tell the class the volunteer has walked the room's perimeter.

Model the Activities

Activity 1 Use 1-cm grid paper to copy the polygon shown. Start in the upper left corner. How many cm long is the top? **4 cm** Write that measurement across the top side. Write the length of each other side moving in a clockwise direction. **5 cm, 7 cm, 3 cm, 11 cm, and 8 cm** What is the total of all the lengths? **38 cm**

Activity 2 Use masking tape on the floor to make a triangle that has three sides each measuring 8 ft. Measure one side with your ruler and write the measurement on the tape. **8 ft** Measure the other two sides and write those measurements. **8 ft, 8 ft** What is the perimeter of the triangle? **24 ft**

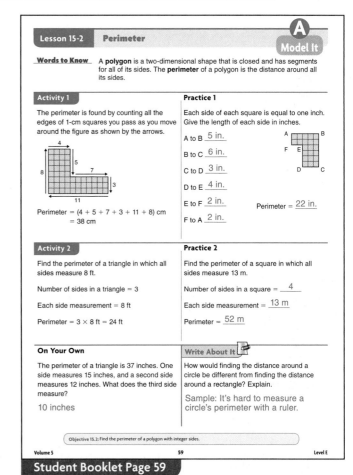

Student Booklet Page 59

Progress Monitoring

The lengths of the sides of a pentagon are 3 in., 9 in., 7 in., 5 in., and 10 in. What is its perimeter? **34 in.**

Error Analysis

Check to see that students are counting each side exactly once and labeling each answer with the correct unit.

Lesson 15-2 Perimeter

Objective 15.2: Find the perimeter of a polygon with integer sides.

Facilitate Student Understanding

Develop Academic Language

Regular polygons have sides which all have the same measure. A square is a regular polygon. A polygon with 5 sides is a pentagon, with 6 sides is a hexagon, and with 8 sides is an octagon. Not all polygons are regular.

Demonstrate the Examples

Example 1 Point out that the lengths have the same measure, as do the widths. Have students first find the perimeter by multiplying the length by 2, multiplying the width by 2, and adding those two products. Then have them add up all the sides by moving around the rectangle. Verify that they get the same perimeter using the two methods.

Example 2 Have students draw an octagon and label each side "16 in." Check to see that each octagon has 8 sides that appear to be the same length. Ask whether the perimeter of all octagons could be found by multiplying the length of one side by 8. **Only if it is a regular octagon**

Computer Tutorial

Some students may benefit from completing a computer tutorial before they attempt the Try It page. A list of the tutorials for each lesson can be found beginning on page x in the front of this book.

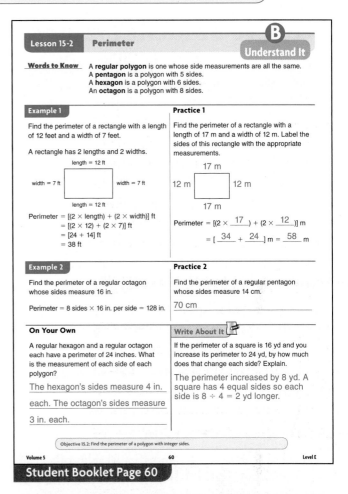

Student Booklet Page 60

⭐ Progress Monitoring

Find the perimeter of a rectangle with a length of 18 cm and width of 11 cm. **58 cm**

Error Analysis

If students have trouble with the concept of perimeter, have them walk around various rooms, hallways, and so on, then use a yardstick or meter stick to measure the distance they walked.

Lesson 15-2 Perimeter

Objective 15.2: Find the perimeter of a polygon with integer sides.

Observe Student Progress

 Error Analysis

Exercise 1 Check to ensure that students are adding each side once and labeling their answers with the correct units.

Exercise 2 If students have difficulty, suggest that they draw each figure and label the measurement of each side.

Exercise 3b Help students to see that only the length is being doubled. The width remains 5 inches.

Exercise 5 Some students may need to be reminded that a regular hexagon has 6 equal sides. Use a pattern block or draw a model.

Exercise 6 Ask what measurement is needed and why. **Perimeter, because the fencing goes around the sides of the garden** Be sure students do not just add 18 + 13 to find the perimeter.

Exercise 8 Remind students that all sides of a square have the same length. **How do you find the perimeter?** 12 + 12 + 12 + 12, or 12 × 4

Student Booklet Page 61

Lesson 15-2 Perimeter — Try It

1. Find the perimeter of each figure.
   a. 29 cm
   b. 24 in.
   c. 41 m

2. Determine the perimeter of each described polygon.
   a. a rectangle with a length of 6 ft and a width of 4 ft
      20 ft
   b. a triangle with sides of length 14 cm, 19 cm, and 23 cm
      56 cm
   c. a regular pentagon with sides of length 7 in.
      35 in.

3. Answer each of the following for a rectangle with a length of 8 in. and a width of 5 in.
   a. What is its perimeter? 26 in.
   b. What would its perimeter be if the length is doubled? 42 in.

4. What is the perimeter of a rectangle with a length of 9 ft and a width of 7 ft? Circle the letter of the correct answer.
   A 16 ft   B 32 in.
   Ⓒ 32 ft   D 63 ft

5. A planting bed is in the shape of a regular hexagon. Each side measures 12 feet. What is the distance you go if you walk all the way around the planting bed?
   72 feet

6. A rectangular garden is 18 feet long and 13 feet wide. How much fencing is required to surround the garden?
   62 feet

7. What is the perimeter of a regular triangle with sides of length 6 in.?
   18 in.

8. What is the perimeter of a square with sides of length 12 feet?
   48 feet

Lesson 15-3 Area

Objective 15.3: Estimate or determine the area of figures by covering them with squares.

Teach the Lesson

Materials
☐ Teaching Aid 4 Centimeter Grid Paper
☐ Teaching Aid 17 Inch Grid Paper

Activate Prior Knowledge

Have students use various square items (carpet squares, tiles, cut out paper squares, and so on) to cover their desks, their math books, and a table in the room. Have them count the number of squares used for each.

Develop Academic Language

The area of a shape is the amount of surface the shape has. Square units can be used to cover a shape's surface. These units can then be counted to find the shape's area.

Model the Activities

Activity 1 We use square units when we measure area. What units are the dimensions of this shape measured in? **Centimeters** So what units will we use to measure the area? **Square centimeters** To help students count all of the squares and not count any square twice, suggest that they color or mark the squares as they count them.

Activity 2 Use Teaching Aid 17 Inch Grid Paper and draw a 3 × 4 rectangle. Start in the top left corner of the rectangle and number that square 1. Then number all the other squares with 2, 3, 4, and so on, until each square has been numbered. **How many squares did you count? 12** Now multiply the number of rows by the number of squares in each row. How many squares is this? **12**

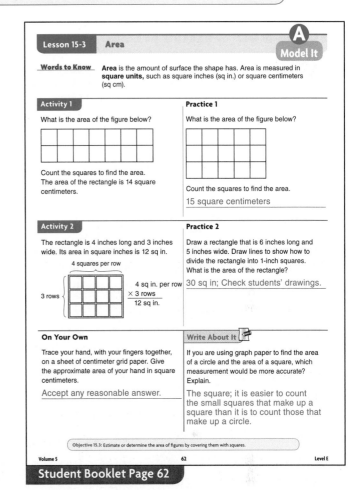

Student Booklet Page 62

 Progress Monitoring

Find the area of a rectangle with a length of 17 m and a width of 6 m. **102 sq m**

Error Analysis

Be sure students appropriately use square and cubic units.

Volume 5   257   Level E

Lesson 15-3 Area

Objective 15.3: Estimate or determine the area of figures by covering them with squares.

Facilitate Student Understanding

Materials
- ☐ Teaching Aid 4 Centimeter Grid Paper
- ☐ Teaching Aid 17 Inch Grid Paper

Demonstrate the Examples

Example 1 Have students use 1-cm grid paper to draw a representation of each side of the box. **Label the six sides as front, back, left, right, top, and bottom. Now find the area of each side and add to find the total area.** Then have them count the total number of squares in all six sides and verify that they get the same answer using these two methods.

Example 2 Discuss why covering a circle with square units to estimate area is less accurate than covering a rectangle with square units. Ask students for ways they might estimate the areas at the edges of the circle that do not get covered by squares.

Computer Tutorial

Some students may benefit from completing a computer tutorial before they attempt the Try It page. A list of the tutorials for each lesson can be found beginning on page x in the front of this book.

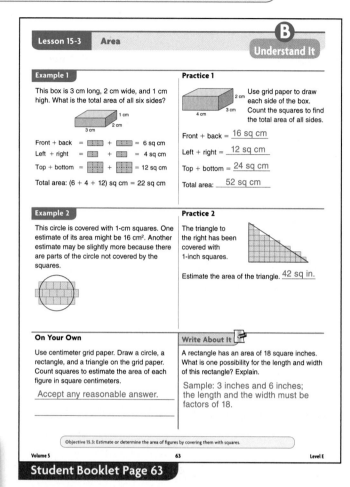

Student Booklet Page 63

Progress Monitoring

Draw a rectangle that is 10 cm long and 8 cm wide. Draw lines to divide it into square centimeters. Find the area of the rectangle.
80 sq cm

Error Analysis

If students are having trouble with Write About It, have them construct a rectangle and then divide it into 18 square units. They can then count to find the number of units that make up the length and the number of units that make up the width.

Lesson 15-3 Area

Objective 15.3: Estimate or determine the area of figures by covering them with squares.

Observe Student Progress

Materials ☐ Teaching Aid 4 Centimeter Grid Paper
☐ Teaching Aid 17 Inch Grid Paper

 Error Analysis

Exercises 1 and 2 Students' answers for Exercise 1b and Exercise 2 may differ as they will have to estimate the partial squares.

Exercise 3 Make sure students understand that area is always in terms of square units.

Exercise 4 If students answer incorrectly, first check to see that they know to find the area, and that area is stated in square units. Then make sure they are adding the areas for all six faces.

Exercise 5 If students have difficulty finding the total area, have them write the area for each of the six faces and then add to find the total area. Students may use Teaching Aid 17 Inch Grid Paper to draw nets to help them find the surface area of each box. Have students use a scale of 1 in. = 1 ft and label each side with the measurements shown in the drawing.

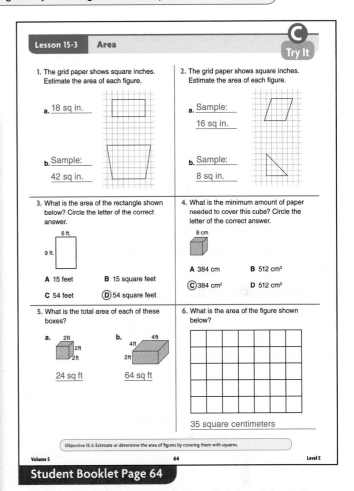

Student Booklet Page 64

ENGLISH LEARNERS Remind students than an *estimate* for area is close to, but not exactly, the area.

Lesson 15-4: Area of Rectangles

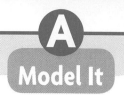

Objective 15.4: Measure the area of rectangular shapes by using appropriate units.

Teach the Lesson

Materials ☐ Teaching Aid 4 Centimeter Grid Paper

Activate Prior Knowledge

Draw a large rectangle on the board about 2 feet by 3 feet. **What can we measure on this rectangle?** The lengths of the sides **How could we measure them?** Use a ruler or a yardstick. Have students measure the lengths in inches, feet, and centimeters. **What other type of measurement can we find?** The area of the rectangle

Develop Academic Language

Have students look at the definitions of *area* and *square unit*. Refer back to the 2-feet-by-3-feet rectangle that you drew on the board. **What kind of square units could we use to measure the area of the rectangle?** Have volunteers draw and label an example of different square units: a square inch, a square centimeter, a square foot, a square meter, a square yard. **Which covers the greatest area?** Square meter

Model the Activities

Activity 1 Provide students with Teaching Aid 4 Centimeter Grid Paper. Have each student draw a 8 cm-by-2 cm rectangle on his or her paper. **What is the area of the rectangle?** 16 square centimeters **What are two ways to find the area?** Count the number of square centimeters inside the rectangle or use the formula. **Does it matter which number you call the length and which one you call the width?** No, the area will be the same. **Why?** Commutative Property of Multiplication

Activity 2 **What units are used to measure the area of the cover of Enrique's math book?** Square inches **Why is it square inches?** The book was measured in inches. We are finding the area so we use square inches for the measurement.

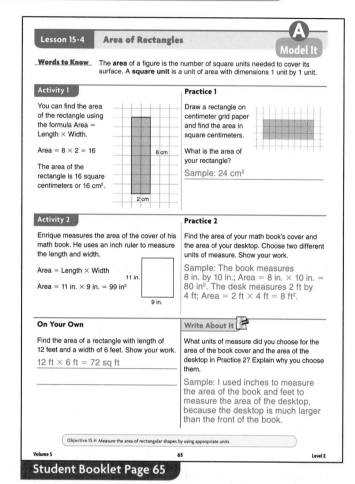

Student Booklet Page 65

Progress Monitoring

If I measure a computer screen in centimeters, what unit should I use to describe the area?
Square centimeters

Error Analysis

Some students might neglect to record the units when finding area. Ask students to underline the units in each answer to help them remember the importance of the units.

Lesson 15-4 Area of Rectangles

Understand It

Objective 15.4: Measure the area of rectangular shapes by using appropriate units.

Facilitate Student Understanding

Activate Prior Knowledge

Compare the sizes of various units of measure. Write on the board units of measure in alphabetical order: centimeter, foot, inch, kilometer, meter, mile, yard. Have students arrange the units in order from smallest to biggest. **Centimeter, inch, foot, yard, meter, kilometer, mile**

Develop Academic Language

Discuss with students the meaning of the word *appropriate*. Students may be familiar with the requirement of school dress codes requiring "appropriate clothing."

ENGLISH LEARNERS You may want to have paper examples of the different units of area measure available and give students the opportunity to touch the different units of measure as they name them.

Demonstrate the Examples

Example 1 As you discuss the example, point out that sometimes it is not so easy to tell what would be the most appropriate unit to use. In this case, since the area is so big it is clear that either miles or kilometers would be best. Tell students that the area of Colorado is over 100,000 square miles. Ask students why a small unit would not be good to measure the area of the state. **The number would be so big that it would hard to understand what it means.**

Example 2 What other units of measure could a tablecloth be measured in? **Feet or centimeters**

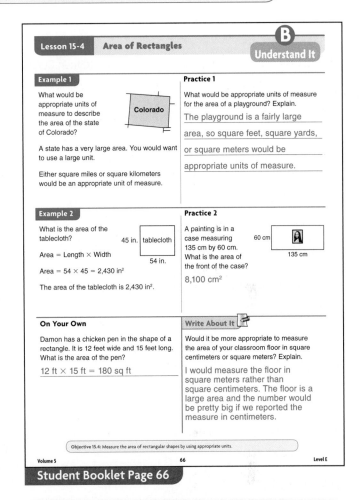

Student Booklet Page 66

Progress Monitoring

Explain how to choose a unit of measure to report the area of a mural. A mural is very large, so I would use square feet.

Error Analysis

If students have difficulty choosing an appropriate measure for various areas, guide them to think of benchmarks they can use to relate the relative sizes of the different measures. A piece of paper is about 1 square foot; a stamp is about 1 square inch; the top of a card table is about 1 square yard. If an item is much greater than the benchmark, then the next larger unit would be a better choice.

Lesson 15-4 | Area of Rectangles

Objective 15.4: Measure the area of rectangular shapes by using appropriate units.

Observe Student Progress

Materials ☐ Teaching Aid 14 Centimeter and Inch Rulers

Computer Tutorial

Some students may benefit from completing a computer tutorial before they attempt the Try It page. A list of the tutorials for each lesson can be found beginning on page x in the front of this book.

Develop Academic Language

ENGLISH LEARNERS Possible words or phrases in this lesson for students to add to their word banks include *square centimeter* (cm^2), *square meter* (m^2), *square kilometer* (km^2), *square inch* (in^2), *square yard* (yd^2), or *square mile* (mi^2), and *appropriate*.

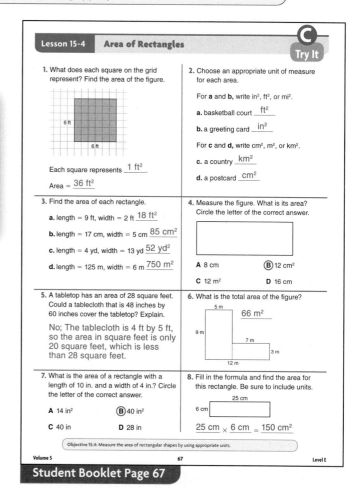

Student Booklet Page 67

⭐ Error Analysis

Exercise 2 If students make multiplication errors finding the area, have them check their answers by estimating and perhaps reversing the order of the factors if they are doing vertical multiplication to find the answers. Remind students that area is found by multiplying the length by the width, and write the formula on the board: Area = Length × Width.

Exercise 4 If students choose answer choices A or D, remind them that the appropriate unit is part of the answer. This question asks for the area, so the answer must be square units. You may want to review the use of abbreviations for square units.

Exercise 5 The question requires students to think about inches versus feet. Point out that the area is given in square feet, but the dimensions of the tablecloth are given in inches. You may want to give them the converted units: 48 inches = 4 feet and 60 inches = 5 feet.

Exercise 6 Some students may need help seeing that the figure is broken into two rectangles whose areas can be added to find the total area. Point out that 5 + 7 = 12, and ask students to find the length of the unlabeled side: 6 m.

Volume 5 — 262 — Level E

Lesson 15-5 Volume

Model It

Objective 15.5: Estimate or determine the volume of solid figures by counting the number of cubes that would fill them.

Teach the Lesson

Materials ☐ Base ten blocks: 24 unit cubes or centimeter cubes per student

Activate Prior Knowledge

Use unit cubes or centimeter cubes to construct a cube 2 centimeters on a side. **What is this figure? a cube What is the area of each face? 4 square centimeters**

Develop Academic Language

Read together the definition of *volume*. Then show a cube made up of 8 centimeter cubes. **How could you find the volume of this cube? Count the cubes.** Have a volunteer dismantle the cube and count them. **How many cubes are there? 8 What is the volume? 8 cubic centimeters**

Have the class read the definition of cubic units. **Why can't we use square units for volume? Sample: Square units are flat, and volume is 3-dimensional, so we need 3-dimensional units.**

ENGLISH LEARNERS You may want to have students write the word *volume* and draw a sketch of the larger cube and the 8 centimeter cubes to help connect the vocabulary to the concept.

Model the Activities

Activity 1 Remind students that the height of the first layer is 1. **How would changing the width to 3 centimeters instead of 2 centimeters change the volume of the first layer? The volume would increase to 9 cubic centimeters.**

Activity 2 Remind students that the number of layers in a rectangular solid is equal to the solid's height. Students can also review how to write cubic units. **How do we abbreviate cubic inches, cubic feet, and cubic meters? In³, ft³, m³**

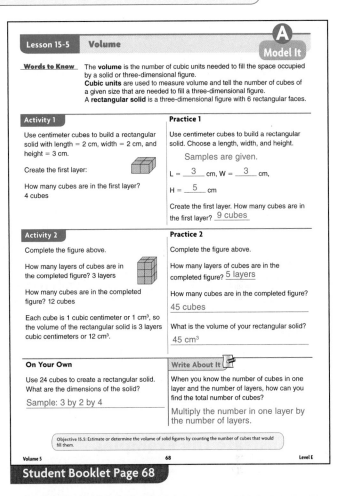
Student Booklet Page 68

⭐ Progress Monitoring

Use centimeter cubes to create a rectangular solid that is 2 centimeters by 4 centimeters by 2 centimeters. What is the volume of the solid? Explain. **16 centimeter cubes; there are 2 layers and each layer has 8 cubes. 2 × 8 = 16**

Error Analysis

Some students may have difficulty visualizing the layers when a rectangular solid is stacked. You may want to have students create the solid and then lay out the layers separately.

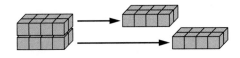

Lesson 15-5 Volume

Objective 15.5: Estimate or determine the volume of solid figures by counting the number of cubes that would fill them.

Facilitate Student Understanding

Materials ☐ Base ten blocks

Develop Academic Language

Discuss all the ways used to write the units for measuring volume. Write on the board: cubic centimeter, cu cm, and cm³. Point out that the same three ways of writing the units for volume can be used with the other measures of length. Mention that the small 3 in cm³ relates to the 3 dimensions of volume, just as the small 2 used in square units relates to the 2 dimensions of area.

ENGLISH LEARNERS Review the three dimensions, showing a line, a square, and then a cube, each time counting the dimensions. Speak specifically about the 3 dimensions of a rectangular solid: the *length*, the *width*, and the *height*.

Demonstrate the Examples

Example 1 Students need to remember that volume is simply the number of cubic units that make up a solid. **What units are used to measure the dimensions of this figure? Inches So what units should we use to measure the volume? Cubic inches**

Example 2 Students having difficulty can draw each layer of the rectangular solid and then count the total number of cubic units. Suggest that students work in pairs to construct the rectangular solid. Have them stack the layers so they can visualize what the entire solid looks like.

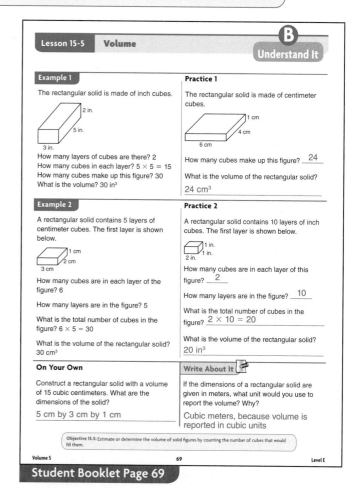

Student Booklet Page 69

✓ Progress Monitoring

Use centimeter cubes to construct a rectangular solid that is 4 centimeters long, 2 centimeters wide, and 3 centimeters tall. What is the volume of the solid? 24 cm³

Error Analysis

Students having difficulty constructing rectangular solids can be reminded that each cube has dimensions of 1 cm on each side. Also, advise students to think of the height of the solid as the number of layers in the solid.

Lesson 15-5 Volume

Objective 15.5: Estimate or determine the volume of solid figures by counting the number of cubes that would fill them.

Observe Student Progress

Materials ☐ Base ten blocks or centimeter cubes

Computer Tutorial
Some students may benefit from completing a computer tutorial before they attempt the Try It page. A list of the tutorials for each lesson can be found beginning on page x in the front of this book.

ENGLISH LEARNERS Discuss the relationship between *cube* and *cubic units*.

Error Analysis

Exercises 1 and 2 If students have difficulty finding the volume from the picture, allow them to use centimeter cubes to create the figure shown.

Exercise 3 Students might need to be reminded that the dimensions of a rectangular solid are the length, the width, and the height. Advise students to count the number of cubes across each dimension.

Exercise 5 Students having difficulty might have added the dimensions instead of multiplying. Advise students to construct the first layer of cubes and to visualize all 10 layers.

Exercise 6 Remind students that the difference between the volumes of the boxes is simply the smaller volume subtracted from the larger volume. Discuss the fact that Box B has twice as many layers and thus it has twice the volume.

Exercise 7 Advise students to try constructing the solid with only one layer. Remind them that the volume of the solid is same as the number of cubic units that makes up the solid.

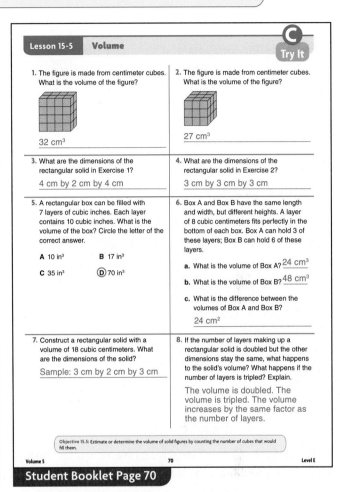

Student Booklet Page 70

Exercise 8 Have students construct simple rectangular solids and then double or triple the height in order to see how the dimensions affect the volume. Discuss what would happen to the volume if the number of layers were multiplied by 4, 5, or 10.

Volume 5: Data, Geometry, and Measurement

Topic 15: Measure Geometric Figures

Topic Summary

Objective: Review measuring geometric figures.

Have students complete the summary page. You may want to have students work in groups of four with each student analyzing a different choice.

Ask students to share their ideas about each answer choice. Be sure they confirm the correct answer for each problem at the end of the discussion.

Answer Evaluation

1. **A** Students used inches instead of square inches.
   B Students found the perimeter instead of the area.
   C This choice is correct.
   D Students added the two measurements instead of multiplying.

2. **A** Students found the area rather than the perimeter.
   B Students multiplied the length by the width.
   C This choice is correct.
   D Students added 1 length + 1 width rather than 2 lengths + 2 widths.

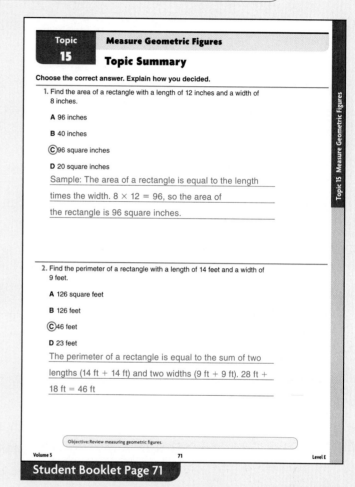

Student Booklet Page 71

 ## Error Analysis

Exercise 1 Suggest that students draw a sketch of the rectangle to help them visualize which operation they need to use to find the area. Make sure they remember to use the correct units.

Exercise 2 For students having difficulty, ask them to draw the rectangle, label the length of each side, then add each measurement as they move around the rectangle in a clockwise direction.

Progress Monitoring

When all assignments for this topic have been completed, assign the corresponding Progress Monitoring page for this topic (Assessment Resources Book, page 15). Be sure students complete the Progress Monitoring page before you administer the final assessment for this volume.

Volume 5: Data, Geometry, and Measurement

Topic 15: Measure Geometric Figures

Mixed Review

Objective: Maintain concepts and skills.

Have students complete the Mixed Review page. Work with each student individually to review results. Identify strengths and weaknesses and correct any misunderstandings.

Error Analysis

Exercise 1 Suggest that students align the numbers vertically before adding or subtracting.

Exercise 2 Students may need to have the following conversions reviewed: 12 inches = 1 foot, 36 inches = 1 yard, 3 feet = 1 yard, 100 cm = 1 m.

Exercise 4 If students choose answer A, have them re-read the directions carefully so that they see that the correct answer will round **up** to 500.

Exercise 5 Students may need to be reminded that they do not show anything in expanded notation for places with the digit 0.

Exercise 6 Remind students that the commutative property of multiplication means that the order of the numbers can be changed without changing the product. The associative property of multiplication means that the way in which the numbers are grouped within parentheses does not affect the product.

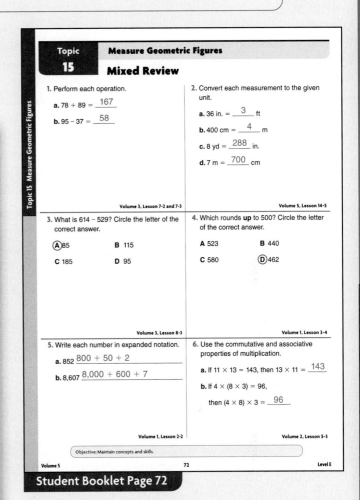
Student Booklet Page 72

Volume 6 — Understand Fractions
Topic 16 — Meaning of Fractions

Topic Introduction

Lesson 16-1 introduces students to the concept of fractions. Lessons 16-2 and 16-3 focus on two important types of fractions: the unit fraction and the whole. Lesson 16-4 gives students practice manipulating two forms of values greater than 1: improper fractions and mixed numbers.

Lesson	Objective	Student Pages	Teacher Pages	Tutorials
Topic 16 Introduction	16.1 Understand that fractions may refer to parts of a set or parts of a whole.	1	269	
	16.2 Recognize, name, and compare unit fractions from $\frac{1}{12}$ to $\frac{1}{2}$.			
	16.3 Know that when all fractional parts are included, such as four fourths, the result is equal to the whole and to one.			
	16.4 Define and manipulate improper fractions and mixed numbers.			
16-1 Basics of Fractions	16.1 Understand that fractions may refer to parts of a set or parts of a whole.	2–4	270–272	16a, 16b
16-2 Unit Fractions	16.2 Recognize, name, and compare unit fractions from $\frac{1}{12}$ to $\frac{1}{2}$.	5–7	273–275	16c, 16d, 16e, 16f
16-3 Parts and the Whole	16.3 Know that when all fractional parts are included, such as four fourths, the result is equal to the whole and to one.	8–10	276–278	16g
16-4 Rename Values Greater than 1	16.4 Define and manipulate improper fractions and mixed numbers.	11–13	279–281	16h, 16i
Topic 16 Summary	Review the meaning of fractions.	14	282	
Topic 16 Mixed Review	Maintain concepts and skills.	15	283	

Computer Tutorial

Some students may benefit from completing the computer tutorial before they attempt the Try It page of each lesson. If you are using the electronic components of *Pinpoint Math,* you will find a complete listing of Tutorial codes and titles when you access them either online or via CD-ROM.

Volume 6: Understand Fractions

Topic 16: Meaning of Fractions

Topic Introduction

Objectives: 16.1 Understand that fractions may refer to parts of a set or parts of a whole. **16.2** Recognize, name, and compare unit fractions from $\frac{1}{12}$ to $\frac{1}{2}$. **16.3** Know that when all fractional parts are included, such as four fourths, the result is equal to the whole and to one. **16.4** Define and manipulate improper fractions and mixed numbers.

Materials ☐ Fraction bars or circles

Write 236 on the board. **In 236, which place is 3 in? Tens Which place is 6 in? Ones Which place is 2 in? Hundreds** Continue this activity with a few more numbers, reviewing the places. Check that students are aware of the difference between *place* and *value*. For instance, in 547, the 4 is located in the *tens* place but its *value* is 40.

Informal Assessment

1. **If you cut an apple into thirds, how many parts make up the whole? 3 Are the parts in the model the same size? Yes. What kind of number names equal parts of a whole? Fraction**

   If you ate 1 of the apple pieces, what fraction names the part of the whole apple you ate? $\frac{1}{3}$

   What part of a fraction tells the number of equal parts that make up the whole? The number below the fraction bar (the denominator)

   What is the fraction in words for the shaded part of the circle? One third What is the fraction in number form? $\frac{1}{3}$

2. **Suppose you want to break 1 graham cracker into fourths. How many parts make up the whole? 4**

   You eat 4 of the 4 parts. What fraction of 1 graham cracker have you eaten? Four-fourths What is the fraction in number form? $\frac{4}{4}$ **Four-fourths make up the whole. One whole is equal to what number? 1**

3. **Which piece is bigger, $\frac{1}{5}$ or $\frac{1}{6}$?** $\frac{1}{5}$ **What is the same about those two fractions? They both have 1 in the numerator. What do you notice about the denominators?** Elicit the idea that the unit fraction with the lesser denominator is the greater fraction.

4. **A mixed number contains a whole number and a fraction. How many parts make up the whole? 4**

   What is the fraction of one part? $\frac{1}{4}$ **How many fourths are there? 15 What is the improper fraction?** $\frac{15}{5}$ **How many wholes are there? 3 What is the fraction of the remaining parts?** $\frac{3}{4}$ **What is the mixed number of the model?** $3\frac{3}{4}$

Another Way Write different fractions on the board and have students model the fractions using fraction bars, circles, or other appropriate material. You may also have students work in pairs. With their partners, students can alternate between drawing fraction models and writing the fraction numbers.

Lesson 16-1 Basics of Fractions

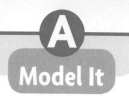

Objective 16.1: Understand that fractions may refer to parts of a set or parts of a whole.

Teach the Lesson

Materials ☐ Fraction Circles Plus

Activate Prior Knowledge

Draw a square and divide it into 4 equal parts. **How many equal parts are there?** 4 Shade 1 of the parts. **How many parts are colored in?** 1 There are 1 out of 4 parts colored in.

Develop Academic Language

Write the fraction $\frac{7}{8}$ on the board. **The top number in a fraction is called the numerator. The numerator tells how many parts are shaded. The bottom number in a fraction is called the denominator. The denominator tells the total number of parts. What is the numerator in this fraction?** 7 **What is the denominator?** 8

Model the Activities

Activity 1 Draw a rectangle on the board and divide it into 4 equal parts. **How many equal parts?** 4 **The total number of equal parts is the denomintor.** Shade 3 of the parts. **How many parts are shaded?** 3 **The number of shaded parts is the numerator. How can you write this as a fraction?** Sample: Write the numerator, 3, above the bar. Write the denominator, 4, below the bar. Write the fraction $\frac{3}{4}$ on the board and read its name.

Activity 2 Draw a fraction circle or a square divided into 8 equal parts. **How can we show $\frac{5}{8}$?** By shading 5 of the 8 parts **Does it make any difference which 5 parts we shade?** No, we can shade any 5 parts. Demonstrate showing $\frac{5}{8}$ in two or three different ways.

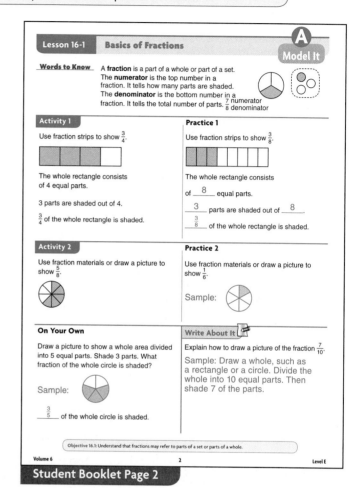

Student Booklet Page 2

Lesson 16-1: Basics of Fractions

Objective 16.1: Understand that fractions may refer to parts of a set or parts of a whole.

Facilitate Student Understanding

Develop Academic Language

In talking about the problems, have students use parts of a set to describe separate items shaded in a group of items rather than parts of a whole.

Demonstrate the Examples

Example 1 Help students distinguish between 5 equal parts of a whole and 5 items in a set. Stress that the fraction compares the number of items shaded to the total number of items.

Example 2 How many equal parts are there? 100 Where does 100 go in the fraction? On the bottom How do you know this? 100 tells you the total number of parts How many parts are shaded? 18 Where does the number of shaded parts go in the fraction? On the top How do you this? 18 is the number of parts that are being shaded. What is the fraction? $\frac{18}{100}$

Student Booklet Page 3

Progress Monitoring

Draw 7 boxes on the board and have students copy onto paper.

Show $\frac{4}{7}$ of this set.

☐ ☐ ☐ ☐ ☐ ☐ ☐

Students should shade 4 boxes.

Error Analysis

If students have difficulty understanding parts of a set, have them work with manipulatives to show $\frac{2}{3}$ of a set, $\frac{4}{7}$ of a set, and so on.

Lesson 16-1 | Basics of Fractions

Objective 16.1: Understand that fractions may refer to parts of a set or parts of a whole.

Observe Student Progress

Computer Tutorial

Some students may benefit from completing a computer tutorial before they attempt the Try It page. A list of the tutorials for each lesson can be found beginning on page x in the front of this book.

ENGLISH LEARNERS Consider a sentence frame students can complete in order to express these concepts. __1__ part of the whole is shaded. $\frac{1}{4}$ of the __set__ of quarters is shaded.

Error Analysis

Exercises 2 and 3 Provide counters for students to model these problems.

Exercise 6 If students have difficulty thinking of an example, ask them to think of things divided into 12 parts, such as an egg carton or a clock.

Exercise 7 Remind students that a model or picture must have equal parts to represent a fraction. Have them use fraction circles to model $\frac{1}{5}$ correctly.

Exercise 8 Check to see that students express and support their solution in a clear and logical manner. Be sure they use the appropriate terms, notations, and symbols and provide evidence to support their solution.

English Learners may benefit from answering this problem orally.

Student Booklet Page 4

Lesson 16-1 | Basics of Fractions | Try It

1. Write a fraction for the shaded part.

   $\frac{25}{100}$

2. Write the fraction.
   a. 2 parts shaded out of 5 $\frac{2}{5}$
   b. 5 parts shaded out of 8 $\frac{5}{8}$
   c. 1 part shaded out of 6 $\frac{1}{6}$

3. Write the fraction of the set.
   a. 4 stars out of 8 $\frac{4}{8}$ or $\frac{1}{2}$
   b. 3 marbles out of 10 $\frac{3}{10}$
   c. 2 hungry dogs out of 9 $\frac{2}{9}$

4. Write a fraction for the shaded part.

   $\frac{2}{7}$

5. Show $\frac{3}{8}$ in two different ways. For one, use 8 objects. For the other, use 1 object.
   Check students' answers for part of a set and part of a whole.

6. Give an example of an everyday situation in which $\frac{1}{12}$ can be represented.
   Sample: one hour on a clock

7. Explain why the picture below does not represent $\frac{1}{5}$.
   Sample: The 5 parts are not equal.

8. On another sheet of paper, explain how to show $\frac{3}{4}$ of a circle. List all of your steps.
   Sample: Draw a circle. Divide it into 4ths with two diameters. Shade 3 of them.

Objective 16.1: Understand that fractions may refer to parts of a set or parts of a whole.

Lesson 16-2: Unit Fractions

Objective 16.2: Recognize, name, and compare unit fractions from $\frac{1}{12}$ to $\frac{1}{2}$.

Teach the Lesson

Materials ☐ Fraction Circles Plus™

Activate Prior Knowledge

Draw two squares on the board. Draw a line through the middle of one square to divide it in half and another line in the second square that divides it into two very unequal pieces. **Which shows $\frac{1}{2}$?** Continue with other shapes and fractions.

Develop Academic Language

On the board draw this chart. Have volunteers complete it for numbers up to 12.

Number of equal parts	Word for one part	Number for one part
2	one-half	$\frac{1}{2}$
3	one-third	$\frac{1}{3}$

Model the Activities

Activity 1 Draw a large circle on the board. **How would you divide this circle into 5 equal parts?** Have students model this with fraction circles. **What is one piece called? One fifth How would you write this in numbers?** $\frac{1}{5}$ Leave this circle on the board to use later.

Draw another circle that is the same size. **How would you divide this into 3 equal parts?** Have a student demonstrate as the rest demonstrate thirds with their fraction circles. **What is this fraction in words? One third In numbers?** $\frac{1}{3}$

Activity 2 Use fraction materials for $\frac{1}{6}$ and $\frac{1}{8}$. **Which is bigger?** $\frac{1}{6}$ **How can you use the fraction materials to compare fractions?** Sample: Place the pieces for the two fractions on top of one another so you can compare.

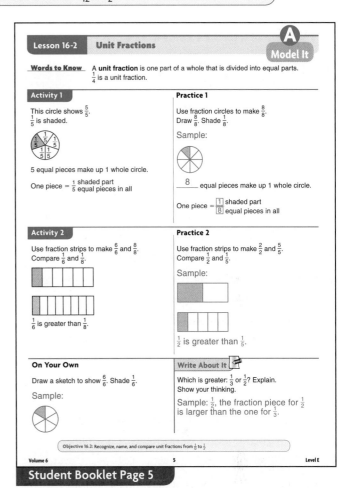

Student Booklet Page 5

✓ Progress Monitoring

Divide a square into six equal pieces. **What are these called? Sixths** Shade one piece. **What fraction is this?** $\frac{1}{6}$

Error Analysis

If students write the fractions incorrectly, remind them that the total number of parts goes on the bottom and the number of shaded parts goes on top.

Lesson 16-2 Unit Fractions

Objective 16.2: Recognize, name, and compare unit fractions from $\frac{1}{12}$ to $\frac{1}{2}$.

Facilitate Student Understanding

Develop Academic Language

Make sure students understand the terms *numerator* and *denominator*.

Demonstrate the Examples

Example 1 You may want to reinforce the example by having students use fraction circles or other materials to look at the relationship.

Tell students to imagine that the circles are pizzas that are the same size. Compare the one divided into 2 equal slices and the one divided into 3. Discuss which pizza has bigger slices. **Sample: The pizza divided into 2 slices. When there are more slices, each is smaller.** Have students think of slices of pizza as fractions. Discuss how they know that if the numerators of two fractions are both 1, then the fraction with the larger denominator is smaller.

Example 2 Write the symbols $>$ and $<$ on the board. Have several students give an example for each. You may want to allow whole number examples to make sure students understand.

Computer Tutorial

Some students may benefit from completing a computer tutorial before they attempt the Try It page. A list of the tutorials for each lesson can be found beginning on page x in the front of this book.

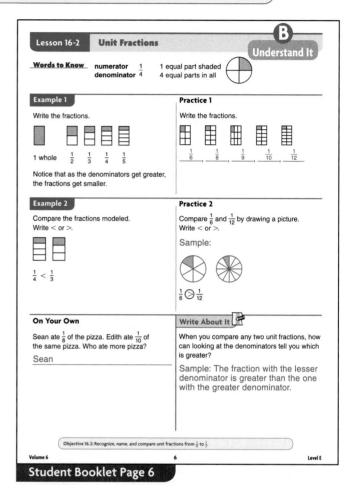

Student Booklet Page 6

Progress Monitoring

Which is greater, $\frac{1}{5}$ or $\frac{1}{8}$? $\frac{1}{5}$ How do you know? **Accept any reasonable answer. Use a symbol to show your answer.** $\frac{1}{5} > \frac{1}{8}$

Error Analysis

If students confuse $>$ and $<$, point out that the open side of the symbol points to the larger number.

Lesson 16-2 Unit Fractions

Objective 16.2: Recognize, name, and compare unit fractions from $\frac{1}{12}$ to $\frac{1}{2}$.

Observe Student Progress

Develop Academic Language

To help students write fractions for word names, discuss how *-th* is added to numbers to make fraction names for numbers greater than 3. Remind them about the irregular fraction names: half, third, fifth, eighth, ninth.

ENGLISH LEARNERS The *-th* sound does not exist in many languages, so some students may need additional pronunciation and listening practice to become proficient at these numbers.

 Error Analysis

Exercises 3–7 If students are having difficulty with these exercises, encourage their use of fraction circles or other fraction materials.

Exercise 8 **ENGLISH LEARNERS** A set of drawings can be a complete answer for some students. Encourage students to number their steps.

Lesson 16-2 Unit Fractions — Try It

1. Shade $\frac{1}{12}$.

2. Shade $\frac{1}{3}$. Shade $\frac{1}{6}$. Which is greater?
   $\frac{1}{3}$ ⊖ $\frac{1}{6}$

3. Write < or > to compare each pair of fractions.
   a. $\frac{1}{3}$ ⊖ $\frac{1}{2}$
   b. $\frac{1}{8}$ ⊖ $\frac{1}{10}$
   c. $\frac{1}{5}$ ⊖ $\frac{1}{12}$
   d. $\frac{1}{6}$ ⊖ $\frac{1}{4}$

4. Which fraction is greater than $\frac{1}{5}$? Circle the letter of the correct answer.
   Ⓐ $\frac{1}{3}$
   B $\frac{1}{6}$
   C $\frac{1}{8}$
   D $\frac{1}{10}$

5. The fraction $\frac{1}{3}$ is greater than $\frac{1}{6}$. Write two other fractions with a numerator of 1 that are greater than $\frac{1}{6}$.
   $\frac{1}{2}$ and $\frac{1}{4}$

6. Pedro says that $\frac{1}{8}$ is greater than $\frac{1}{6}$ because 8 is greater than 6. Cindy says $\frac{1}{6}$ is greater than $\frac{1}{8}$. Who is right? Explain.
   Cindy

7. Write these fractions in order from greatest to least.
   $\frac{1}{10}, \frac{1}{2}, \frac{1}{6}, \frac{1}{8}, \frac{1}{5}$
   $\frac{1}{2}, \frac{1}{5}, \frac{1}{6}, \frac{1}{8}, \frac{1}{10}$

8. On another sheet of paper, describe how to show $\frac{1}{8}$ of a square. Show all of your steps.
   Sample: First divide the square both horizontally and vertically to make 4 equal parts. Then from the center of the square divide each part diagonally to make 8 equal parts. Then shade one part.

Student Booklet Page 7

| Lesson 16-3 | Parts and the Whole |

Model It

Objective 16.3: Know that when all fractional parts are included, such as four fourths, the result is equal to the whole and to one.

Teach the Lesson

Materials ☐ Fraction Circles Plus

Activate Prior Knowledge

Draw a square on the board and divide it into 5 equal parts. **How many equal parts? 5 How many parts in the whole? 5 If we divided the square into 10 equal parts, how many parts would be in the whole? 10** Use gestures to reinforce parts and whole.

Develop Academic Language

Be sure students use the word *whole* correctly to describe all the parts together. They can also use 1 to describe the whole.

ENGLISH LEARNERS Since the words *whole* and *hole* are homophones with different meanings, some students might find it confusing to listen to discussions of problems with wholes. Ask them to devise a signal that lets everyone know when they are referring to the whole.

Model the Activities

Activity 1 Draw a rectangle on the board, divide it into 6 equal parts, and shade all of the parts. **How many equal parts are there? 6 How many parts are shaded? 6 What fraction is shaded? $\frac{6}{6}$ What is another name for $\frac{6}{6}$? Whole, or 1**

Draw a line across the middle of the rectangle. **How does this change the problem? There are now 12 equal parts and 12 are shaded. What fraction can you write for this problem? $\frac{12}{12}$ What is another name for the shaded part? Whole, or 1**

Activity 2 If students think that only 8 of the 9 fenceposts are painted, remind them to read the problem carefully. **Tami and her 8 friends means how many people all together? 9**

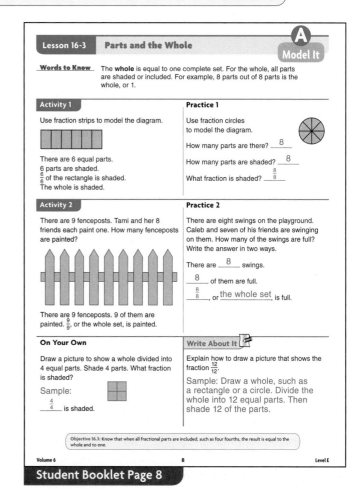
Student Booklet Page 8

✓ Progress Monitoring

Have students draw a square and divide it into 8 equal parts. Then have them shade all 8 parts and describe what is shaded in three different ways. $\frac{8}{8}$, whole, or 1

Error Analysis

Students might have difficulty equating all the parts shaded to the whole being shaded. To help make the connection, use fraction circles or other fraction materials and ask students to identify the whole in each case.

Lesson 16-3 Parts and the Whole

Objective 16.3: Know that when all fractional parts are included, such as four fourths, the result is equal to the whole and to one.

Facilitate Student Understanding

Develop Academic Language

Continue to have students read fractions such as $\frac{8}{8}$ *eight eighths* and use the word *whole* to describe such fractions.

ENGLISH LEARNERS Remind students to listen carefully for the fractional parts that end with *-th*. Model clear pronunciation of these words.

Demonstrate the Examples

Example 1 Explain to students that finding fractions of a set works the same way as finding fractions of a whole. **How many parts are there?** 4 **How many are shaded?** 4 **What is the fraction?** $\frac{4}{4}$ If students have trouble seeing this, let them use fraction materials to show that 4 fourths can make one square when they are put together, or a set of 4 triangles when they're broken apart. Help students realize that when the numerator and denominator of a fraction are the same, the fraction is equal to 1, which is another name for a whole. Thus they should be able to write 1 for a fraction with the same numerator and denominator or provide the missing numerator or denominator when a fraction is equal to 1.

Example 2 It may help students to use fraction materials to model the problem. Again, stress that all the parts out of a number of parts gives a fraction equal to 1.

Computer Tutorial

Some students may benefit from completing a computer tutorial before they attempt the Try It page. A list of the tutorials for each lesson can be found beginning on page x in the front of this book.

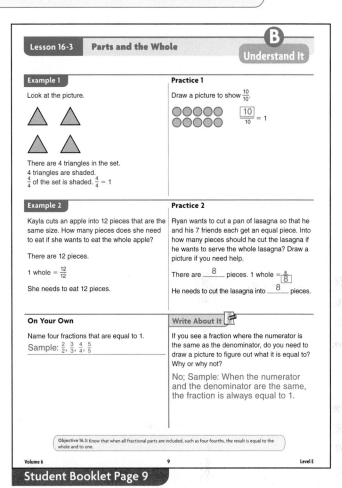

Student Booklet Page 9

Progress Monitoring

Draw 6 boxes on the board and have students copy onto paper.

Show $\frac{6}{6}$ of this set.

Students should shade 6 boxes.

Error Analysis

If students have difficulty completing problems such as $\frac{\square}{10} = 1$, have them work with fraction circles, identify the whole, and then count the number of parts in the whole.

Lesson 16-3 Parts and the Whole

Objective 16.3: Know that when all fractional parts are included, such as four fourths, the result is equal to the whole and to one.

Observe Student Progress

Develop Academic Language

Exercise 7 Encourage students to use $\frac{4}{4}$, 1, and one whole to describe the answer.

 Error Analysis

Exercise 3b Have students draw a picture divided into 8 parts, then shade the whole. **Now count how many sections are shaded to find the answer.**

Exercise 4 If the student does not select C as the answer, use models to help show how each of the fractions can be represented.

Exercise 5 Make sure students have drawn equal parts. If the parts are unequal, suggest that they trace fraction materials.

Exercise 6 It may help students to model this problem if they are not sure how to find the solution.

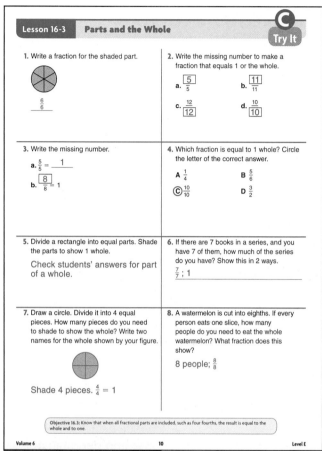

Student Booklet Page 10

Lesson 16-4: Rename Values Greater Than 1

Objective 16.4: Define and manipulate improper fractions and mixed numbers.

Teach the Lesson

Materials ☐ Teaching Aid 12 Fraction Strips

Activate Prior Knowledge

Have students use the fraction strips to model the fractions $\frac{2}{3}$, $\frac{3}{8}$, and $\frac{3}{4}$. **How do you know which fraction strips to use?** The denominator of the fraction tells which fraction strips to use. **How do you know how many fraction strips to use?** The numerator of the fraction tells how many fraction strips to use.

Develop Academic Language

The numerator of a fraction is above the fraction bar, and the denominator of a fraction is below the fraction bar. Write the fraction $\frac{3}{4}$ on the board. **What is the numerator in this fraction?** 3 **What is the denominator?** 4

Model the Activities

Activity 1 **How many one-third fraction strips do you need to model the improper fraction $\frac{10}{3}$?** 10 Model the fraction. **How many one-thirds are there in one whole?** 3 **How do you know?** The fraction $\frac{3}{3}$ equals 1. Now divide your fraction strips into groups of 3. **How many whole groups of 3 do you have?** 3 **Are there any one-thirds left over?** Yes. **How many?** 1 Use the fraction strips to rename the improper fraction as a mixed number. **What mixed number has 3 wholes and one-third?** $3\frac{1}{3}$

Activity 2 **How many fourths do you need to model one whole?** 4 Model $2\frac{1}{4}$ using the fraction strips. **How many fourths did you use?** 9 **What is the numerator of the improper fraction?** 9 **What is the denominator?** 4 **So, what improper fraction is equivalent to $2\frac{1}{4}$?** $\frac{9}{4}$

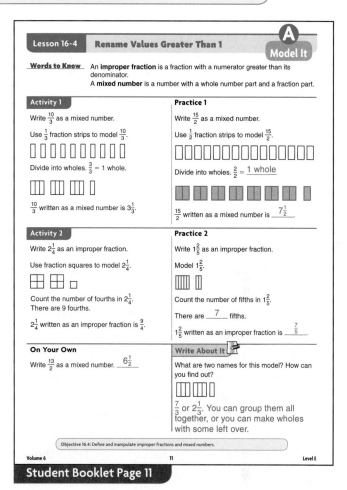

Student Booklet Page 11

✓ Progress Monitoring

Rename $\frac{8}{5}$ as a mixed number. **What fraction blocks will you use?** Fifths **How many of them do you need?** 8 Separate the blocks to show the mixed number. Check students' blocks. **What is $\frac{8}{5}$ as a mixed number?** $1\frac{3}{5}$

Error Analysis

Remind students to write the whole number part of a mixed number larger than the numbers in the numerator and denominator of the fraction part. This will help to keep the whole number and the fraction separate.

| Lesson 16-4 | Rename Values Greater Than 1 |  Understand It |

Objective 16.4: Define and manipulate improper fractions and mixed numbers.

Facilitate Student Understanding

Develop Academic Language

Discuss the term *improper* with students. **What is the everyday meaning of improper?** Wrong, incorrect An improper fraction is not wrong, but because its numerator is greater than its denominator, it isn't in the usual form.

Demonstrate the Examples

Example 1 When you used fraction pieces to change improper fractions into mixed numbers, you were using division. You divided the numerator into as many equal groups of the denominator as possible. Now we'll do the same thing without the fraction pieces. Divide the numerator by the denominator. **What do you get?** $20 \div 3 \rightarrow 6$ R2 The 6 is the whole number part of the mixed number. The remainder is the numerator of the fraction part of the mixed number. The denominator from the improper fraction is the denominator for the mixed number. **What is the mixed number?** $6\frac{2}{3}$

Example 2 To change a mixed number into an improper fraction, we need to do the opposite of changing an improper fraction to a mixed number. **What is the inverse operation of division?** Multiplication So for this problem, we need to multiply the whole number by the denominator. **What do you get?** $3 \times 5 = 15$ Now we need to add that to the original numerator. **What do you get?** $15 + 2 = 17$ Then we put that numerator over the original denominator. **What is the improper fraction?** $\frac{17}{3}$

Student Booklet Page 12

ENGLISH LEARNERS To help students remember the terms whole number, numerator, and denominator, create a poster with all parts of a mixed number labeled.

 **Progress Monitoring**

Change $\frac{9}{4}$ into a mixed number. $9 \div 4 \rightarrow 2$ R1. 2 is the whole number. 1 is the numerator. 4 is the denominator. $2\frac{1}{4}$

Error Analysis

Students might forget to add the numerator when converting a mixed number to an improper fraction. Tell them to change their improper fraction back to a mixed number and see if they get the same mixed number they started with.

Lesson 16-4: Rename Values Greater Than 1

Objective 16.4: Define and manipulate improper fractions and mixed numbers.

Observe Student Progress

Computer Tutorial

Some students may benefit from completing the computer tutorial before they attempt the Try It page. A list of the tutorials for each lesson can be found beginning of page x in the front of this book.

Error Analysis

Exercises 1 and 2 Make sure students are writing mixed numbers properly, with the numerator clearly above the denominator and the whole number part of mixed numbers larger than the numerator and denominator. Show students that using a horizontal fraction bar is less confusing than a diagonal fraction bar when writing mixed numbers.

Exercise 3 If students choose answer D, they are only using the whole number part of the quotient, and not dealing with the remainder.

Exercise 4 If students choose answer D, they are multiplying the numerator by the whole number instead of multiplying the denominator by the whole number. Encourage students to check their answers by converting the improper fraction back to a mixed number.

Exercise 6 If students have trouble with the division involved in writing improper fractions as mixed numbers, encourage them to show the division problem instead of computing it mentally.

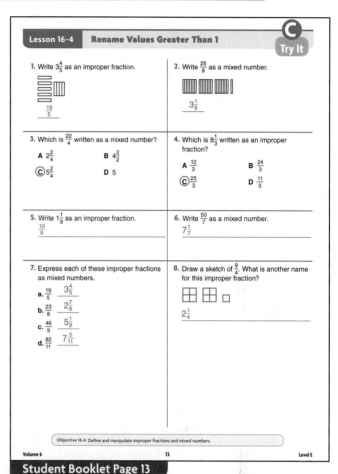

Student Booklet Page 13

Volume 6: Understand Fractions

Topic 16: Meaning of Fractions

Topic Summary

Objective: Review the meanings of fractions.

Have students complete the student summary page. You may want to have students work in groups of four with each student analyzing a different choice. Ask students to share their ideas about each answer choice. Be sure they confirm the correct answer for each problem at the end of the discussion.

Answer Evaluation

1. **A** Students do not know how to compare fractions.

   B Students do not know how to compare fractions.

   C Students think a fraction with a larger denominator is greater.

   D This choice is correct.

2. **A** Students gave the fraction of red pens.

   B This choice is correct.

   C Students gave a fraction of all the pens.

   D Students confused the numerator and the denominator.

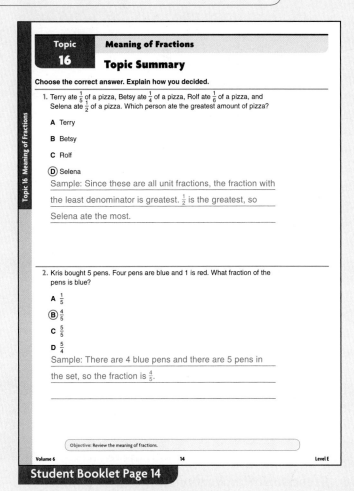

Student Booklet Page 14

 ## Error Analysis

Exercise 1 Have students compare fractions using fraction strips. Remind students that unit fractions with the least denominator are the greatest, and unit fractions with greater denominators are the least.

Exercise 2 If students confuse the numerator and denominator, have them find the denominator (a set or whole), then the numerator (number of parts in a set or whole).

Progress Monitoring

When all assignments for this topic have been completed, assign the corresponding Progress Monitoring page for this topic (Assessment Resources Book, page 16). Be sure students complete the Progress Monitoring page before you administer the final assessment for this volume.

Volume 6 — Understand Fractions
Topic 16 — Meaning of Fractions
Mixed Review

Objective: Maintain concepts and skills.

Have students complete the Mixed Review page. Work with each student individually to review results. Identify strengths and weaknesses and correct any misunderstandings.

 Error Analysis

Exercise 1 If students record numbers in incorrect places, review place values. Have students practice reading numbers and writing the word names.

Exercises 2 and 3 If students make arithmetic mistakes, continue to review addition and multiplication facts.

Exercise 4 Have students check their work by multiplying the quotient and divisor and adding the remainder. Also remind students that the remainder is never larger than the divisor.

Exercise 5 Recommend that students use a place-value chart to be sure they are adding the correct values.

Student Booklet Page 15

Topic 16 — Meaning of Fractions
Mixed Review

1. Write the following number in standard form.

   three million, eight hundred five thousand, two hundred twelve

   3,805,212

2. Find each sum or difference.

   a. 38 + 96   134
   b. 73 − 27   46
   c. 328 + 273 601
   d. 525 − 261 264

3. Find each product.

   a. 23 × 4   92
   b. 4,298 × 6   25,788
   c. 75 × 8   600
   d. 7,802 × 4   31,208

4. Find each quotient.

   a. 340 ÷ 3   113 R1
   b. 646 ÷ 4   161 R2
   c. 89 ÷ 7   12 R5
   d. 525 ÷ 5   105

5. Write the expanded form for seven thousand, three hundred four.

   7,000 + 300 + 4

6. How many different addition facts have a sum of 9? What are they?

   5: 0 + 9, 1 + 8, 2 + 7, 3 + 6, 4 + 5

Volume 6: Understand Fractions

Topic 17: Equivalence of Fractions

Topic Introduction

Lesson 17-1 gives students their first exposure to comparing fractions by using visual or tactile models. Lesson 17-2 provides students with a review of factoring whole numbers, an important skill to master before they move on to finding equivalent fractions in Lesson 17-3. Lesson 17-4 teaches students how to use models, multiplication, and division to put fractions in lowest terms. Finally, Lesson 17-5 gives students an overview of some of the different ways to interpret fractions.

Lesson	Objective	Student Pages	Teacher Pages	Tutorials
Topic 17 Introduction	**17.1** Compare fractions represented by drawings, number lines, or concrete materials. **17.3** Multiply and divide by forms of 1 to write equivalent fractions. **17.4** Rewrite fractions in lowest terms.	16	285	
17-1 Compare Fractions	**17.1** Compare fractions represented by drawings, number lines, or concrete materials.	17–19	286–288	17a, 17b, 17c, 17d
17-2 Factor Whole Numbers	**17.2** Find the factors of whole numbers.	20–22	289–291	17e, 17f, 17g
17-3 Equivalent Fractions	**17.3** Multiply and divide by forms of 1 to write equivalent fractions.	23–25	292–294	17h, 17d, 17i
17-4 Fractions in Lowest Terms	**17.4** Rewrite fractions in lowest terms.	26–28	295–297	17j, 17k
17-5 Interpretations of Fractions	**17.5** Explain different interpretations of fractions, such as parts of a whole, parts of a set, and division of whole numbers by whole numbers.	29–31	298–300	17l, 17m, 17n, 17o
Topic 17 Summary	Review finding equivalent fractions.	32	301	
Topic 17 Mixed Review	Maintain concepts and skills.	33	302	

Computer Tutorial

Some students may benefit from completing the computer tutorial before they attempt the Try It page of each lesson. If you are using the electronic components of *Pinpoint Math*, you will find a complete listing of tutorial codes and titles when you access them either online or via CD-ROM.

Volume 6 — Understand Fractions

Topic 17 — Equivalence of Fractions

Topic Introduction

Objectives: 17.1 Compare fractions represented by drawings, number lines, or concrete materials. 17.3 Multiply and divide by forms of 1 to write equivalent fractions. 17.4 Rewrite fractions in lowest terms.

Materials ☐ Fraction Circles Plus

Have students practice finding common multiples and common factors for pairs of whole numbers.

Informal Assessment

1. **For Exercise 1, what do you notice about the three sections in this circle?** They are not the same size; they are marked with fractions. You can have students use fraction circles to verify that the fractions named form 1 whole. **Which two fractions are named in the top half?** $\frac{1}{6}$ and $\frac{1}{3}$ **What can you conclude about the sum of $\frac{1}{6}$ and $\frac{1}{3}$?** $\frac{1}{6} + \frac{1}{3} = \frac{1}{2}$ Have students work in pairs or groups to complete Exercise 1.

2. **How many sections is the circle divided into?** 4 **Are they equal parts?** No. **Which section of this circle is the largest?** $\frac{1}{2}$ **Which section of this circle is the smallest?** $\frac{1}{12}$ **Which is greater, $\frac{1}{4}$ or $\frac{1}{6}$?** $\frac{1}{4}$ Have students complete the inequalities.

3. **Look at the fraction $\frac{2}{4}$. What numbers divide evenly into both 2 and 4?** 1 and 2 **Which of these numbers is greater than the other?** 2 is greater than 1. **If you divide the numerator and denominator of $\frac{2}{4}$ by 2, what will the new fraction be?** $\frac{1}{2}$

4. **Use $\frac{4}{8}$ and $\frac{3}{6}$ as an example. What is the greatest number that divides into 4 and 8?** 4 **If you divide both numbers by 4, what will the new fraction be?** $\frac{1}{2}$ **Find the greatest number that divides into 3 and 6.** 3 **Tell me the new fraction.** $\frac{1}{2}$ **Therefore, $\frac{4}{8}$ and $\frac{3}{6}$ are equivalent.** Have students rewrite all the fractions in lowest terms and then find the matches.

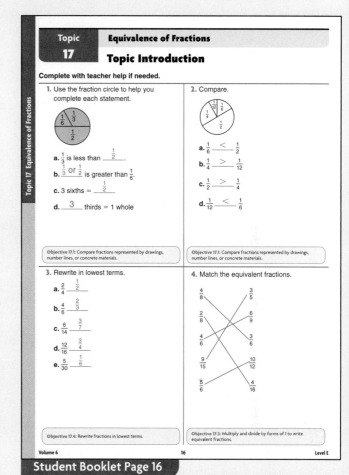

Student Booklet Page 16

Another Way If students have a difficult time comparing fractions, have them use Fraction Circles Plus.

Lesson 17-1 Compare Fractions

Objective 17.1: Compare fractions represented by drawings, number lines, or concrete materials.

Teach the Lesson

Materials ☐ Fraction Circles Plus

Activate Prior Knowledge

Provide fraction circles. **Find the green pieces and use them to make 1 whole circle. How many pieces did you use?** 3 **What fractional parts are shown, thirds or fourths?** Thirds **How do you know?** There are 3 parts in 1 whole. Have a volunteer draw the model on the board. Write $\frac{1}{3}$ on each part and *thirds* beneath. Repeat for other fractions in the set.

Develop Academic Language

List the names of fractional parts from halves through twelfths on the board. Have students look for and underline base words. <u>Four</u>ths, <u>six</u>ths, ...

ENGLISH LEARNERS Not all languages have the *-th* sound at the end of words, so the sound may be hard to pronounce and perceive. Enunciate carefully and have students practice saying these words.

Model the Activities

Activity 1 Have students work in pairs with fraction circles. **How can you show $\frac{2}{3}$?** Put together 2 of the thirds pieces. **Now gather the sixths pieces. How can you find out how many sixths cover the same part of a circle as $\frac{2}{3}$?** Lay sixths over the model until it is covered by sixths without overlap. **How many sixths do you need?** 4 $\frac{2}{3} = \frac{4}{6}$.

Activity 2 **Which fractional parts does the model on the left show?** Thirds **The model on the right?** Fourths **How will you shade to show the fractions you need to compare?** Shade 2 thirds on the left model and 3 fourths on the right model. **How can you tell which is greater?** The one that takes up more of the whole Emphasize that the wholes must be the same size for this method to work.

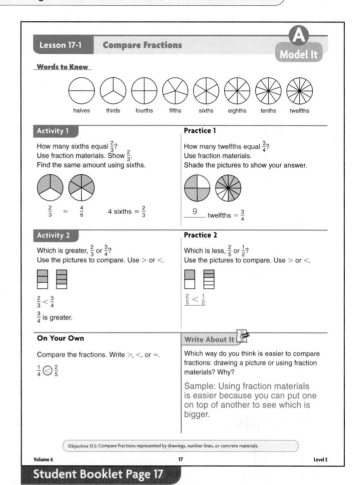

Student Booklet Page 17

Progress Monitoring

Make circle models to compare $\frac{1}{4}$ to $\frac{1}{3}$. If students draw, make sure they draw two circles of the same size. **How can you decide which is greater?** The one that takes up more of the whole circle is greater: $\frac{1}{3}$.

Error Analysis

Some students may not be able to compare two fraction drawings shown side-by-side. Have them model the fractions and put one model on top of the other. Emphasize that the wholes must be the same size, and give an example of a misleading drawing with wholes of very different sizes.

Volume 6   286   Level E

| Lesson 17-1 | Compare Fractions |

Objective 17.1: Compare fractions represented by drawings, number lines, or concrete materials.

Facilitate Student Understanding

Materials ☐ Teaching Aid 11 Fraction Number Lines

Develop Academic Language

Draw this number line on the board.

Ask students what the fraction $\frac{2}{5}$ represents on this model. **The distance from 0 to the $\frac{2}{5}$ mark** Draw a circle model of fifths. **How are these models the same? Both have 5 equal parts, but the number line shows lengths, while the circle model shows areas.** Repeat with number line models for other fractions.

Demonstrate the Examples

Example 1 Distribute Teaching Aid 11 and draw a number line from 0 to 1 on the board. **How can I mark this number line to show eighths? Divide it in 8 equal sections.** Without drawing any more tick marks, how can we show fourths on the same line? **Group the 8 spaces in 4 equal groups.** Help students see that other fractions such as thirds, fifths, sevenths, or ninths could not be plotted on the same tick marks used for the eighths. **Label the fourths.** If the process confuses students, show how to start instead with a number line showing the lesser denominator, fourths, and then how to divide each space to show eighths. **Which of the eighths fractions is on the same tick mark as $\frac{3}{4}$? $\frac{6}{8}$ Then those fractions are equal.** Have students find the other pairs of equal fractions. $\frac{4}{8} = \frac{2}{4}, \frac{6}{8} = \frac{3}{4}$

Example 2 To compare fractions, both should have the same denominator. Find an eighths fraction equal to $\frac{3}{4}$. $\frac{6}{8}$ **How can you tell whether this is greater or less than $\frac{5}{8}$? Since $\frac{6}{8}$ is farther right, it is greater. $\frac{3}{4}$ is the same as $\frac{6}{8}$, so $\frac{3}{4} > \frac{5}{8}$.**

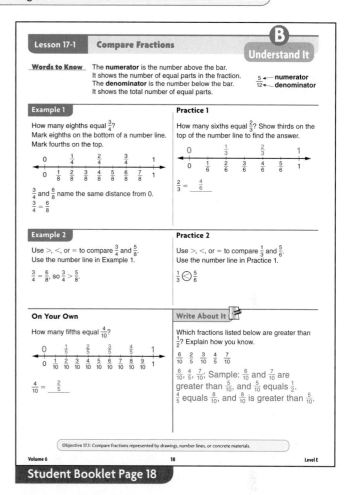

Student Booklet Page 18

ENGLISH LEARNERS Be sure that students hear the *-th* at the end of twelfths. With 5 consonants in a row, it can be difficult to sort out.

★ Progress Monitoring

Draw a number line from 0 to 1 that shows sixteenths. Circle the fraction that equals $\frac{1}{2}$. $\frac{8}{16}$ Draw a box around the fraction that equals $\frac{1}{4}$. $\frac{4}{16}$

Error Analysis

In Example 2 and Practice 2, students may note that the fractions can be compared visually, without giving both the same denominator. However, when both scales on the number line aren't given, as in Exercise 3 on the next page, students will need to proceed as for Activity 1.

Lesson 17-1 Compare Fractions

Objective 17.1: Compare fractions represented by drawings, number lines, or concrete materials.

Observe Student Progress

Computer Tutorial

Some students may benefit from completing a computer tutorial before they attempt the Try It page. A list of the tutorials for each lesson can be found beginning on page x in the front of this book.

Error Analysis

Exercise 1 Ask what fractional parts the two models show. **Fourths, fifths** Some students may need to use fraction circles.

Exercise 2 If students need help getting started, remind them that there are 10 spaces to be divided in 5 equal parts. Students could also start with a number line showing fifths and then divide the spaces to show tenths.

Exercise 3 Help students identify sixths as the fractional parts they need to show along the top of the number line. Remind students that the numbers become greater from left to right.

Exercise 4 Have students model each answer choice with fraction circles, and then use one of the half pieces to compare. Students should place the half on top of each model and see whether the fraction underneath is greater.

Exercise 6 Give students several unit fractions to compare. Allow students to use fraction circles, and then help them write a general rule about fractions that have the same numerator.

Exercise 8 Suggest that students begin by using $\frac{1}{2}$ as a benchmark. Alternatively, students could make a number line showing halves and eighths, and a separate number line showing halves and tenths.

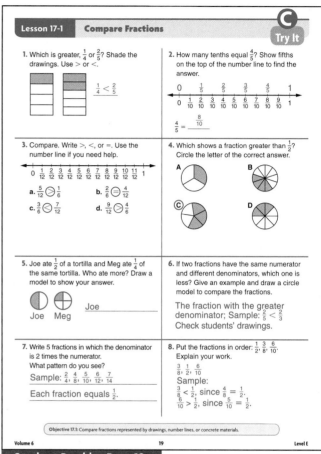

Student Booklet Page 19

Lesson 17-2 Factor Whole Numbers

Objective 17.2: Find the factors of whole numbers.

Teach the Lesson

Materials ☐ MathFlaps

Activate Prior Knowledge

Name one multiplication fact with a product of 18. Sample: $2 \times 9 = 18$ Name the rest of the facts with the product 18. $3 \times 6 = 18, 1 \times 9 = 18$

Develop Academic Language

Draw this chart on the board. Have students give the factors and product for each factorization that you supply. Do several others besides the ones illustrated.

Factorization	Factors	Product
$3 \times 7 = 21$	3, 7	21
$12 \times 5 = 60$	12, 5	60

ENGLISH LEARNERS Make sure that students understand that the term *factor* can be both a noun and a verb, and that the meanings are related.

Model the Activities

Activity 1 Have students use MathFlaps. **Model any multiplication fact that has a product of 10.** They may model 2 rows of 5 or 5 rows of 2. **What factors have we modeled?** $2 \times 5 = 10$ or $5 \times 2 = 10$ Point out that each is called a factorization of 10 but they are not considered to be distinct because they involve the same two factors. Show that any number can be factored as 1 times itself; for this page, however, you may want to ask that students factor each product without using 1 as a factor.

Activity 2 **What is the total number of plants?** 15 **Use 15 MathFlaps. How can you arrange them in equal rows?** In 3 rows, 5 rows, 15 rows, or 1 row **When we're making equal rows, what operation are we modeling?** Multiplication **What is one multiplication fact with a product of 15?** $3 \times 5 = 15$

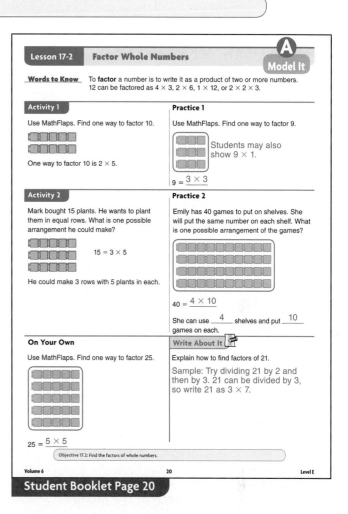

Student Booklet Page 20

Write About It

Point out that since 21 is an odd number, you know that you can't divide by 2. Therefore, 2 is not a factor.

Progress Monitoring

Give two ways to factor 49. $1 \times 49, 7 \times 7$

Error Analysis

For students who have difficulty finding factors, suggest they start with 2. After testing 2 as a factor, try 3, 4, and so on. When students use this orderly method, they can stop factoring as soon as they reach a factor that has already appeared in the list.

Lesson 17-2 — Factor Whole Numbers

Objective 17.2: Find the factors of whole numbers.

Facilitate Student Understanding

Develop Academic Language

Continue to encourage students to use the words *factorization* and *factor* as they work the problems.

Demonstrate the Examples

Example 1 Help students see that the "try 1, 2, 3 ..." method requires division by those numbers. Students can factor by recalling multiplication facts, but that will become more difficult as students encounter larger numbers.

Point out that when students factor 16 as 2×8, they can tell there are other possible ways to factor 16 because 8 can be factored further. It may help students to find the possibilities if they start by breaking down the numbers as far as possible ($16 = 2 \times 2 \times 2 \times 2$) and then see how many different ways those factors can be combined.

Example 2 Here, you need only 2 factors in each factorization. What will each of the factors stand for? *The number of rows and the number of tiles in each row* Guide students to factor. In this example, students are asked to go beyond factoring to find the number of arrangements. Each factorization shows two possible arrangements. Have students demonstrate that while 5×8 and 8×5 are not different factorizations, they do represent different arrangements. The number of arrangements is twice the number of factorizations, unless any of the factorizations are "doubles" facts.

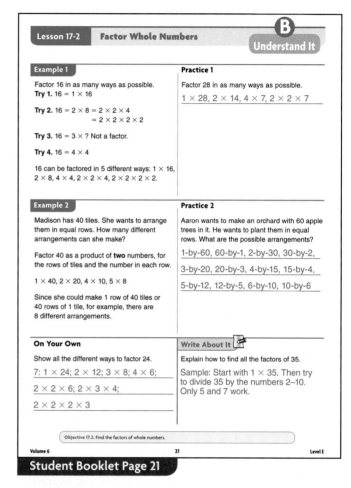

Student Booklet Page 21

 Progress Monitoring

In how many different ways can you factor 42? 5: $1 \times 42, 6 \times 7, 2 \times 3 \times 7, 2 \times 21, 3 \times 14$ **Explain how you found the factorizations.** Sample: I factored 42 into 6×7. Then I factored 6 into 2×3. I combined the factors 2, 3, and 7 in as many ways as possible to make 42.

Error Analysis

If students have difficulty finding all the possible factorizations, encourage them to begin with 1 times the number, then to list any facts with that product, and then to break the factors down and recombine them in as many ways as possible.

Lesson 17-2 Factor Whole Numbers

Objective 17.2: Find the factors of whole numbers.

Observe Student Progress

Computer Tutorial

Some students may benefit from completing a computer tutorial before they attempt the Try It page. A list of the tutorials for each lesson can be found beginning on page x in the front of this book.

Develop Academic Language

Have students explain the difference between the noun *factor* and the verb *factor*. As a noun, it refers to a number multiplied by another number to give a product; as a verb, it means to write a number as a product of factors.

Exercise 2 Here, *single expression* indicates that students should write a single multiplication problem for each product.

Student Booklet Page 22

Lesson 17-2 Factor Whole Numbers

1. One way to factor 20 is shown. What is another way?

   1×20, 2×10, or $2 \times 2 \times 5$

2. Factor into a single expression with as many factors as possible.
   a. 36 $\underline{2 \times 2 \times 3 \times 3}$
   b. 60 $\underline{2 \times 2 \times 3 \times 5}$
   c. 100 $\underline{2 \times 2 \times 5 \times 5}$
   d. 72 $\underline{2 \times 2 \times 2 \times 3 \times 3}$

3. Factor each number in two different ways as a product of two whole numbers.
   a. 63 $\underline{1 \times 63, 3 \times 21, \text{ or } 7 \times 9}$
   b. 36 $\underline{1 \times 36, 2 \times 18, 3 \times 12,}$
      $\underline{4 \times 9, \text{ or } 6 \times 6}$
   c. 28 $\underline{1 \times 28, 2 \times 14, \text{ or } 4 \times 7}$
   d. 70 $\underline{1 \times 70, 2 \times 35, 5 \times 14, \text{ or}}$
      $\underline{7 \times 10}$

4. Which of the following is **not** a way to factor 88?
   A 8×11
   B $2 \times 2 \times 2 \times 11$
   Ⓒ 4×44
   D 1×88

5. List all the different ways to factor 80 as a product of 2 whole numbers.
   $1 \times 80, 2 \times 40, 4 \times 20,$
   $5 \times 16, 8 \times 10$

6. Natasha says the only six ways to factor 90 are 1×90, 2×45, 3×30, 5×18, 6×15, and 9×10. What is her mistake?
   She forgot factorizations with more than 2 factors: $9 \times 2 \times 5$, $2 \times 3 \times 15$, $5 \times 3 \times 6$, $2 \times 3 \times 3 \times 5$.

Error Analysis

Exercise 3 If students cannot find more than one way to factor a number, continue to model testing different numbers systematically by division to see whether they are factors. Point out that for this exercise, only factorizations with 2 factors should be part of the answer.

Exercise 4 Suggest that students perform each multiplication to find the one that does not equal 88.

Exercise 5 Remind students that simply reversing the order of the factors does not give you another way to factor. 1×80 is the same as 80×1.

Exercise 6 Remind students that if a pair of factors contains a number that can be factored further, there are more ways to factor the number. Refer to Exercise 2 as an example. Have students find the factorizations not included in Natasha's list.

Lesson 17-3 Equivalent Fractions

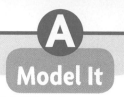

Objective 17.3: Multiply and divide by forms of 1 to write equivalent fractions.

Teach the Lesson

Materials ☐ Fraction Circles Plus

Activate Prior Knowledge

What do you get when you multiply a number by 1? **The number itself** What do you get when you divide a number by 1? **The number itself** How can you write 1 as a fraction? **Put the same number in the numerator and denominator: for example, $\frac{3}{3} = 1$.**

Develop Academic Language

How do you find a *factor* of a number? **Find a whole number that divides evenly into the number.** List all the factors of 36. **1, 2, 3, 4, 6, 9, 12, 18, 36** List all the factors of 9. **1, 3, 9** What are the *common factors* of 36 and 9? **1, 3, 9**

Model the Activities

Activity 1 Model $\frac{2}{3}$. To find an equivalent fraction, we can choose any number, and multiply both the numerator and denominator by that number. If the way this is shown on the page is confusing to students who haven't multiplied fractions yet, you can rewrite the equation as $\frac{2 \times 4}{3 \times 4} = \frac{8}{12}$. You may want to begin building the understanding that multiplying the numerator and denominator by the same number is the same as multiplying by 1. Let's try multiplying 2 and 3 by 4. What is 2 × 4? **8** What is 3 × 4? **12** Try covering your $\frac{2}{3}$ model with $\frac{8}{12}$. Does it fit exactly? **Yes.** So $\frac{2}{3}$ is equivalent to $\frac{8}{12}$.

Activity 2 Model $\frac{3}{12}$. Another way to find an equivalent fraction is to divide both the numerator and denominator by the same number. Can we divide both numbers evenly by 5? **No.** What is a good way to pick a number that can divide 3 and 12? **Find common factors of 3 and 12.** What factor goes into both 3 and 12? **3** Then divide the numerator and denominator by 3. $\frac{1}{4}$ Try covering your $\frac{3}{12}$ model with $\frac{1}{4}$. Does it match? **Yes.** So $\frac{3}{12}$ and $\frac{1}{4}$ are equivalent fractions.

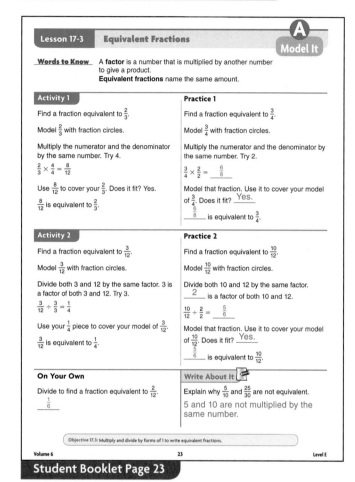

Student Booklet Page 23

✓ Progress Monitoring

Write $\frac{4}{10}$ on the board. How can we multiply to find an equivalent fraction? **Multiply 4 and 10 by the same number.** If we multiplied both by 3, what fraction would we have? $\frac{12}{30}$ How can we use factors to find an equivalent fraction? **Divide 4 and 10 by the same factor.** What is a common factor of 4 and 10? **2** What fraction do we get if we divide by $\frac{2}{2}$? $\frac{2}{5}$

Error Analysis

Make sure that students understand that you have to multiply or divide the numerator and denominator by the same number in order for the fractions to be equivalent.

Lesson 17-3 Equivalent Fractions

Objective 17.3: Multiply and divide by forms of 1 to write equivalent fractions.

Facilitate Student Understanding

Develop Academic Language

Even though two *equivalent* fractions may look different, they name the same amount. Draw two same-size squares on the board. Divide one in halves and one in quarters. Shade $\frac{1}{2}$ of the first and $\frac{2}{4}$ of the second. $\frac{1}{2}$ is equivalent to $\frac{2}{4}$.

ENGLISH LEARNERS Relate *equivalent* and *equal* to help students remember the definition.

Demonstrate the Examples

Example 1 Write $\frac{8}{15}$ on the board. **How do you create an equivalent fraction using multiplication?** Multiply the numerator and denominator by the same number. **This is the same as multiplying by a fraction equal to 1. Multiply by $\frac{2}{2}$.** Show the multiplication on the board. **What is the resulting equivalent fraction?** $\frac{16}{30}$

Example 2 Write $\frac{15}{20}$ on the board. **How do you create an equivalent fraction using division?** Divide the numerator and denominator by the same number. **How do you know that this won't change the value of the fraction?** This is the same as dividing by a fraction equal to 1. **Could you divide by $\frac{2}{2}$ in this problem?** No. 15 isn't divisible by 2. **How do you find a fraction equal to 1 that you can divide by?** Look for a common factor of the numerator and denominator. **What number is a factor of both 5 and 20?** 5 **Divide the fraction by $\frac{5}{5}$. What is the resulting equivalent fraction?** $\frac{3}{4}$

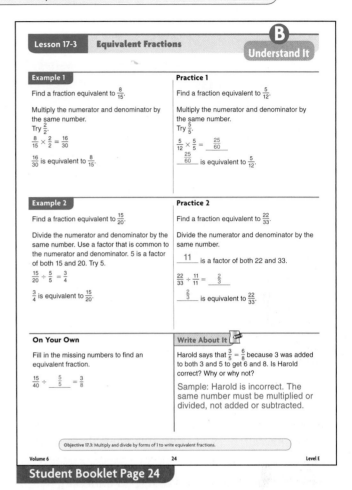

Student Booklet Page 24

✓ Progress Monitoring

Find another fraction that is equivalent to $\frac{15}{20}$.
Sample: $\frac{30}{40}$

Error Analysis

If students try to find equivalent fractions by adding or subtracting a number or by any other incorrect method, ask them to use models to prove their answers.

Lesson 17-3 Equivalent Fractions

Objective 17.3: Multiply and divide by forms of 1 to write equivalent fractions.

Observe Student Progress

Computer Tutorial

Some students may benefit from completing a computer tutorial before they attempt the Try It page. A list of the tutorials for each lesson can be found beginning on page x in the front of this book.

 Error Analysis

Exercise 1 Students who choose D are adding the same number to 6 and 10 instead of multiplying or dividing by the same number. Encourage them to use models.

Exercise 2 If students find an equivalent fraction by multiplying instead of dividing, ask them to read the directions again.

Exercise 3 You can rewrite the problem as $\frac{1}{8} \times \frac{}{} = \frac{}{32}$.

Exercise 5 Encourage students to underline or highlight the word *not* in Exercise 5 so they know that they are looking for the fraction that doesn't equal $\frac{4}{6}$.

Exercise 6 If students think that the reciprocals in Exercise 6a are equivalent, remind them of the rules for finding equivalent fractions or have students model both fractions.

Exercise 7 Students who use fraction materials to check that the fractions are equivalent should also refer to dividing by $\frac{2}{2}$ in the explanation.

Exercise 8 If students do not initially understand this question, ask them to start making a list of all of the equivalent fractions.

ENGLISH LEARNERS Rephrase the question: **Can *anyone* list *all* the fractions equivalent to $\frac{2}{3}$?**

Student Booklet Page 25

Lesson 17-3 Equivalent Fractions — Try It

1. Which fraction is equivalent to $\frac{6}{10}$?
   A $\frac{2}{5}$ B $\frac{12}{10}$
   C $\frac{3}{5}$ D $\frac{8}{12}$

2. Divide to find a fraction that is equivalent to $\frac{10}{25}$.
   $\frac{2}{5}$

3. Find the value of *x* that makes the statement true.
   $\frac{1}{8} = \frac{x}{32}$
   4

4. Find the value of *x* that makes the statement true.
   $\frac{10}{15} = \frac{2}{x}$
   3

5. Which fraction is *not* equivalent to $\frac{4}{6}$?
   A $\frac{2}{3}$ B $\frac{8}{12}$
   C $\frac{16}{24}$ D $\frac{7}{12}$

6. Write *yes* if the fractions are equivalent, and *no* if they are not.
   a. $\frac{15}{7}$ and $\frac{7}{15}$ No.
   b. $\frac{15}{6}$ and $\frac{5}{2}$ Yes.

7. Is $\frac{8}{12}$ equivalent to $\frac{4}{6}$? Explain.
   They are equivalent. 8 and 12 were both divided by 2 to get 4 and 6.

8. Is it possible to list every single fraction that is equivalent to $\frac{2}{3}$? Explain.
   Sample: No, because 2 and 3 can be multiplied by an infinite number of numbers.

Objective 17.3: Multiply and divide by forms of 1 to write equivalent fractions.

Lesson 17-4 Fractions in Lowest Terms

Objective 17.4: Rewrite fractions in lowest terms.

Teach the Lesson

Materials ☐ Teaching Aid 12 Fraction Strips

Activate Prior Knowledge

What are the factors of 12? 1, 2, 3, 4, 6, 12 **What are the factors of 32?** 1, 2, 4, 8, 16, 32 Write the factors on the board. **How do you find the greatest common factor of 12 and 32?** Find the greatest of all the common factors listed: 4.

Develop Academic Language

Write $\frac{2}{4}$ and $\frac{3}{8}$ on the board. **A fraction is in *lowest terms* when the only common factor of the numerator and denominator is 1. Do 2 and 4 have any common factors?** Yes; 2. **Is $\frac{2}{4}$ in lowest terms?** No, it can be reduced to $\frac{1}{2}$. **Is $\frac{3}{8}$ in lowest terms?** Yes.

Model the Activities

Activity 1 Use fraction strips to model $\frac{8}{12}$ as the picture shows. **Which fraction strips did you use?** Twelfths Now line up other equal parts below. Try 2 halves. **Does this group of equal parts equal the length of your model?** No. Try another group of equal parts. Guide students to continue until they see that two thirds requires the fewest equal parts to equal the model of eight twelfths. **What fraction does your second model show?** $\frac{2}{3}$ **So, what is $\frac{8}{12}$ in lowest terms?** $\frac{2}{3}$

Activity 2 **What kind of number is $2\frac{5}{10}$?** Mixed number **What is a mixed number?** A number with a whole number part and a fraction part To write a mixed number in lowest terms, write the fraction in lowest terms and leave the whole number part as it is. **What is the whole number part of $2\frac{5}{10}$?** 2 **What is the fraction part?** $\frac{5}{10}$ Have students complete the activity by finding the smallest number of equal parts that will cover the same length as the model of $\frac{5}{10}$.

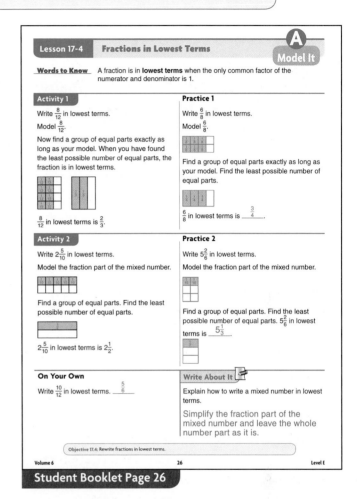

Student Booklet Page 26

ENGLISH LEARNERS Model the process of trying out halves, then thirds, then fourths, and so on, to clarify the meaning of "the least possible number of equal parts."

✓ Progress Monitoring

Write $\frac{9}{15}$ on the board. **How do you know whether or not the fraction is in lowest terms?** If 9 and 15 have no common factors other than 1, then the fraction is in lowest terms. **Is $\frac{9}{15}$ in lowest terms?** No.

Error Analysis

Remind students that writing a fraction in *lowest terms* is the same as writing a fraction in *simplest form*.

Lesson 17-4 Fractions in Lowest Terms

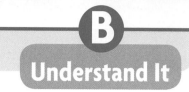

Objective 17.4: Rewrite fractions in lowest terms.

Facilitate Student Understanding

Develop Academic Language

Write the number 42 on the board. **A prime number is a whole number greater than 1 that is divisible by only itself and 1. Is 42 a prime number?** No. **What are two factors of 42?** Sample: 6 and 7 **Continue to factor the factors until they are all primes. What is the prime factorization of 42?** $2 \times 3 \times 7$

Demonstrate the Examples

Example 1 You can write a fraction in lowest terms by finding the greatest common factor of the numerator and denominator. **Find factors of 16 and of 34. What factors do they have in common?** 1 and 2 **What is the greatest common factor of 16 and 34?** 2 **Divide the numerator and denominator by 2. What is $\frac{16}{34}$ in lowest terms?** $\frac{8}{17}$

Example 2 Another way to write a fraction in lowest terms is to find the prime factorization of the numerator and the denominator and cancel common terms. **Find the prime factorization of 18 and 24.** Check students' factors. Write the prime factorizations in place of the numerator and denominator. Remember that a factor in the numerator cancels an equal factor in the denominator. **What factors cancel?** A 2 and a 3 **What is $\frac{18}{24}$ in lowest terms?** $\frac{3}{4}$

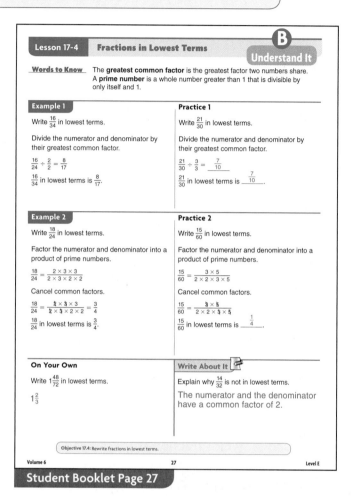

Student Booklet Page 27

Progress Monitoring

What is the fraction $\frac{12}{12}$ written in lowest terms? $\frac{1}{1}$ or 1

Error Analysis

Remind students that they must find the *greatest* common factor to express the fraction in lowest terms. Students should always check to see whether the numerator and denominator of the answer have a common factor.

Lesson 17-4: Fractions in Lowest Terms

Objective 17.4: Rewrite fractions in lowest terms.

Observe Student Progress

Computer Tutorial

Some students may benefit from completing a computer tutorial before they attempt the Try It page. A list of the tutorials for each lesson can be found beginning on page x in the front of this book.

Error Analysis

Exercise 1 Allow students to use fraction strips to model the problem.

Exercises 3 and 4 If students aren't sure how to proceed with these exercises, tell them to think, 33 divided by what number equals 11? 3 Then remind them that the numerator and the denominator must both be divided by the same number in order for the fractions to be equivalent.

Exercise 5 Students who choose A or C have simplified the fraction but have not reduced it to lowest terms. Remind students to find the greatest common factor of the numerator and denominator or to use prime factorization to reduce the fraction to lowest terms.

Exercise 6 Students may be confused because the fraction $\frac{9}{35}$ is already in lowest terms. Have them list all the factors of both numbers so they can see that there are no common factors of 9 and 35.

Student Booklet Page 28

Lesson 17-4 Fractions in Lowest Terms — Try It

1. Write $\frac{4}{12}$ in lowest terms.
   $\frac{1}{3}$

2. Write $\frac{9}{24}$ in lowest terms.
   $\frac{3}{8}$

3. Find the value of x that makes the statement true.
   $\frac{9}{33} = \frac{x}{11}$
   3

4. Find the value of x that makes the statement true.
   $\frac{25}{100} = \frac{1}{x}$
   4

5. Which is $\frac{18}{72}$ in lowest terms?
   A $\frac{9}{36}$ **B** $\frac{1}{4}$
   C $\frac{3}{12}$ D $\frac{1}{3}$

6. Which is $\frac{9}{35}$ in lowest terms?
   A $\frac{9}{35}$ B $\frac{3}{17}$
   C $\frac{1}{4}$ D $\frac{3}{8}$

7. Write the steps for putting $\frac{12}{20}$ in lowest terms.
   First choose the greatest common factor, 4. Then divide both the numerator and the denominator by 4. You get $\frac{3}{5}$.

8. Explain why $\frac{18}{36}$ in lowest terms is **not** $\frac{0}{3}$.
   18 is the greatest common factor of 18 and 36. 18 ÷ 18 = 1, not 0.

Objective 17.4: Rewrite fractions in lowest terms.

ENGLISH LEARNERS Tell students that another name for *lowest terms* is *simplest form*.

Lesson 17-5: Interpretations of Fractions

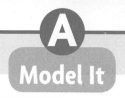

Objective 17.5: Explain different interpretations of fractions, such as parts of a whole, parts of a set, and division of whole numbers by whole numbers.

Teach the Lesson

Materials
- ☐ Fraction Circles Plus
- ☐ Two-color counters

Activate Prior Knowledge

Provide students with two-color counters. **Show 8 red counters. Turn over 3 of them. What fraction of the group is red?** $\frac{5}{8}$ **What fraction is yellow?** $\frac{3}{8}$ Have students work in pairs to model different fractions.

Develop Academic Language

Use Fraction Circles Plus to model $2\frac{1}{4}$. Write $2\frac{1}{4} = \frac{9}{4}$ on the board. Point to $2\frac{1}{4}$. **This is a *mixed number*** because it has a whole-number part and a fraction part. Point to $\frac{9}{4}$. You need 9 fourths to make $2\frac{1}{4}$. This is an *improper fraction* because the numerator is greater than the denominator.

Model the Activities

Activity 1 Have students work in small groups. **Use fifths to make a model of one whole. Put stickers on 2 of the pieces. How many parts are in the whole circle?** 5 **What fraction of the whole has stickers?** $\frac{2}{5}$ **Now pull the pieces of the circle apart. What is the fraction now?** The same Provide each group with two-color counters. **Make another model using 2 yellow counters and 3 red counters. How many parts are in the group?** 5 **What fraction of the group is yellow?** $\frac{2}{5}$

Activity 2 **Look at the model. How many whole stacks of blue squares are there?** 3 Write 3 on the board. **What is the fraction of the blue squares in the last stack?** $\frac{1}{5}$ Write $+ \frac{1}{5}$ next to 3. **What is the total number of blue *stacks*?** $3\frac{1}{5}$ Write $= 3\frac{1}{5}$ next to $\frac{1}{5}$. **What is the total number of blue *squares*?** 16 **What fraction does each square represent?** $\frac{1}{5}$ There are 16 fifths. **How can you write the number of blue squares as an improper fraction?** $\frac{16}{5}$

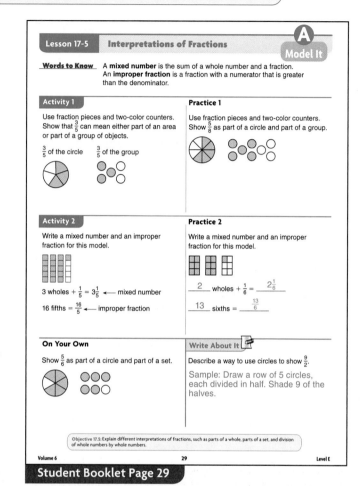

Student Booklet Page 29

Progress Monitoring

Write on the board $\frac{5}{3}, \frac{3}{8}, \frac{2}{6}, \frac{9}{2}, \frac{10}{4}$. **Which of these are equal to mixed numbers?** $\frac{5}{3}, \frac{9}{2}, \frac{10}{4}$ **How do you know?** The numerator is greater than the denominator. The fraction equals more than 1 whole.

Error Analysis

Provide fraction materials for students to model fractions such as $\frac{4}{3}$ and $\frac{5}{4}$ in which there is 1 whole and only 1 extra fractional piece.

Lesson 17-5: Interpretations of Fractions

Objective 17.5: Explain different interpretations of fractions, such as parts of a whole, parts of a set, and division of whole numbers by whole numbers.

Facilitate Student Understanding

Materials ☐ Fraction Circles Plus

Develop Academic Language

Write on the board 8 ÷ 4 = 2. Write *divisor* and *quotient* on the board. Ask students to identify the divisor and quotient. **4, 2** Have students tell you their definitions. **The divisor is the number you divide by. The quotient is the answer.**

Demonstrate the Examples

Example 1 Draw a number line on the board labeled from 0 to 5. **How many jumps do we have to make to divide this line into 8 equal parts? 8 Will each jump cover 1 unit, more than 1 unit, or less than 1 unit? less Count the spaces from 0 to 5. How many are there? 40 How can we divide 40 spaces in 8 equal groups?** Include 5 spaces in each group. Invite a student to draw the division on the number line on the board. **Where does the first jump end?** $\frac{5}{8}$ **If 5 units are divided into 8 equal parts, how long is each part, in units?** $\frac{5}{8}$

Example 2 Provide students with fraction circles. Have them put out 3 wholes. Ask how 8 students can share 3 squares. **Start with 1 square. How could 8 people share 1 square?** Each square can be cut into 8 pieces. Have students replace the 3 wholes with 24 eighths. Separating these into 8 equal groups shows that each student gets $\frac{3}{8}$ of a square. **The fraction bar is the same as a division sign telling us to divide the numerator by the denominator.** Emphasize that when the divisor is greater than the dividend, the answer will be a fraction.

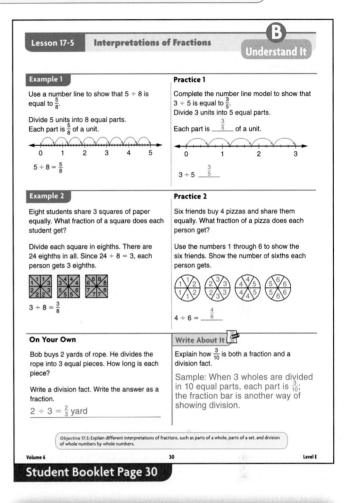

Student Booklet Page 30

Progress Monitoring

When is the quotient in a division problem equal to a fraction less than 1 whole? Give an example.

When you are dividing by a greater number; Sample: $3 \div 10 = \frac{3}{10}$; 10 is greater than 3, so the quotient is less than 1.

Error Analysis

The relationship between fractions and division is most obvious using improper fractions such as $\frac{6}{3}$ or $\frac{10}{2}$. Have students use fraction materials to model these numbers and see that $\frac{6}{3} = 6 \div 3$ and $\frac{10}{2} = 10 \div 2$.

Lesson 17-5 Interpretations of Fractions

Objective 17.5: Explain different interpretations of fractions, such as parts of a whole, parts of a set, and division of whole numbers by whole numbers.

Observe Student Progress

Materials
- ☐ Fraction Circles Plus
- ☐ Two-color counters

Computer Tutorial

Some students may benefit from completing a computer tutorial before they attempt the Try It page. A list of the tutorials for each lesson can be found beginning on page x in the front of this book.

Error Analysis

Exercise 2 Remind students to assign each friend a number. Students can divide each melon separately and then count the pieces that have the same number, or they can proceed as shown.

Exercise 3 Point out that students are writing two different names for the same quantity. For the mixed number, they count the whole circles **3** and then find the fraction shaded in the circle on the right. $\frac{4}{10}$ For the improper fraction, they count or compute the total number of blue pieces; for example, 3 times 10 is 30, plus 4 more is 34 tenths.

Exercise 5 Have students use fraction circles to model the problem. **How many wholes are formed?** 1 whole **What fraction is left over?** $\frac{2}{3}$ You can also help students write the fraction as the sum of a fraction equal to 1 and another fraction.

Exercise 6 If students need a visual model for this problem, they can draw 7 circles and divide each into fourths.

Exercise 7 Have students use fraction circles or two-color counters to help model the problem.

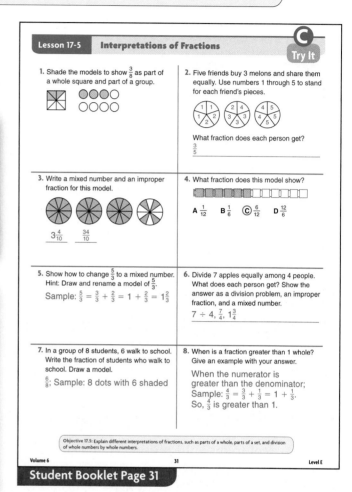

Student Booklet Page 31

Volume 6: Understand Fractions
Topic 17: Equivalence of Fractions
Topic Summary

Objective: Review finding equivalent fractions.

Have students complete the student summary page. You may want to have students work in groups of four with each student analyzing a different choice.

Ask students to share their ideas about each answer choice. Be sure they confirm the correct answer for each problem at the end of the discussion.

Answer Evaluation

1. **A** This choice is correct.
   B Students divided incorrectly.
   C Students divided incorrectly.
   D Students divided incorrectly.

2. **A** Students omitted the whole number.
   B Students counted the wholes incorrectly.
   C Students did not put the answer in lowest terms.
   D This choice is correct.

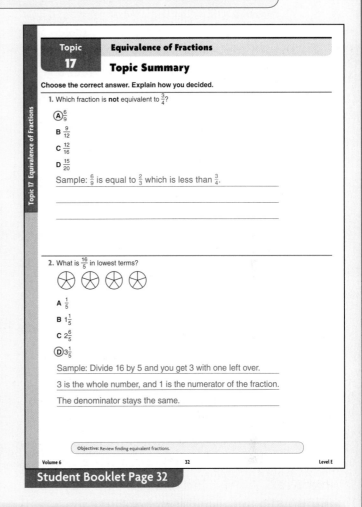

Student Booklet Page 32

 Error Analysis

Exercise 1 Have students put each fraction in lowest terms and compare to three fourths.

Exercise 2 Tell students to shade in 16 sections of the diagram to help change the improper fraction into a mixed number in lowest terms.

Progress Monitoring

When all assignments for this topic have been completed, assign the corresponding Progress Monitoring page for this topic (Assessment Resources Book, page 17). Be sure students complete the Progress Monitoring page before you administer the final assessment for this volume.

Volume 6: Understand Fractions

Topic 17: Equivalence of Fractions

Mixed Review

Objective: Maintain concepts and skills.

Have students complete the Mixed Review page. Work with each student individually to review results. Identify strengths and weaknesses and correct any misunderstandings.

Error Analysis

Exercise 1 Put a place-value chart on the board for students to use for reference.

Thousands	Hundreds	Tens	Ones

Exercise 3 Students who continue to make errors in basic facts should practice with addition and subtraction flash cards. Have students work in pairs for 10- or 15-minute practice sessions.

Exercise 4 Remind students to look at the ones digit when rounding to the nearest ten. If the digit is 5 or greater, they round up. Otherwise, they round down. Drawing a number line from 460 to 470 can help students see that 467 is closer to 470 than it is to 460.

Exercise 6 Ask students to explain how they can solve the problem by first breaking it into simpler parts. **Sample: List the multiples of 2 from 0 to 30. List the multiples of 5 from 0 to 30. Find the numbers in both lists and then look for a pattern in the ones digits.**

Exercise 7 Have students list the numbers they multiply to get 24. Then have them list the factors of those numbers as well. Afterward, they can combine the lists.

Exercise 8 Remind students that after they find the whole number, they must still put the fraction into lowest terms.

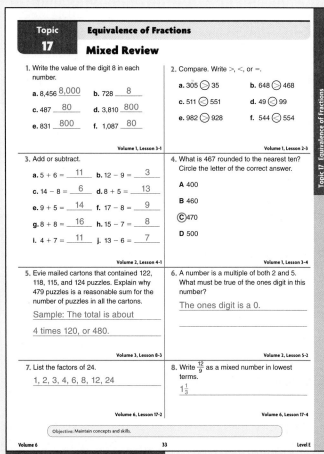

Student Booklet Page 33

Volume 6: Understand Fractions
Topic 18: Addition and Subtraction of Fractions

Topic Introduction

Lesson 18-1 starts students on adding and subtracting fractions with like denominators. Then Lessons 18-2 and 18-3 move on to adding and subtracting fractions with unlike denominators. Lessons 18-4 and 18-5 build on the previous lessons by having students add and subtract mixed numbers.

Lesson	Objective	Student Pages	Teacher Pages	Tutorials
Topic 18 Introduction	18.1 Add and subtract fractions with like denominators. 18.4 Add mixed numbers with like and unlike denominators. 18.5 Subtract mixed numbers with like and unlike denominators.	34	304	
18-1 Add and Subtract Fractions	18.1 Add and subtract fractions with like denominators.	35–37	305–307	18a, 18b, 18c
18-2 Unlike Denominators: Add	18.2 Add fractions having unlike denominators and change answers to mixed numbers in lowest terms when appropriate.	38–40	308–310	18d, 18e
18-3 Unlike Denominators: Subtract	18.3 Subtract and simplify fractions with unlike denominators.	41–43	311–313	18f, 18g
18-4 Add Mixed Numbers	18.4 Add mixed numbers with like and unlike denominators.	44–46	314–316	18c, 18h
18-5 Subtract Mixed Numbers	18.5 Subtract mixed numbers with like and unlike denominators.	47–49	317–319	18g
Topic 18 Summary	Review adding and subtracting fractions.	50	320	
Topic 18 Mixed Review	Maintain concepts and skills.	51	321	

Computer Tutorial

Some students may benefit from completing the computer tutorial before they attempt the Try It page of each lesson. If you are using the electronic components of *Pinpoint Math*, you will find a complete listing of Tutorial codes and titles when you access them either online or via CD-ROM.

Volume 6 — Understand Fractions

Topic 18: Addition and Subtraction of Fractions

Topic Introduction

Objectives: 18.1 Add and subtract fractions with like denominators. **18.4** Add mixed numbers with like and unlike denominators. **18.5** Subtract mixed numbers with like and unlike denominators.

$1\frac{4}{5}$ is how many total fifths? **9** How do you know? **There are 5 fifths in one whole, and 4 fifths in the fraction, so 4 fifths + 5 fifths = 9 fifths.** Let's write $\frac{20}{3}$ as a mixed number. How many groups of 3 are in 20? **6** How many pieces are left over? **2** How do we write this answer as a mixed number? $6\frac{2}{3}$

Informal Assessment

1. What is the first fraction? $\frac{1}{3}$ How do you know? **1 part out of 3 parts is shaded.** What is the second fraction? $\frac{2}{3}$ If you combined the shaded parts, how many would you have? **3** What is the top number of a fraction called? **Numerator** What is the bottom number called? **Denominator** If the denominators are the same, how do you add two fractions? **Add the numerators.**

2. Why can't you just take away the 3 parts in the second fraction bar? **The 3 parts are 3 eighths, not 3 fourths.** What do you need to do before you can perform the subtraction? **Change both fractions to a common denominator, eight.** How many eighths are equal to $\frac{3}{4}$? **6** How many eighths are in $1\frac{3}{8}$? **11** How could you show the subtraction of $\frac{6}{8}$? **Cross out 6 of the eighths.** How many eighths are left? **5** How many eighths are in $\frac{1}{4}$? **2** What is the sum of $\frac{2}{8}$ and $\frac{3}{8}$? $\frac{5}{8}$ What is $\frac{5}{8} + 1$? $1\frac{5}{8}$ Is this already in lowest terms? **Yes.** How do you know? **The only common factor that 5 and 8 have is 1.**

3. How can we show $1\frac{1}{4}$ on these fraction models? **Shade the first whole rectangle; that's 1. Shade 2 eighths in the second rectangle. $\frac{2}{8}$ is equivalent to $\frac{1}{4}$.**

4. Repeat the procedure from Exercise 2. If students are having difficulty, provide fraction materials for them to model the problems with.

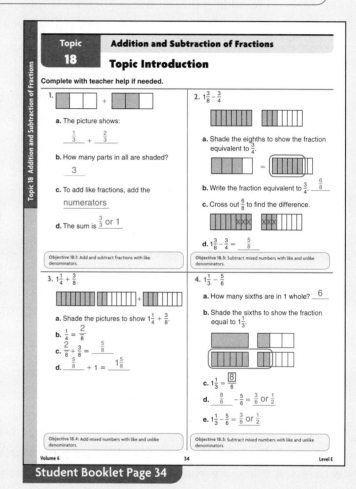

Student Booklet Page 34

Another Way For Exercise 3, suggest that students change the mixed fraction to an improper fraction before adding. After they get the sum, they can put the fraction in lowest terms.

Lesson 18-1 | Add and Subtract Fractions

Objective 18.1: Add and subtract fractions with like denominators.

Teach the Lesson

Materials
- ☐ Fraction Circles Plus or other fraction materials
- ☐ Teaching Aid 11 Fraction Number Lines

Develop Academic Language

Writing fractions in the form *3 eighths* (rather than $\frac{3}{8}$) can help to emphasize the method of adding and subtracting like fractions. In the same way that 3 apples plus 2 apples equals 5 apples, 3 eighths plus 2 eighths equals 5 eighths.

ENGLISH LEARNERS You might want to make a chart of halves, thirds, fourths, fifths, and so on, to help students spell them and know the ones that don't fit the model of simply adding *-th* to the number word.

Model the Activities

Activity 1 Provide copies of Teaching Aid 11 Fraction Number Lines. Have students find the number line for sixths and mark the location of $\frac{1}{6}$. Ask for ideas on how to show $\frac{1}{6}$ plus $\frac{3}{6}$. Accept any suggestion, provided that students include the idea of moving to the right. Have a few volunteers draw their models for $\frac{1}{6} + \frac{3}{6}$. Ask where they land on the number line. Point out that the model shows that $\frac{4}{6}$ is the sum of $\frac{1}{6}$ and $\frac{3}{6}$.

Activity 2 Repeat the procedure above for subtraction by having students start at $\frac{7}{8}$ and go left 4 units. Remind them that on this line each unit is 1 eighth. Students will land on $\frac{3}{8}$, the difference between $\frac{7}{8}$ and $\frac{4}{8}$. **When you show addition on a number line, which way do you move?** To the right **Why?** The numbers increase from left to right. **What does it mean when you move to the left?** You are subtracting, taking away, or decreasing. Encourage students to give examples to show that they understand how to use the number line models.

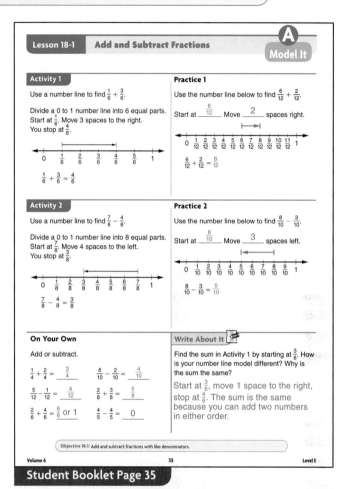

Student Booklet Page 35

⭐ Progress Monitoring

Provide students with copies of Teaching Aid 11. Have them choose the number line for twelfths and find $\frac{7}{12} - \frac{4}{12}$. $\frac{3}{12}$

Error Analysis

Check whether students know how to count the spaces on a fraction number line. They use the numerator of the fraction they are adding or subtracting. For example, to find $\frac{7}{12} - \frac{4}{12}$, start on $\frac{7}{12}$ and count 4 spaces to the left to land on $\frac{3}{12}$.

Lesson 18-1 Add and Subtract Fractions

Objective 18.1: Add and subtract fractions with like denominators.

Facilitate Student Understanding

Activate Prior Knowledge

How do you make equivalent fractions? You multiply or divide a fraction by a form of 1, that is, by a fraction with the same number in the numerator and denominator.

Develop Academic Language

Write a fraction, for example, $\frac{3}{5}$, on the board. Point to the 3 as you say **This number is up on top so we say it is up. This number** (pointing to the 5) **is down on the bottom so we say it is down.** Write *up*, *down*, and then *numerator* and *denominator*. Underline the *d* in *down* and *denominator*. **The denominator is down and the numerator is up.** To help students remember which number in a fraction is the numerator and which is the denominator, point out that the word *down* starts with the letter *d* as does *denominator*.

Demonstrate the Examples

Example 1 Have students use fraction circles to find $\frac{3}{10} + \frac{2}{10}$. **What is the sum?** $\frac{5}{10}$ **How can you show this sum with just one fraction piece?** Trade $\frac{5}{10}$ for $\frac{1}{2}$. Encourage students to find simplest forms of their sums and differences. Remind them about dividing by a form of 1, which doesn't change the value of a number.

Example 2 Check that students understand that the Xs are used to show pieces taken away.

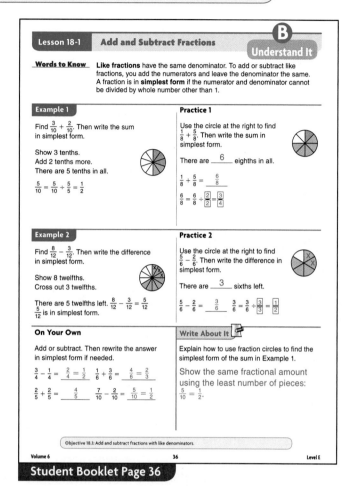

Student Booklet Page 36

Progress Monitoring

Have students find the sum of $\frac{2}{5}$ and $\frac{1}{5}$ with fraction circles. $\frac{3}{5}$ Write $\frac{2}{5} + \frac{1}{5} = \frac{3}{5}$ on the board. Ask students to state a general rule for adding two like fractions. **Sample: Add the numerators and use the same denominator.**

Error Analysis

Check that students are not making the mistake of adding the denominators and writing, for example, $\frac{2}{5} + \frac{1}{5} = \frac{3}{10}$.

Lesson 18-1: Add and Subtract Fractions

Objective 18.1: Add and subtract fractions with like denominators.

Observe Student Progress

Computer Tutorial

Some students may benefit from completing a computer tutorial before they attempt the Try It page. A list of the tutorials for each lesson can be found beginning on page x in the front of this book.

Develop Academic Language

Exercise 8 Review the definition of *like fractions* by asking why two like fractions are alike. **They have the same denominator.**

Error Analysis

Exercise 1 Students can shade 1 tenth more to find $\frac{6}{10} + \frac{1}{10}$. They can erase this shading and mark 2 Xs to show $\frac{6}{10} - \frac{2}{10}$. Provide fraction circles if students cannot do the exercises using the picture model.

Exercise 2 Remind students to start on the first number in each exercise.

Exercise 3 Students who are making errors, such as adding the denominators, need further work with fraction circles or number lines.

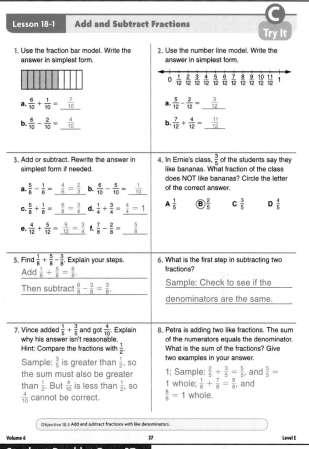

Student Booklet Page 37

Lesson 18-2 Unlike Denominators: Add

Objective 18.2: Add fractions having unlike denominators and change answers to mixed numbers in lowest terms when appropriate.

Teach the Lesson

Materials
- ☐ Fraction Circles Plus or other fraction materials
- ☐ Teaching Aid 11 Fraction Number Lines
- ☐ Teaching Aid 12 Fraction Strips

Activate Prior Knowledge

Have available Teaching Aid 12 Fraction Strips. Write on the board: $\frac{1}{6} + \frac{3}{6} = ?$

How do you find the sum? Add the numerators, 1 + 3, then write 4 over 6, $\frac{4}{6}$. **Is that fraction in simplest form?** No. **How do you find an equivalent fraction?** Divide both the numerator and the denominator by 2: $\frac{4}{6} = \frac{2}{3}$.

Develop Academic Language

What does it mean when fractions have like denominators? The denominators are the same. When we put "un" before a word it means "not." *Unhappy* means "not happy". **What do you think *unlike denominators* means?** The denominators are not the same.

Model the Activities

Activity 1 Provide students with copies of Teaching Aid 12 Fraction Strips. **How could you show $\frac{3}{10} + \frac{2}{5}$?** Use 3 tenths and 2 fifths. **How can you show the same amount using all tenths?** Replace $\frac{2}{5}$ with $\frac{4}{10}$. Now both fractions have like denominators. **How many tenths do you have in all?** 7 **What is the sum?** $\frac{7}{10}$

Activity 2 Draw a number line labeled 0, 1, and 2 on the board. Have 2 volunteers to mark the number line in halves and in fourths. Mark $\frac{1}{2}$ on the number line. Start at $\frac{1}{2}$. **What do you have to add?** $1\frac{1}{4}$ **How many fourths are in $1\frac{1}{4}$?** 5 Show how to count 5 fourths on the number line. Because this is an addition, the value increases. **Which way do you move?** Right Draw an arrow from $\frac{1}{2}$ to $1\frac{3}{4}$ while counting the fourths. **What is the sum of $\frac{1}{2}$ and $1\frac{1}{4}$?** $1\frac{3}{4}$

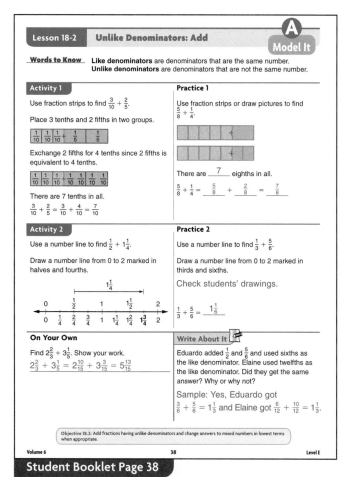

Student Booklet Page 38

Progress Monitoring

Have students use fraction strips to demonstrate $\frac{3}{4} + \frac{1}{8}$. Replace $\frac{3}{4}$ with $\frac{6}{8}$, so then there are $\frac{7}{8}$.

Error Analysis

Check that students make an accurate replacement of the fractions when replacing with equivalent fractions. Ask them to demonstrate that the two fractions are equivalent.

Volume 6   308   Level E

Lesson 18-2 Unlike Denominators: Add

Understand It

Objective 18.2: Add fractions having unlike denominators and change answers to mixed numbers in lowest terms when appropriate.

Facilitate Student Understanding

Develop Academic Language

Discuss the meaning of the word *common*. Tell students that *common* has several meanings in everyday language. One of those meanings is "shared", as in, "Their common language is Japanese." In mathematics, the meaning is similar. It means "belonging equally to two or more quantities." So, a *common denominator* is one that can be used to rewrite two or more fractions with the same denominator.

Demonstrate the Examples

Example 1 What is the first step to solve this addition problem? **Find the common denominator.** Ask students to explain how to find the common denominator. **Multiply 8 × 5 = 40.** What is the second step? **Write equivalent fractions.** Ask students to explain how to write the equivalent fractions. $\frac{7}{8} \times \frac{5}{5}, \frac{2}{5} \times \frac{8}{8}$ What is the third step? **Add the numerators.** Show $\frac{35}{40} + \frac{16}{40} = \frac{51}{40}$ on the board. To change $\frac{51}{40}$ to a mixed number, write $\frac{51}{40}$ as a sum of a fraction equal to 1 and another fraction. Write $\frac{51}{40} = \frac{40}{40} + \frac{11}{40}$ on the board. What is $\frac{51}{40}$ written as a mixed number? $1\frac{11}{40}$ Is $1\frac{11}{40}$ in simplest form? Why? **Yes; 11 and 40 have no common multiples.**

Example 2 Have the students read the problem aloud and identify the operation required to solve this problem. **Addition** Guide them step by step to add the fractions and change the improper fraction to a mixed number.

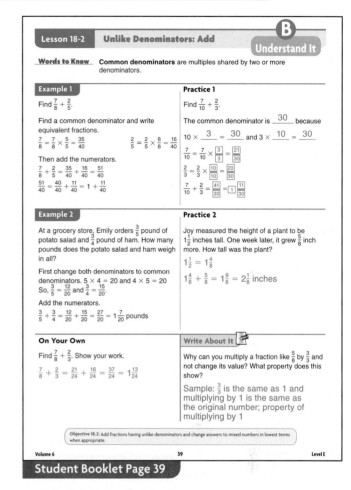
Student Booklet Page 39

Progress Monitoring

Find $\frac{3}{4} + \frac{1}{3}$. $1\frac{1}{12}$

Error Analysis

Check that students are not making mistakes because their work is too cramped. Encourage them to use more space on the paper for each problem rather than crowding their calculations.

Lesson 18-2 — Unlike Denominators: Add

Objective 18.2: Add fractions having unlike denominators and change answers to mixed numbers in lowest terms when appropriate.

Observe Student Progress

Materials
- ☐ Fraction Circles Plus or other fraction materials
- ☐ Teaching Aid 12 Fraction Strips

Computer Tutorial

Some students may benefit from completing a computer tutorial before they attempt the Try It page. A list of the tutorials for each lesson can be found beginning on page x in the front of this book.

Develop Academic Language

Review the meaning of *simplest form* as no common multiples other than 1 for the numerator and the denominator. The fraction $\frac{2}{4}$ is not in simplest form because 2 is a common multiple of both 2 and 4.

ENGLISH LEARNERS For Exercise 6 check that students understand the word *gained*. Tell students that the phrase "the puppy gained $\frac{3}{8}$ pound" means that "the puppy weighs $\frac{3}{8}$ pound more."

Student Booklet Page 40

Lesson 18-2 — Unlike Denominators: Add — Try It

1. Add $\frac{2}{3} + \frac{5}{6}$. Use the number line.

   $1\frac{3}{6}$, or $1\frac{1}{2}$

2. Add.
   a. $\frac{1}{4} + \frac{1}{3}$ $\frac{7}{12}$
   b. $\frac{3}{5} + \frac{2}{3}$ $1\frac{4}{15}$
   c. $\frac{7}{8} + \frac{1}{2}$ $1\frac{3}{8}$
   d. $\frac{4}{5} + \frac{3}{4}$ $1\frac{11}{20}$

3. Add. Rewrite in simplest form if needed.
   a. $\frac{21}{4} + 1\frac{1}{2} = 6\frac{3}{4}$
   b. $3\frac{3}{5} + 1\frac{9}{10} = 5\frac{1}{2}$
   c. $1\frac{1}{3} + 1\frac{5}{8} = 2\frac{23}{24}$
   d. $1\frac{3}{4} + \frac{3}{4} = 2\frac{1}{2}$

4. What is the simplest form for the sum of $3\frac{4}{5}$ and $1\frac{3}{4}$? Circle the letter of the correct answer.
   - Ⓐ $5\frac{11}{20}$
   - B $5\frac{1}{4}$
   - C $4\frac{1}{20}$
   - D $3\frac{11}{20}$

5. Enrique has $\frac{3}{4}$ pound of flour to bake bread. He adds $\frac{5}{8}$ pound more to make extra bread. How many pounds of flour does he use all together?
   $1\frac{3}{8}$ pounds

6. Micah's puppy weighed $3\frac{1}{4}$ pounds last week. This week the puppy gained $\frac{3}{8}$ pound. How much does the puppy weigh now?
   $3\frac{5}{8}$ pounds

7. When Rosa added $\frac{5}{12}$ and $\frac{1}{4}$ she got $\frac{2}{3}$. When her friend Melissa did the same problem she got $\frac{32}{48}$. Who was right? Explain.
   Both answers are correct. Rosa's answer is in simplest form.

8. The sum of two fractions is $\frac{5}{6}$. One of the fractions is $\frac{1}{2}$. What is the other fraction in simplest form?
   $\frac{1}{3}$

Objective 18.2: Add fractions having unlike denominators and change answers to mixed numbers in lowest terms when appropriate.

★ Error Analysis

Exercise 2 If students make mistakes renaming fractions you may want to have them rework this exercises using Teaching Aid 12 Fraction Strips to model each problem.

Exercise 3 Working the problems vertically might be difficult. Encourage students to rewrite the problems, perhaps using a number line or fraction pieces to model.

Exercise 7 If students neglect to write an explanation, have them explain it orally and then ask them to write the explanation that they just shared with you.

Exercise 8 If students do not know how to approach this problem, point out that it is a guess-and-check type problem. There are two conditions to meet.

1. The sum of the fractions is $\frac{5}{6}$.
2. One of the fractions is $\frac{1}{2}$.

Some students may find it helpful to use fraction strips or fraction materials.

$\frac{3}{6} = \frac{1}{2}$ $\frac{2}{6} = \frac{1}{3}$

Lesson 18-3 | Unlike Denominators: Subtract

Objective 18.3: Subtract and simplify fractions with unlike denominators.

Teach the Lesson

Materials ☐ Teaching Aid 12 Fraction Strips

Activate Prior Knowledge

Count forward and back with fraction strips or a number line to model addition and subtraction of eighths. Then present the problem $\frac{3}{8} - \frac{1}{4}$, and show that there is no $\frac{1}{4}$ piece to be taken away. Tell students they will solve this type of problem by finding a common denominator. Model the process with addition of $\frac{3}{8}$ and $\frac{1}{4}$.

Develop Academic Language

Count by 3s. These numbers are *multiples* of 3. Three is a *factor* of each number. Have students list more multiples.

Model the Activities

Activity 1 Model $\frac{3}{4}$. Are there any sixths in our model to take away? No. To subtract sixths from fourths, we need to find equivalent fractions with a common denominator, or the same denominator. Place 3 of the fourths strips on your desk. How many twelfths strips are equal to that length? 9 $\frac{9}{12}$ is equivalent to $\frac{3}{4}$. Repeat for $\frac{1}{6}$. Now we have a common denominator and we can subtract.

Activity 2 You need to find a common denominator to subtract. First, see if one denominator is a multiple of the other. How do we know 9 is a multiple of 3? 3 × 3 = 9 So 9 is a common denominator. Draw a frame: $\frac{2}{3} \times \frac{\Box}{\Box} = \frac{\Box}{9}$. What do we multiply by 3 to get a denominator of 9? 3 If you do the same thing to the numerator, then you are multiplying the fraction by $\frac{3}{3}$, which is the same as multiplying by 1.

Student Booklet Page 41

Progress Monitoring

Solve $\frac{3}{8} - \frac{1}{4}$. $\frac{3}{8} - \frac{1}{4} \times \frac{2}{2} = \frac{3}{8} - \frac{2}{8} = \frac{1}{8}$

Error Analysis

Students who have difficulty with Practice 2 should continue using fraction strips.

ENGLISH LEARNERS Relate *common* to the meaning "shared," as in having something in common.

Lesson 18-3 — Unlike Denominators: Subtract

Objective 18.3: Subtract and simplify fractions with unlike denominators.

Facilitate Student Understanding

Develop Academic Language

Review *lowest terms*, also known as *simplest form*. Use fraction strips to show that $\frac{4}{8}$ can be represented with fewer equal pieces as $\frac{1}{2}$. Show the division, and point out that it is the same as dividing by 1. Start with $\frac{42}{48}$ and show that repeated division may be necessary. **When the numerator and the denominator can no longer be divided by the same number, then the fraction is in lowest terms.**

Demonstrate the Examples

Example 1 Why is 15 the common denominator? It is a multiple of 5. Complete the subtraction. How do we know that $\frac{5}{15}$ isn't in simplest terms? We can divide the numerator and denominator by 5 to find an equivalent fraction.

Example 2 Look at the greater denominator, 15. Is it a multiple of 6? No. One way to find a common denominator is to multiply the denominators. What is 15 × 6? 90 If we use 90 as our common denominator, we'll have to work with greater numbers when we add and when we reduce the answer to lowest terms. Another way is to look for the least common denominator, or the smallest multiple that 15 and 6 have in common. We list multiples of each number until the same number appears in both lists.

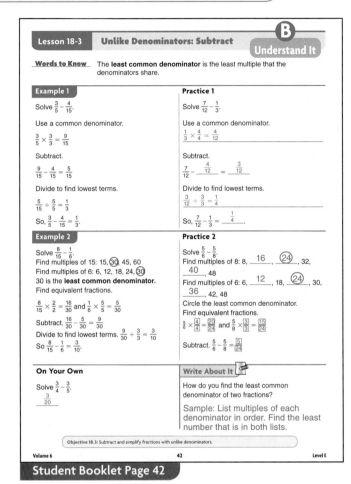

Student Booklet Page 42

⭐ Progress Monitoring

Solve and reduce the answer to lowest terms.

$\frac{3}{7} - \frac{1}{3}$ $\frac{2}{21}$ $\frac{2}{3} - \frac{5}{12}$ $\frac{1}{4}$

Error Analysis

Allow students to use fraction strips to support their understanding of the calculations. You may want to help students make a poster showing a decision tree for subtraction of fractions; start with "Are the denominators the same?" and continue through "Is this fraction in lowest terms?"

Volume 6 — 312 — Level E

Lesson 18-3 — Unlike Denominators: Subtract

Objective 18.3: Subtract and simplify fractions with unlike denominators.

Observe Student Progress

Computer Tutorial

Some students may benefit from completing the computer tutorial before they attempt the Try It page of each lesson. A list of the tutorials for each lesson can be found beginning on page x in the front of this book.

Error Analysis

Exercises 1 and 2 Allow students to use fraction strips to support their calculations.

Exercise 3 Help students recognize that subtraction is required. Use fraction strips to show $\frac{1}{4}$. **This is how much she has.** Use fraction strips to model $\frac{2}{3}$. **This is how much she needs in all. Take away the part she already has.**

Exercise 5 If students select answer A, remind them to read all the answers before choosing one. Answer A is a correct answer to the subtraction, but the problem specifies simplest terms, so it is an incorrect answer to the question.

Exercise 7 If students do not understand how to answer in ounces, have them write their fractional answers first. Then help them use the hint.

Exercise 8 Have students solve the problem to help them see that both answers are equivalent. Point out that although $\frac{8}{12}$ isn't incorrect, answers should be given in lowest terms unless otherwise noted.

Student Booklet Page 43

Lesson 18-3 — Unlike Denominators: Subtract — Try It

1. Solve $\frac{5}{9} - \frac{1}{3}$.
   $\frac{2}{9}$

2. Solve $\frac{5}{12} - \frac{1}{4}$.
   Give your answer in lowest terms.
   $\frac{1}{6}$

3. Jen has $\frac{1}{4}$ bottle of lemonade. She needs $\frac{2}{3}$ bottle to serve to her friends. How much more does she need?
   $\frac{5}{12}$ bottle

4. Kenneth bought $\frac{5}{6}$ pound of cheese. Lina bought $\frac{1}{2}$ pound of cheese. How much more did Kenneth buy? Remember to give your answer in lowest terms.
   $\frac{1}{3}$ pound

5. Which of the following shows the difference $\frac{12}{15} - \frac{1}{5}$ in lowest terms? Circle the letter of the correct answer.
   A $\frac{9}{15}$ B $\frac{12}{30}$
   C $\frac{2}{3}$ **D** $\frac{3}{5}$

6. Elias cut $\frac{1}{2}$ yard of paper from a roll that still had $\frac{5}{8}$ yard on it. How long was the paper he had left?
   $\frac{1}{8}$ yard

7. When Collin bought his pet rat, it weighed $\frac{3}{8}$ pound. Now it weighs $\frac{9}{16}$ pound. How many ounces did it gain? Circle the letter of the correct answer.
   Hint: $\frac{1}{16}$ pound = 1 ounce
   A 3   B 6
   C 8   D 12

8. Helene subtracted $\frac{3}{4} - \frac{1}{12}$ and got $\frac{2}{3}$. Julia did the same subtraction and got $\frac{8}{12}$. Who is correct? Explain.
   Sample: They are both correct, but Helene's answer is in lowest terms. $\frac{2}{3}$ is equivalent to $\frac{8}{12}$, but $\frac{2}{3}$ is in lowest terms.

Objective 18.3: Subtract and simplify fractions with unlike denominators.

ENGLISH LEARNERS Review the meanings of the words in the subtraction questions. Exercises 3 and 4 use *how much more*. Exercise 7 uses *gain*, which is another way of saying *how much more*. Exercise 6 uses *had left*, which indicates taking one amount away from another.

Lesson 18-4 Add Mixed Numbers

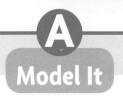

Objective 18.4: Add mixed numbers with like and unlike denominators.

Teach the Lesson

Materials ☐ Teaching Aid 12 Fraction Strips

Activate Prior Knowledge

Write the fraction $4\frac{2}{3}$ on the board. **What kind of number is $4\frac{2}{3}$?** Mixed number **How do you use fraction strips to model $4\frac{2}{3}$?** Use 4 whole numbers and 2 one-thirds. **Model $4\frac{2}{3}$.** Check that students have modeled the number correctly.

Develop Academic Language

Write $\frac{3}{8} + \frac{3}{4}$ on the board. **What do you need to do to add the two fractions?** Find the least common denominator of the two fractions. **What is the LCD of two or more fractions?** The least common multiple of their denominators **What is the LCD of $\frac{3}{8}$ and $\frac{3}{4}$?** 8

Model the Activities

Activity 1 Model the fractions $1\frac{1}{3}$ and $2\frac{2}{3}$. **How many whole fraction strips did you use?** 3 **How many one-third fraction strips did you use?** 3 Group all the whole fraction strips together, then group all the fractional parts together. **What did adding $\frac{1}{3}$ and $\frac{2}{3}$ form?** 1 whole Count the fraction strips. **How many whole fraction strips are there?** 4 **How many fractions are there?** 0 **What is the sum of $1\frac{1}{3}$ and $2\frac{2}{3}$?** 4

Activity 2 **How is this problem different than the one in Activity 1?** The fractions do not have the same denominator. **How can you find the LCD of 6 and 4?** List multiples of 6 and multiples of 4. List the first 5 multiples of 6 and the first 5 multiples of 4 on the board: 6, 12, 18, 24, 30 and 4, 8, 12, 16, 24. **What is the LCD of 6 and 4?** 12 Have students complete the activity.

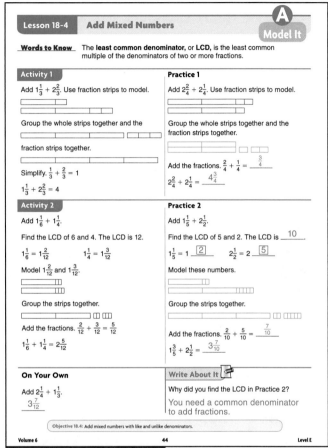

Student Booklet Page 44

⭐ Progress Monitoring

Write $3\frac{1}{8} + 5\frac{1}{2}$ on the board. **What is the first step when modeling this addition problem with fraction strips?** Find the LCD of 8 and 2.

Error Analysis

Students sometimes get so involved with finding the LCD that they "lose" the whole number part of the mixed number. Encourage students to always write both parts of the mixed number.

ENGLISH LEARNERS Students might confuse LCD and LCM. Discuss the LCD of two fractions and the LCM of two numbers.

Lesson 18-4 Add Mixed Numbers

Objective 18.4: Add mixed numbers with like and unlike denominators.

Facilitate Student Understanding

Develop Academic Language

Write $\frac{10}{12}$ on the board. **I can simplify this fraction so that the numerator and denominator are in the lowest terms. What number is a factor of both 10 and 12?** 2 **What is $\frac{10}{12}$ in simplest form?** $\frac{5}{6}$

Demonstrate the Examples

Example 1 To add two mixed numbers, you add the whole number parts, add the fraction parts, and simplify. **What are the whole number parts of the two mixed numbers?** 3 and 12 **What are the fraction parts?** $\frac{7}{9}$ and $\frac{1}{9}$ **What is the sum of the whole number parts?** 15 **What is the sum of the fraction parts?** $\frac{8}{9}$ **Is $15\frac{8}{9}$ in simplest form? Why or why not?** Yes, $\frac{8}{9}$ cannot be reduced any further.

Example 2 **What are the whole number parts of the two mixed numbers?** 6 and 9 **What are the fraction parts?** $\frac{1}{2}$ and $\frac{1}{6}$ **What is the sum of the whole number parts?** 15 **What do you need to do to find the sum of the fraction parts?** Find the LCD of 2 and 6. **What is the LCD?** 6 Have students complete the activity.

Student Booklet Page 45

 Progress Monitoring

Write $4\frac{4}{6}$ on the board. **Is this mixed number in simplest form? Why or why not?** No. $\frac{4}{6}$ can be simplified. **What is $4\frac{4}{6}$ written in simplest form?** $4\frac{2}{3}$

Error Analysis

Students might forget to simplify the final answer. Remind students of all the steps used to add mixed numbers. Write the steps on the board if necessary.

Step 1: Find the LCD of the fractions, if necessary.
Step 2: Rewrite each fraction using the LCD, if necessary.
Step 3: Add the whole numbers.
Step 4: Add the fractions.
Step 5: Simplify the fraction, if necessary.

Lesson 18-4 | Add Mixed Numbers

Objective 18.4: Add mixed numbers with like and unlike denominators.

Observe Student Progress

Computer Tutorial

Some students may benefit from completing a computer tutorial before they attempt the Try It page. A list of the tutorials for each lesson can be found beginning on page x in the front of this book.

 Error Analysis

Exercises 1 and 2 Provide fraction blocks for students to model these exercises before they make their sketches.

Exercise 3 Students might add the fractions and not simplify. Remind them that $\frac{7}{8} + \frac{1}{8}$ is $\frac{8}{8}$, which reduces to 1, and should be added to the whole numbers. There should be no fractional part to this answer.

Exercise 5 Students might add the numerators and denominators and choose answer A. Reread the rules for adding mixed numbers on the board.

Exercise 6 Have students compute the answer, then use fraction strips to model it.

Exercise 7 Students might not remember to simplify the answer. Remind students of the steps used when adding mixed numbers.

Lesson 18-4 | Add Mixed Numbers — Try It

1. Add $2\frac{2}{5} + 3\frac{1}{5}$. Sketch fraction strips to help you find the answer.
   $5\frac{3}{5}$

2. Add $3\frac{1}{3} + 1\frac{1}{2}$. Find the LCD and sketch fraction strips to help you find the answer.
   $4\frac{5}{6}$

3. Add $5\frac{7}{8} + 2\frac{1}{8}$.
   8

4. Add $12\frac{1}{4} + 5\frac{2}{6}$.
   $17\frac{7}{12}$

5. Which is $8\frac{1}{4} + 4\frac{2}{4}$? Circle the letter of the correct answer.
   A $12\frac{3}{8}$ B $12\frac{2}{4}$
   C $12\frac{3}{4}$ D $13\frac{3}{4}$

6. Which is $1\frac{5}{9} + 1\frac{1}{3}$? Circle the letter of the correct answer.
   A $1\frac{7}{12}$ B $2\frac{2}{9}$
   C $2\frac{8}{9}$ D $3\frac{2}{9}$

7. Add $9\frac{2}{6} + 15\frac{4}{12}$.
   $24\frac{2}{3}$

8. Add $4\frac{1}{3} + 10\frac{2}{7}$. Explain how you solved it.
   Sample: I found the LCD for 3 and 7 which is 21. I then renamed the fractions to $\frac{7}{21}$ and $\frac{6}{21}$. I added the whole numbers and got 14. I then added the fractions and got $\frac{13}{21}$.
   $4\frac{1}{3} + 10\frac{2}{7} = 14\frac{13}{21}$

Objective 18.4: Add mixed numbers with like and unlike denominators.

Student Booklet Page 46

Lesson 18-5: Subtract Mixed Numbers

Objective 18.5: Subtract mixed numbers with like and unlike denominators.

Teach the Lesson

Materials
- ☐ Teaching Aid 12 Fraction Strips
- ☐ Index cards

Activate Prior Knowledge

Write $\frac{7}{8} - \frac{2}{8}$ on the board. **How do I subtract fractions that have like denominators?** Subtract the numerators. **What is $\frac{7}{8} - \frac{2}{8}$?** $\frac{5}{8}$ **How do I subtract fractions that have unlike denominators?** Find the LCD and then subtract.

Develop Academic Language

Refer to the fractions in Activate Prior Knowledge. Review the meaning of *numerator* and *denominator*. Point to each fraction and have students identify the numerator and denominator.

Model the Activities

Activity 1 Model the fraction $3\frac{1}{2}$. **How many whole fraction strips did you use?** 3 **How many one-half fraction strips did you use?** 1 **What mixed number are you subtracting from $3\frac{1}{2}$?** $1\frac{1}{2}$ Remove 1 whole and $\frac{1}{2}$ fraction strips. **What number remains?** 2 **What is $3\frac{1}{2} - 1\frac{1}{2}$?** 2

Activity 2 **How is this problem different than the one in Activity 1?** The fractions do not have the same denominator. **How can you find the LCD of 6 and 3?** List multiples of 6 and multiples of 3. List the first 5 multiples of 6 and the first 5 multiples of 3 on the board: 6, 12, 18, 24, 30 and 3, 6, 9, 12, 15 **What is the LCD of 6 and 3?** 6 Have students complete the Activity.

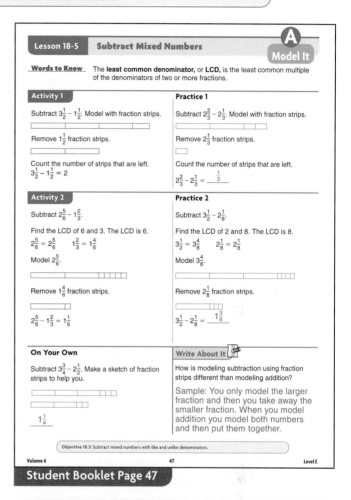

Student Booklet Page 47

Progress Monitoring

Write $4\frac{3}{5} - 2\frac{1}{10}$ on the board. **What is the first step when doing this subtraction?** Find the LCD of 10 and 5.

Error Analysis

Students might try and model both fractions with fraction strips. Remind students to look at the operation before they begin modeling the problem.

Lesson 18-5 | **Subtract Mixed Numbers**

Understand It

Objective 18.5: Subtract mixed numbers with like and unlike denominators.

Facilitate Student Understanding

Develop Academic Language

Write the whole number 1 on the board. **How do you rename 1 as a fraction?** $\frac{1}{1}$ **Name a fraction that is equivalent to $\frac{1}{1}$. Sample:** $\frac{3}{3}$ **How do you name a fraction equivalent to $\frac{1}{1}$ with a denominator of 9?** $\frac{9}{9}$

Demonstrate the Examples

Example 1 **When you subtract fractions, what must be true about the denominators of the fractions?** They must be like. **What are the denominators of the fractional parts of the mixed numbers in Example 1?** 10 To subtract mixed numbers, subtract the whole numbers, then subtract the fractions. Simplify the mixed number if possible. Subtract the mixed numbers. **Is $3\frac{3}{10}$ is lowest terms?** Yes.

Example 2 **What is different about this problem compared to Example 1?** The denominators are different. **What is the LCD of 3 and 12?** 12 **What is $5\frac{2}{3}$ renamed as?** $5\frac{8}{12}$ **What is $5\frac{8}{12} - 4\frac{5}{12}$?** $1\frac{3}{12}$ **Is $1\frac{3}{12}$ in simplest from?** No. **How do you write $1\frac{3}{12}$ in simplest form?** Divide the numerator and denominator by 3. **What is $5\frac{2}{3} - 4\frac{5}{12}$ in simplest form?** $1\frac{1}{4}$

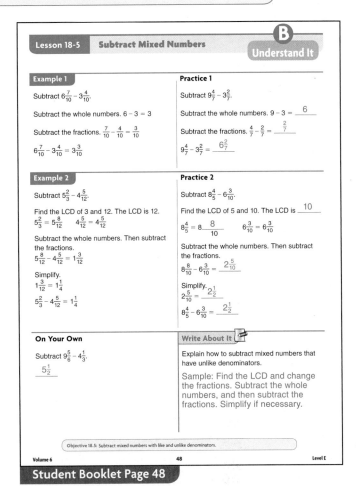

Student Booklet Page 48

ENGLISH LEARNERS Review the differences among fractions, improper fractions, and mixed numbers by writing examples of each on index cards. Have students sort the cards into the three groups.

Progress Monitoring

How do you know if the answer is in simplest form? Look at the numerator and denominator and see if they could be divided by the same number. If so, the fraction needs to be simplified. If not, it is already in simplest form.

Error Analysis

Students might forget to see if their answers could be simplified. Remind students to look at each answer carefully to determine if it can be simplified.

Lesson 18-5 Subtract Mixed Numbers

Objective 18.5: Subtract mixed numbers with like and unlike denominators.

Observe Student Progress

Computer Tutorial

Some students may benefit from completing a computer tutorial before they attempt the Try It page. A list of the tutorials for each lesson can be found beginning on page x in the front of this book.

 Error Analysis

Exercises 1 and 2 Provide fraction strips for students to use before drawing sketches for these problems.

Exercise 3 Students who pick answer A subtracted the numerators and denominators. Establish that the denominators must be the same in order to subtract, and the numerators and the whole numbers are the only numbers that get subtracted.

Exercise 4 This is the first problem which has not had a whole number part to the subtrahend. Remind students that 0 can be used to hold a place for the whole number part to help them in subtracting.

Exercise 5 Have students use fraction strips to model the problem. Make sure students simplify their answers after subtracting the fractions.

Exercise 8 If students have difficulty thinking of similarities, have them write the steps for adding mixed numbers, then the steps for subtracting mixed numbers, and see what steps happen in each procedure.

Student Booklet Page 49

Volume 6 — **Understand Fractions**

Topic 18 — **Addition and Subtraction of Fractions**

Topic Summary

Objective: Review adding and subtracting fractions.

Have students complete the student summary page. You may want to have students work in groups of four with each student analyzing a different choice.

Ask students to share their ideas about each answer choice. Be sure they confirm the correct answer for each problem at the end of the discussion.

Answer Evaluation

1. **A** Students subtracted instead of added.
   B Students changed the mixed fractions into incorrect improper fractions, then added.
   C Students only added the whole numbers.
   D This choice is correct.

2. **A** Students left the whole number out of the equation.
   B Students changed the mixed fractions into incorrect improper fractions, then subtracted.
   C This choice is correct.
   D Students added instead of subtracted.

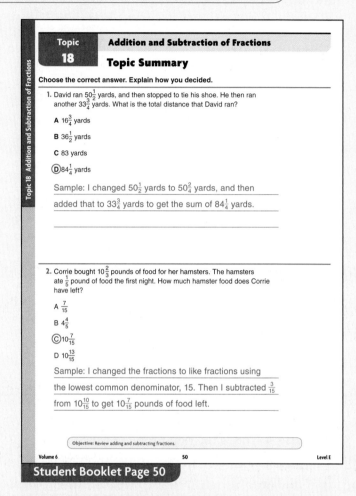

Student Booklet Page 50

 Error Analysis

Exercise 1 When students add $\frac{1}{2}$ and $\frac{3}{4}$ and get the sum $1\frac{1}{4}$, remind them to add the 1 to the sum of the whole numbers.

Exercise 2 If students change $10\frac{2}{3}$ to an improper fraction, remind them to multiply 10 by 3 and then to add the product to 2.

Progress Monitoring

When all assignments for this topic have been completed, assign the corresponding Progress Monitoring page for this topic (Assessment Resources Book, page 18). Be sure students complete the Progress Monitoring page before you administer the final assessment for this volume.

Volume 6 — Understand Fractions
Topic 18 — Addition and Subtraction of Fractions
Mixed Review

Objective: Maintain concepts and skills.

Have students complete the Mixed Review page. Work with each student individually to review results. Identify strengths and weaknesses and correct any misunderstandings.

✓ Error Analysis

Exercise 2 Have students check their answers by solving each side of the equation.

Exercise 3 If students have difficulty comparing the fractions while in lowest terms, tell them to give the fractions common denominators.

Exercise 4 Students might divide the numerator and denominator by a different common factor which will give an equivalent fraction. However, remind students to use the greatest common factor so the fractions will be in lowest terms.

Exercise 6 Be sure that students understand that the pizza was completely eaten by the two boys, so Ben's part + Michael's part = 1 whole pizza.

Exercise 8 Have students write an equation for the problem, such as 5(2 + 2), before trying to solve it.

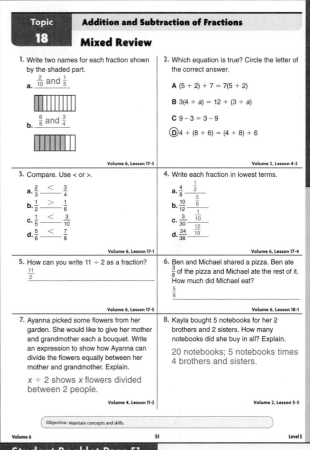

Student Booklet Page 51

Volume 6: Understand Decimals

Topic 19: Decimals and Money

Topic Introduction

Lesson 19-1 teaches students to understand and use the dollar and cent signs for money. Lesson 19-2 introduces place-value charts to model decimals to hundredths. Lesson 19-3 brings in students' previous knowledge of comparing and ordering whole numbers and extends this knowledge to decimals. Lessons 19-4 and 19-5 return to concepts of money and introduce decimal notation for money. Lesson 19-6 has students work with money as a model for fractions and decimals. Finally, 19-7 teaches students to find unit costs.

Lesson	Objective	Student Pages	Teacher Pages	Tutorials
Topic 19 Introduction	19.1 Know, understand, and use the dollar and cent signs for money. 19.2 Write tenths and hundredths in decimal notation. 19.3 Order and compare whole numbers and decimals to two decimal places. 19.7 Determine the unit cost when given the total cost and number of units.	52	323	
19-1 Money	19.1 Know, understand, and use the dollar and cent signs for money.	53–55	324–326	19a, 19b
19-2 Tenths and Hundredths	19.2 Write tenths and hundredths in decimal notation.	56–58	327–329	19b
19-3 Compare and Order Decimals	19.3 Order and compare whole numbers and decimals to two decimal places.	59–61	330–332	19f, 19g
19-4 Use Coins and Bills	19.4 Count combinations of coins or bills.	62–64	333–335	19c, 19d, 19e
19-5 Decimal Notation for Money	19.5 Know and use the decimal notation and the dollar and cent signs for money.	65–67	336–338	19a
19-6 Money in Fractions and Decimals	19.6 Know and understand that fractions and decimals are two different representations of the same concept.	68–70	339–341	19a, 19b
19-7 Unit Costs	19.7 Determine the unit cost when given the total cost and number of units.	71–73	342–344	19f, 19g, 19h, 19i, 19b, 19j
Topic 19 Summary	Review the relationship between decimals and money.	74	345	19k, 19l
Topic 19 Mixed Review	Maintain concepts and skills.	75	346	

Computer Tutorial

Some students may benefit from completing the computer tutorial before they attempt the Try It page of each lesson. If you are using the electronic components of *Pinpoint Math,* you will find a complete listing of Tutorial codes and titles when you access them either online or via CD-ROM.

Volume 6 — Understand Decimals
Topic 19: Decimals and Money
Topic Introduction

Objectives: 19.2 Write tenths and hundredths in decimal notation. **19.3** Order and compare whole numbers and decimals to two decimal places. **19.4** Know and understand that fractions and decimals are two different representations of the same concept. **19.5** Know and use the decimal notation and the dollar and cent signs for money.

Materials ☐ Play coins and bills

Distribute the bills and coins to pairs of students. Begin with the coins and review the value of each coin. Create different coin amounts and have students recreate them with their partners and find the values. Introduce bills and again create money amounts allowing students to name the values. Reinforce writing the value in decimal form including the dollar or cent sign.

Informal Assessment

1. **How much is each quarter worth?** 25 cents **How much are 3 quarters worth?** 75 cents **How much are 3 dimes worth?** 30 cents **How much is 1 nickel worth?** 5 cents **How will you find the total number of cents?** Add.

   You may want to suggest that students write the numbers for each value in cents in a column to make adding easier. **Where does the decimal point go to show dollars and cents?** After the first 1

2. **How many squares is the grid divided into?** 100 **How many are shaded?** 53 **What is the word name for this number?** Fifty-three hundredths Have students write the decimal and fraction forms for fifty-three hundredths.

3. **How do you say the first decimal?** Eight and one tenth **How do you say the second decimal?** Eight and ten hundredths **What do you know about tenths and hundredths?** One tenth is equal to ten hundredths.

4. Remind students that they can write these fractions over a denominator of 100, because they all end in the hundredths place, but some of the fractions can also be simplified.

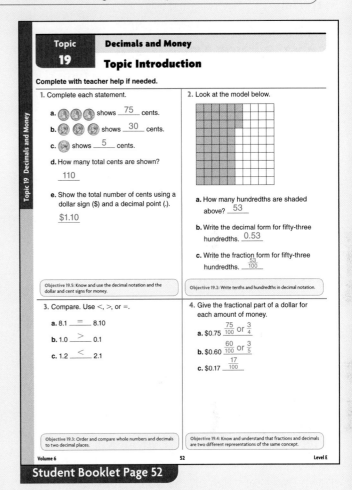

Student Booklet Page 52

Another Way Have students use base ten blocks to represent each decimal in Exercise 3. This will allow students to easily determine which decimal is larger.

Lesson 19-1 Money

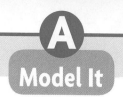

Objective 19.1: Know, understand, and use the dollar and cent signs for money.

Teach the Lesson

Materials ☐ Play coins and bills

Activate Prior Knowledge

Write 5 on the board. Have students take turns skip counting by 5s and write the numbers on the board. Repeat for 10s and 25s.

Divide students in small groups. Distribute a set of play coins and bills to each group. Ask them to name each coin shown and give its value in cents. Remind students that the size of the coin does not indicate value, and that a single coin can be worth more than several other coins.

Develop Academic Language

Remind students that the cent sign, ¢, is placed after a number that indicates the cent value of an amount of money as we would say the word *cents* after the number.

Model the Activities

Activity 1 Remind students that a paper dollar bill represents 100 cents. Have volunteers to write out the name and value of each dollar bill and coin shown using cent sign on the board. 100¢, 25¢, 10¢, 5¢, 1¢ **How can you find the total amount of money shown?** Add: 100¢ + 25¢ + 10¢ + 5¢ + 1¢. **What is the total value?** 141¢

Activity 2 **What is the value of a nickel?** 5¢ **How can you find the number of nickels?** Skip count by 5s. Have students skip count by 5s while writing the sequence of numbers on the board until they reach 45. **How many 5s did you count to get 45?** 9 This represents the number of nickels it takes to make 45¢.

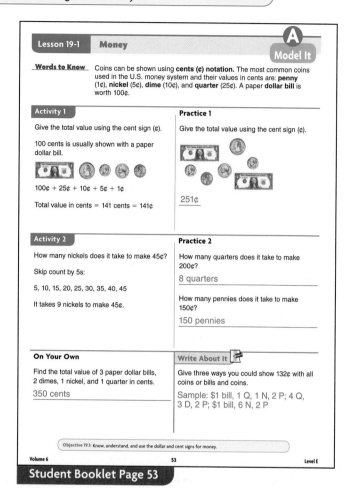

Student Booklet Page 53

⭐ Progress Monitoring

Display several different combinations of bills and coins. **What is the total number of cents shown? How would you write the value using a cent sign?**

Error Analysis

If students have difficulty counting the total amount of money shown, suggest they begin with the dollar bill or largest coin value, skip count by that amount and then begin skip counting by the next-greatest coin value, continuing until all coins have been counted.

Lesson 19-1   Money

Objective 19.1: Know, understand, and use the dollar and cent signs for money.

Facilitate Student Understanding

Develop Academic Language

A whole number of dollars can be shown with a dollar sign, as in $5. Watch for students who place the dollar sign behind the amount of money. This can result from the manner in which we say the word *dollars* **after** the number is stated, "five *dollars*."

ENGLISH LEARNERS Be aware that if students have attended school outside the United States, they may be accustomed to writing a cent sign with two lines through it.

Demonstrate the Examples

Example 1 Provide play bills for students to model the problems with as they count. **Count the ten-dollar bills.** $10, $20 **Now continue and count the five-dollar bill.** $25 Make sure students do not repeat $20 and also that they do not mistakenly skip count by $10 again and say $30. **Now continue and count the one-dollar bill. What is the total value?** $26

Example 2 Repeat the procedure from Example 1. Make sure that students are confident in the values of the coins and in the method of skip counting.

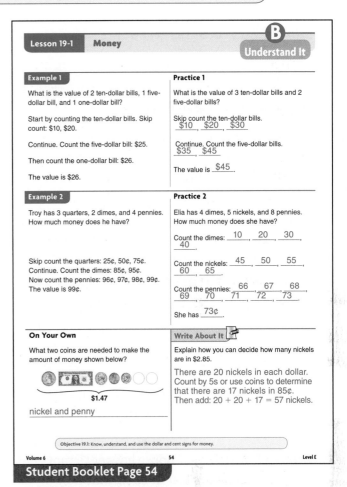

Student Booklet Page 54

⭐ Progress Monitoring

Draw various numbers of coins and bills on the board and have students give the total values.

Error Analysis

Allow students to continue to use play coins and bills or to draw and label sketches of coins and bills to model the problems.

Volume 6   325   Level E

Lesson 19-1 Money

Objective 19.1: Know, understand, and use the dollar and cent signs for money.

Observe Student Progress

Materials ☐ Play Coins and Bills

Computer Tutorial

Some students may benefit from completing a computer tutorial before they attempt the Try It page. A list of the tutorials for each lesson can be found beginning on page x in the front of this book.

★ Error Analysis

Exercise 1 For students having difficulty, suggest that they write the value of each set of similar bills and coins, and then find the total amount of money.

Exercise 2 Check to ensure that students are correctly showing the cent sign to the right of the number.

Exercise 3 If students choose answer C, they are confusing the number of coins with the value of the coins. Review the names and values of the coins with students.

Exercise 4 If students choose answer D, they are repeating numbers when skip counting. Suggest that they model the problem with play bills to keep track of what they have already counted.

Exercise 5 If students answer incorrectly, have them use play money coins and skip count until they reach the value shown, then count the number of coins that were required.

Student Booklet Page 55

Lesson 19-1 Money

1. Write the value of the money shown below using a cent sign.

   478¢

2. Write each amount of money with a cent sign.
   a. 127 cents 127¢
   b. 691 cents 691¢
   c. 58 cents   58¢
   d. 4 cents   4¢

3. What is the value of 1 quarter, 3 dimes, and 2 pennies? Circle the letter of the correct answer.
   A 5¢   B 123¢
   (C) 57¢   D 46¢

4. What is the value of 3 ten-dollar bills, 3 five-dollar bills, and 6 one-dollar bills? Circle the letter of the correct answer.
   A $45   C $12
   B $336   (D) $51

5. Tell how many of each coin it takes to make the given value.
   a. $1.45 = 29 nickels
   b. $0.67 = 67 pennies
   c. $2.50 = 25 dimes
   d. $3.75 = 15 quarters

6. Find a combination of 6 coins with a value of 95¢. Then write the total value of the coins using a cent sign.

   3 quarters, 1 dime, 2 nickels; $0.95

Objective 19.1: Know, understand, and use the dollar and cent signs for money.

ENGLISH LEARNERS Point out that a quarter is a coin that is worth 25¢. **Think about what is a quarter of 100¢.** Point out that the word *cent* is worth 100. **Think about what other things have the word *cent* in them.** Sample: centipede, century, centimeter

Lesson 19-2 Tenths and Hundredths

Objective 19.2: Write tenths and hundredths in decimal notation.

Teach the Lesson

Materials ☐ Teaching Aid 9 Decimal Models

Activate Prior Knowledge
Review writing and reading fractions with denominators of 10 or 100.

Develop Academic Language
Use a place-value chart to relate the names of the first two decimal places, tenths and hundredths, to the whole-number place values tens and hundreds.

Explain that *equivalent* forms of a number may look different but have the same value. Use an example of expanded form and standard form of a number.

ENGLISH LEARNERS Not all languages have words that end in -*th*, so articulate these words clearly, and ask students to listen carefully.

Model the Activities

Activity 1 Provide copies of Teaching Aid 9 Decimal Models. Draw a tenths grid. **How many parts is it possible to shade?** 10 **So what fractions could we show on this grid?** Tenths **How do we show six tenths?** Shade 6 parts. Have students write the fraction. **To write the decimal, you'll write the numerator. The denominator tells you what place that number should end in. What place does the denominator tell us?** Tenths Write 0.____. **The tenths place is right after the decimal point. What do we put in the tenths place?** 6 Guide students to fill in the chart with attention to the places' names.

Activity 2 Repeat the process from Activity 1. Help students use the model or the place-value chart to understand that 0.47 has 4 tenths and 7 hundredths.

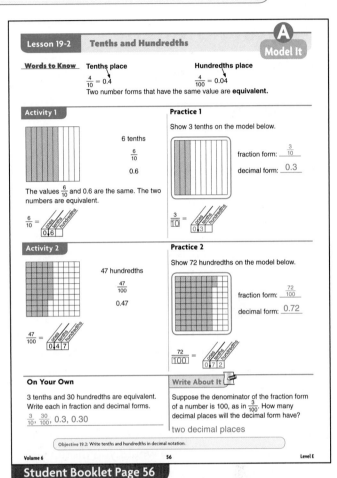

Student Booklet Page 56

★ Progress Monitoring

Shade 34 hundredths on a grid on the board. **What fraction does this show?** $\frac{34}{100}$ **What decimal does this show?** 0.34

Error Analysis

Watch for students who confuse tenths and hundredths. Show models of 0.3 and 0.03 side by side. Have them write the correct decimals below each model. Then write the equivalent fractions.

Lesson 19-2: Tenths and Hundredths

Objective 19.2: Write tenths and hundredths in decimal notation.

Facilitate Student Understanding

Develop Academic Language

To reinforce the names and order of the decimal places, teach students to make and label their own place-value charts to use when none are provided.

Demonstrate the Examples

Example 1 Guide students to read the examples given. **Look at $\frac{1}{10}$ through $\frac{9}{10}$. What do all of the fractions have in common?** 10 in the denominator **What do all of the decimals have in common?** There is a digit in the first place to the right of the decimal. **Do you have to memorize these, or is there a way to figure out the decimal when a fraction has a denominator of 10?** You can write the numerator in the tenths place. Allow students to draw models if necessary, but emphasize that simply knowing the names of the decimal place values is enough. **Now look at the fraction $\frac{10}{10}$. How would you model $\frac{10}{10}$ on a grid?** Shade 10 of 10 bars. **Since you would color the entire grid, what is another way to state the value of $\frac{10}{10}$?** 1 Notice that the decimal also is 1 whole. You may want to point out that the method of writing the numerator to end in the place named by the denominator works here; 10 is written to end in the tens place, so the decimal falls between the 1 and the 0.

Example 2 **First write the whole number part. How many wholes are shaded?** 1 Then add the decimal part. **What part of the second grid is shaded?** Five tenths One and five-tenths is shaded. Write the decimal point between the whole number part and the fractional part, where you hear the word *and*. 1.5 Ask students to relate the whole number part and decimal part to the parts of the equivalent mixed number shown.

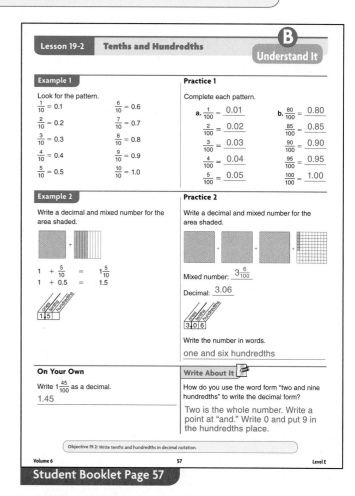

Student Booklet Page 57

Progress Monitoring

Shade $3\frac{7}{10}$ on tenths grids. **What mixed number does this show?** $3\frac{7}{10}$ **How can that be written as a decimal?** 3.7

Error Analysis

Remind students that when writing the decimal form for a mixed number, the whole number part always goes to the left of the decimal point; the tenths or hundredths go to the right.

Lesson 19-2 — Tenths and Hundredths

Objective 19.2: Write tenths and hundredths in decimal notation.

Observe Student Progress

Computer Tutorial

Some students may benefit from completing a computer tutorial before they attempt the Try It page. A list of the tutorials for each lesson can be found beginning on page x in the front of this book.

Error Analysis

Exercise 1 Make sure students don't start in the hundredths place for 1b and 1c.

Exercise 2 For students having difficulty writing the two equivalent fractions, remind them to write one with a denominator of 10 and the other with a denominator of 100. Beginning with the fractions may help some students write the decimal.

Exercise 4 Have students write the word name for each answer option. Then find the match for $7\frac{39}{100}$.

Exercise 5 Remind students that *and* signals the beginning of the decimal part.

Exercise 6 Have students place the labels and the decimal point before filling in any numbers.

Lesson 19-2 — Tenths and Hundredths — Try It

1. Write each of the following in decimal and fraction notation.

   a. nine tenths $0.9, \frac{9}{10}$

   b. sixty-eight hundredths $0.68, \frac{68}{100}$

   c. three hundredths 0.03

   d. one tenth $0.1, \frac{1}{10}$

2. Rewrite each of the following decimals showing hundredths. Then write two equivalent fractions.

   a. 0.7 $0.70, \frac{7}{10}, \frac{70}{100}$

   b. 0.3 $0.30, \frac{3}{10}, \frac{30}{100}$

   c. 0.6 $0.60, \frac{6}{10}, \frac{60}{100}$

   d. 0.9 $0.90, \frac{9}{10}, \frac{90}{100}$

3. Write two equivalent numbers for the model below: one mixed number and one decimal.

   $3\frac{27}{100}, 3.27$

4. What is the decimal form for $7\frac{39}{100}$? Circle the letter of the correct answer.

   A 7.039 B 73.9
   Ⓒ 7.39 D 70.39

5. Write each number in decimal form.

   a. forty and four hundredths 40.04

   b. forty and four tenths 40.4

   c. forty and forty-four hundredths 40.44

6. Write a number with 6 hundredths, 7 tenths, 9 ones, and 4 tens. Write the places' names and a decimal point in the chart to help you.

tens	ones	tenths	hundredths
4	9	7	6

Objective 19.2: Write tenths and hundredths in decimal notation.

Student Booklet Page 58

Lesson 19-3 Compare and Order Decimals

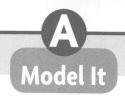

Objective 19.3: Order and compare whole numbers and decimals to two decimal places.

Teach the Lesson

Materials
- ☐ Teaching Aid 8 Place-Value Charts
- ☐ Index cards

Activate Prior Knowledge

Have students compare pairs of two- and three-digit numbers using the < or > symbol.

Discuss tenths and hundredths as decimal place values. Post a place-value chart. Point out that values decrease as we extend the chart to the right.

Develop Academic Language

What does it mean to compare two numbers?
Sample: *to find the greater number* Discuss using and reading the symbols < and >.

Model the Activities

Activity 1 Have students copy the numbers into a place-value chart. Some students may find it helpful to write zeros in empty decimal places at the end of the number until the numbers are equal in length. **What is the value of the 4 in 49.25?** 4 tens; 40 Continue for all places. **Which place has the greatest value?** Tens Start with the tens. **Is one of these digits greater than the other?** No. **Move to the ones. Is one of these digits greater than the other?** No. **What place to we move to next?** Tenths **Is one of these digits greater than the other?** 7 **Which number is greater?** 49.7

Activity 2 Have students write the numbers in a place-value chart and add a terminal zero to 357.5. Justify the addition of the zero by showing that 5 tenths and 50 hundredths are equivalent, and point out that zeros *cannot* be added between the decimal point and any digit. **Compare the digits in the place with the greatest value. Move right and continue to compare. Which place has digits that are different?** Hundredths **What are the digits?** 5 and 0 **Which is greater?** 5 **Which of the numbers is greater?** 357.55 This number has five tenths *and* five hundredths. The other number only has five tenths.

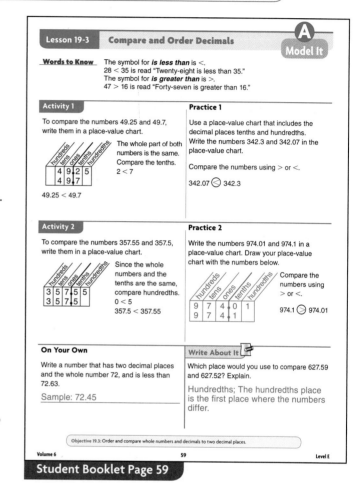

Student Booklet Page 59

Progress Monitoring

Distribute a set of index cards with various digits and one with a decimal point to each student. Have students create numbers, pair up, and compare their numbers. Then have them change partners and compare again.

Error Analysis

Point out that the comparison symbol is smaller next to the smaller number and gets larger next to the larger number. It always points to the smaller number.

Students can shade squares on Teaching Aid 8 Place-Value Charts to compare decimals when whole numbers are the same.

Lesson 19-3: Compare and Order Decimals

Objective 19.3: Order and compare whole numbers and decimals to two decimal places.

Facilitate Student Understanding

Develop Academic Language

Clarify to students that when they are asked to order numbers, they are to show them from least to greatest unless otherwise stated.

ENGLISH LEARNERS Make sure that students know that *ordering* means putting numbers in order.

You may want to review the words *least* and *greatest*. **How would you write these five numbers from least to greatest: 92, 47, 78, 53, 24?** 24, 47, 53, 78, 92 Create a model showing objects arranged from least to greatest in size.

Demonstrate the Examples

Example 1 Ask students to begin from the left and tell whether the digits in each place are the same or different. When a difference occurs, have them determine which is greater. The greater number will be the last number in the list and can be crossed off. Continue comparing successive place values until students have determined the order of the numbers from least to greatest.

Example 2 What decimals are shown on this number line? Tenths **Starting with a number line from 0 to 1, how could you make a number line like this?** Draw 9 tick marks to divide the space into 10 equal parts; each tick mark will be labeled with a tenth in order from one tenth to nine tenths. **Find two tenths and seven tenths. How does the number line show which is greater?** 0.7 comes after 0.2, so 0.7 is greater. You may want to have students write a zero at the end of each decimal number and then label the unlabeled tick marks.

ENGLISH LEARNERS Students may need a discussion of the comparative and superlative forms in English, such as *less* and *lesser* versus *least*.

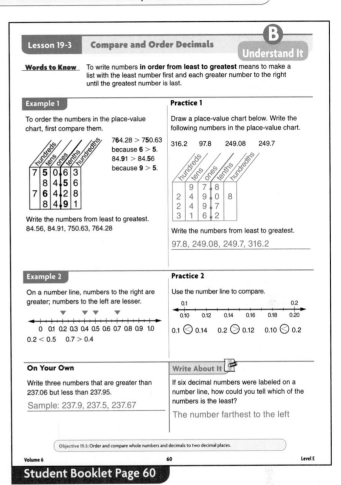

Student Booklet Page 60

Progress Monitoring

Write the numbers 347, 326.01, 30.92, and 304.07 on the board. **Write these numbers in order from least to greatest.** 30.92, 304.07, 326.01, 347

Error Analysis

Suggest that students create their list from right to left by finding the greatest number, comparing again to find the next greatest number, and so on.

Lesson 19-3: Compare and Order Decimals

Objective 19.3: Order and compare whole numbers and decimals to two decimal places.

Observe Student Progress

Computer Tutorial

Some students may benefit from completing a computer tutorial before they attempt the Try It page. A list of the tutorials for each lesson can be found beginning on page x in the front of this book.

Error Analysis

Exercises 1 and 2 Suggest that students arrange the numbers in a vertical list to compare place values more easily. For those still having difficulty, encourage them to write the numbers in a place-value chart and use terminal zeros if necessary.

Exercise 3 You may want to suggest that students assign their best estimate of the number each letter is pointing to before answering the question. Changing all instances of .5 to .50 may help students label the unlabeled tick marks.

Exercise 5 Have students begin by writing each pair of letters in order from lesser to greater: B, A; A, C; and D, B. Then students can merge the second and third pairs into the first.

Exercise 6 Have students write the first number in a place-value chart. Then help them try digits 0 through 9 in the tenths place to form the second number and test whether each number formed is greater than the first number.

Student Booklet Page 61

Lesson 19-3 — Compare and Order Decimals — Try It

1. Write < or > in each ◯.
   a. 0.63 > 0.51
   b. 89.14 > 89.1
   c. 352.1 > 351.6
   d. 1.12 > 1
   e. 70.34 < 73.04

2. Write each set of numbers in order, least to greatest.
   a. 28.43   37   21.8   28   31.2
      21.8, 28, 28.43, 31.2, 37
   b. 3.04   3.14   3.01   3.41   3.31
      3.01, 3.04, 3.14, 3.31, 3.41

3. Which letter shows where 43.12 would be located on the number line below? C
   A   B   C   D
   42 42.5 43 43.5 44 44.5

4. Circle the letter of the least number below.
   A 87.03   B 87.3
   C 873   (D) 80.73

5. If A > B, C > A, and D < B, show the order of A, B, C, and D from least to greatest on the number line below.
   D   B   A   C

6. Write a number with 2 decimal places. Write a number with 1 decimal place that is **greater** than the other number.
   Sample:
   0. _5_ _9_
   0. _6_

7. Below is the answer a student gave when comparing 34.2 and 30.7.
   Since there are 7 tenths in 30.7 and only 2 tenths in 34.2, 30.7 is greater.
   What mistake did the student make? Explain and then write the correct response.
   No; 34.2 has 4 ones and 30.7 has 0 ones, so 34.2 is greater. The correct response is 34.2 is greater than 30.7.

Objective 16.3: Order and compare whole numbers and decimals to two decimal places.

Lesson 19-4   Use Coins and Bills

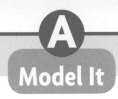

Objective 19.4: Count combinations of coins or bills.

Teach the Lesson

Materials ☐ Teaching Aid 10 Coins and Bills

Activate Prior Knowledge

Hold up different coins and ask students to name each one. Show a penny, nickel, dime, and quarter. Also have them identify a $1 bill, a $5 bill, and a $10 bill.

Develop Academic Language

As students handle the coins, be sure they use the names of the coins correctly.

Model the Activities

Activity 1 Have students model the group of coins and name each coin shown. **One way to find the total value is to add the coins' values. What is the first addend?** 5¢ **Why?** The first coin is a nickel. **The second addend?** 5¢ **Why?** This coin is also a nickel. **The third addend?** 10¢ **How do you know?** This coin is a dime. **Add.** 20¢

Activity 2 Have students model the group of coins, name each coin, and give each coin's value. **To count money, line the coins up in order from greatest value to least value. Which coin in this group has the greatest value?** Half dollar **Next?** Quarter **Next?** Dime **Next?** Nickel **Next?** Penny **Start with the value of the half dollar.** 50¢ **Count on by the value of the next coin. Start at 50¢, and count on 25¢ for the quarter.** 75¢ **How much will you count on for the next coin?** 10¢ **Start at 75¢, and count on 10¢ for the dime.** 85¢ **Repeat for the nickel and penny. What is the total value?** 91¢

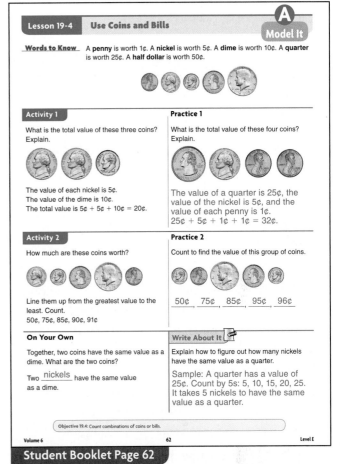

Student Booklet Page 62

Progress Monitoring

What is the value of these two coins? 26¢

Error Analysis

If students have difficulty remembering the values of the coins, suggest they make a chart showing each coin, its name, and its value.

You may want to supply a hundred char to help students count on. Another counting strategy is to replace each dime, quarter, or half dollrs with nickels, and then count by fives.

Volume 6   333   Level E

Lesson 19-4: Use Coins and Bills

Objective 19.4: Count combinations of coins or bills.

Facilitate Student Understanding

Materials ☐ Teaching Aid 10 Coins and Bills

Develop Academic Language

Use expressions such as *a five-cent coin* or *a quarter of a dollar* to describe coins and have students supply their standard names. **Nickel, quarter**

Demonstrate the Examples

Example 1 You may want to practice a few coin trades before beginning. First find the total value. **Line the coins up by value.** Guide students to count. **46¢ To show 46¢ another way, you can replace two nickels with 1 dime. Is the value still the same?** Have students count to verify. **Yes. Can you find a different way to make 46¢?** Sample: **Replace the dimes with 4 nickels.** Continue. Make sure students practice replacing a coin of greater value with coins of lesser value, and vice versa.

Example 2 Have students model the group of bills. **What is the value of each bill?** Guide students to read the value printed on each bill. **How can we arrange these bills in order to count them more easily? In order from greatest to least value** Guide students to count as they move each bill to the side. **$20, $40, $50, $55, $60, $61 What is the total value? $61**

Computer Tutorial

Some students may benefit from completing a computer tutorial before they attempt the Try It page. A list of the tutorials for each lesson can be found beginning on page x in the front of this book.

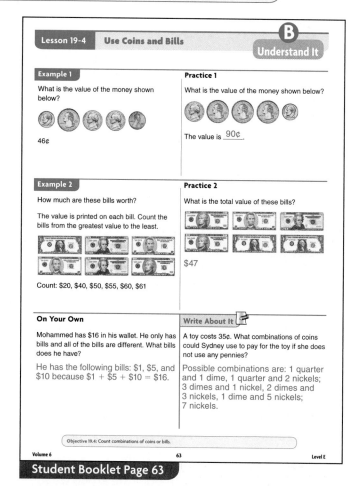

Student Booklet Page 63

Progress Monitoring

Show one way to make 72 cents.

Error Analysis

If students continue to have difficulty finding total values, continue to use a chart of coins. Add a 1-dollar bill, a 5-dollar bill, and a 10-dollar bill and their values to the chart.

Lesson 19-4: Use Coins and Bills

Objective 19.4: Count combinations of coins or bills.

Observe Student Progress

Develop Academic Language

Exercise 2 To help students learn the names of coins, have a student pick two coins at random and hold up each one for the other students to name.

 Error Analysis

Exercise 1 Have students model or draw and label Ed's coins.

Exercise 3 In 3a, the coins are not in order from greatest to least value. Remind students that they can reorder the coins to help them count.

Exercise 4 If students select item B, they do not fully understand the meaning of *greater than*. Review *greater than* and remind students that if two quantities are equal, one is not greater than the other.

ENGLISH LEARNERS Remind students that *greater than* means "more than."

Exercise 6 If students choose incorrect coins and bills, have them use models to see which of the coins and bills Irena might have. Students should find the greatest possible amount first, find the least possible amount next, and then subtract.

Student Booklet Page 64

Lesson 19-4 Use Coins and Bills

1. Ed has 2 quarters, 3 dimes, and 2 nickels. How much money does he have?

   90¢

2. Give the value of each coin.
   a. quarter 25¢
   b. nickel 5¢
   c. dime 10¢
   d. penny 1¢

3. Give the total value of the coins.
   a. 1 half dollar, 2 nickels, 1 dime 70¢
   b. 2 quarters, 3 nickels 65¢
   c. 1 dime, 3 nickels, 8 pennies 33¢
   d. 1 quarter, 3 dimes, 1 nickel, 4 pennies 64¢

4. Which group of coins has a value greater than 50¢? Circle the letter of the correct answer.
   (A) 3 quarters
   B 5 dimes
   C 4 dimes, 1 nickel, 1 penny
   D 8 nickels

5. Melba has 4 coins, all with different values. If she has no dollar coins, what is the greatest amount of money she could have?

   90¢

6. Irena has 3 bills with the same value. Each bill is less than $50. What is the difference between the greatest and least amounts she could have?

   $60 − $3 = $57

7. Thom has 2 nickels and 3 dimes. Show another group of coins with the same value as Thom's money.

   Check that students' models show 40¢.

8. Chris has $50 in bills. List 3 different combinations of bills he might have.

   1 dollar and 1 quarter; 1 dollar, 2 dimes, and 1 nickel; 1 dollar, 1 dime, and 3 nickels; 8 dimes, 1 quarter, and 4 nickels

Lesson 19-5 Decimal Notation for Money

Objective 19.5: Know and use the decimal notation and the dollar and cent signs for money.

Teach the Lesson

Materials ☐ Teaching Aid 10 Coins and Bills

Activate Prior Knowledge

Hold up a $1 bill and a $5 bill. **What is the value of these bills?** $6 Repeat with other combinations of bills. Have students use decimal notation to write each amount. Then practice trading coins for $1 bills.

Develop Academic Language

Have students read $1.80 as "one dollar and eighty cents" rather than as "a dollar eighty" or "one eighty" so they learn correct decimal notation. *And* indicates the location of the decimal point.

Model the Activities

Activity 1 Have students work with quarters. **How can you count to find the value of the quarters?** Record the count in cents notation on the board. **Count together in cents: 25¢, 50¢, 75¢, 100¢. What is another way of saying the value 100¢?** 1 dollar Let's count again, from 25 cents through 1 dollar. Record the count in decimal notation. **$0.25, $0.50, $0.75, $1.00. How is $1.00 written differently from the other amounts before it?** There is a nonzero digit to the left of the decimal point. Show that dollars are recorded to the left of the decimal point and that there are two decimal places for cents.

Activity 2 Have students work with nickels and count by 5s. **How many nickels do you have to count to get to $1?** 20 Make as many groups of 20 as possible. **How many groups can you make?** 2 **How can you use the dollar sign and decimal point to write the value of each group?** $1.00 **What is the value of these two groups together?** $2.00 Look at the coins that you couldn't use to form a dollar. **How can we add the value of these coins to the value we've already found?** Count them and add to $2.00, or count on by 5s from $2.00. Guide students through both methods.

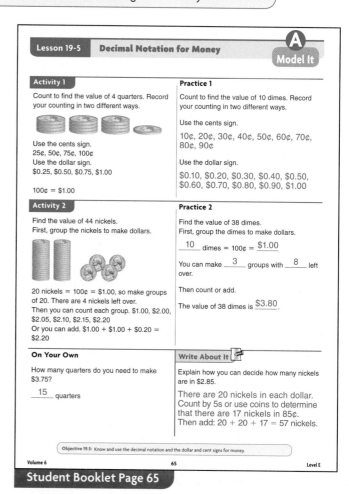

Student Booklet Page 65

✓ Progress Monitoring

Write 125¢ and $1.25 on the board. **What do these amounts represent?** 1 dollar and 25 cents Have students model $1.25 with quarters.

Error Analysis

If students have difficulty determining the number of a given coin in a dollar amount, review the coins and their values, and then practice counting by 5s, 10s, or 25s.

To help students record values correctly in decimal notation, provide a frame: $____.__ includes unlimited space before the decimal point but only 2 places after the decimal point.

Lesson 19-5 Decimal Notation for Money

Objective 19.5: Know and use the decimal notation and the dollar and cent symbols for money.

Facilitate Student Understanding

Develop Academic Language

Be sure students read $326, $3.26, and 326¢ correctly. The only amount where you say *and* is "three dollars and twenty-six cents."

ENGLISH LEARNERS In some countries, a comma is used in place of the decimal point. Discuss the use of the comma and decimal point in money amounts in the United States.

Demonstrate the Examples

Example 1 We need to find the number of cents in $1.79. Which coin is worth 1 cent? **Penny** Then find the number of pennies in $1.79. How many pennies are in one dollar? **100** How many pennies are in 79 cents? **79** What is the sum of 100 pennies and 79 pennies? **179 pennies** As students work these and similar problems, encourage them to generalize: To write dollars and cents as cents, move the decimal point two places to the right, drop the dollar sign, and add the cents sign. To write cents as dollars and cents, do the opposite. You can explain to students that since there are 100 cents in 1 dollar, you either divide by 100 to change cents to dollars, or multiply by 100 to change dollars to cents.

Example 2 Have students describe the two steps needed to solve this problem. **First add to find the total number of cents and then write the number of cents as dollars and cents.** Provide additional practice by having students make up similar problems for others to solve.

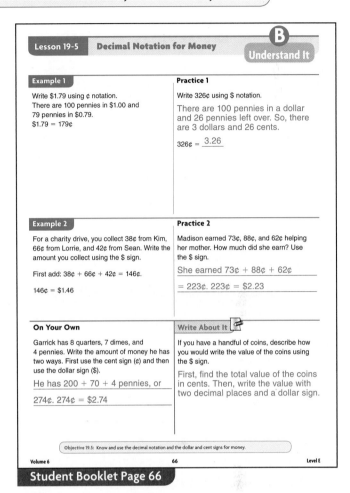

Student Booklet Page 66

Progress Monitoring

Find the sum of 48¢, 67¢, and 98¢. Write the sum using the ¢ and $ signs. **213¢, $2.13**

Error Analysis

If students continue to have difficulty writing a number of cents using the dollar sign, allow them to work with manipulatives. For example, they can count 122 pennies and isolate 100 of them to show that 122 pennies represent 1 dollar and 22¢, or $1.22.

Lesson 19-5 Decimal Notation for Money

Objective 19.5: Know and use the decimal notation and the dollar and cent symbols for money.

Observe Student Progress

Computer Tutorial

Some students may benefit from completing a computer tutorial before they attempt the Try It page. A list of the tutorials for each lesson can be found beginning on page x in the front of this book.

Develop Academic Language

Exercises 2 and 3 To help students learn to read the notation correctly, have them read both the given amounts and their answers aloud.

Student Booklet Page 67

Lesson 19-5 Decimal Notation for Money — Try It

1. You have 48 dimes. Write the amount of money using first the cent sign and then the dollar sign.
   480¢ = $4.80

2. Write each amount as cents.
   a. $2.37 237¢
   b. $1.99 199¢
   c. $0.88 88¢
   d. $5.30 530¢

3. Write each amount as dollars.
   a. 422¢ $4.22
   b. 108¢ $1.08
   c. 170¢ $1.70
   d. 98¢ $0.98

4. What is the value of 88¢ + 45¢? Circle the letter of the correct answer.
   A $0.33
   B $1.33
   C 1.33¢
   D $133

5. Write the value of a dime using the $ symbol.
   $0.10

6. Jamil has 1 half dollar, 3 quarters, 4 dimes, 8 nickels, and 7 pennies. Write the amount of money he has using the ¢ sign and the $ sign.
   212¢, $2.12

7. Suppose you have one coin of each type less than a half dollar. How much do you have? Write the amount using the ¢ sign and the $ sign.
   41¢, $0.41

8. Find a combination of 6 coins with a value of 95¢. Then write the value of the coins using the $ sign.
   3 quarters, 1 dime, 2 nickels; $0.95

Objective 19.5: Know and use the decimal notation and the dollar and cent signs for money.

Error Analysis

Exercise 1 Have students find the value of 48 dimes first. They can begin by finding the number of dimes in $1.00, and they can determine the number of equal groups worth $1.00 can be made. Then they can count on or add the value of the remaining 8 dimes.

Exercise 2 Remind students that they can think of pennies to find the number of cents. If students are confused, begin with 2c. **How many pennies are in 88 cents?** 88 Repeat for 2b and then for 2a.

Exercise 3 If students have trouble with 3d, ask them if any dollars can be made from 98¢. Then ask them to look in Exercise 2 for the amount that has no dollars and to explain how they can tell. **There is a 0 to the left of the decimal point.**

Exercise 4 If students choose either C or D, they do not understand how to write 133 cents using either the $ or ¢ sign. Have students read C and D and remind them that the decimal point is never used with the ¢ sign, but it does need to be used with the $ sign if there are cents.

Exercise 6 Students can first count the cents or add the values of the groups. Have students use one method to check the other before they write the answer in decimal notation.

Lesson 19-6: Money in Fractions and Decimals

Objective 19.6: Know and understand that fractions and decimals are two different representations of the same concept.

Teach the Lesson

Materials ☐ Teaching Aid 10 Coins and Bills

Activate Prior Knowledge
Review writing and naming various fractions and decimals.

Develop Academic Language
Remind students that in a fraction, the denominator tells how many equal parts a whole is divided into. The numerator tells how many of those equal parts are being considered. Show a labeled example.

In a decimal, the place value of the last digit tells how many equal parts the whole is divided into. The decimal 0.6 shows a 6 in the tenths place which means that the whole is divided into 10 equal parts and 6 of those are represented by 0.6.

Model the Activities

Activity 1 Cut a dollar from Teaching Aid 10 into 10 equal parts. **What coin is each of these parts of a dollar equal to?** 1 dime Put a dime on top of each strip. **One dime is what part of a dollar?** $\frac{1}{10}$ One dime represents 10 out of 100 cents. The fractions $\frac{1}{10}$ and $\frac{10}{100}$ are equivalent; they both show that a dime is one tenth of a dollar. **How do we write $\frac{10}{100}$ as a decimal?** 0.10

Activity 2 Cut a dollar from Teaching Aid 10 into fourths. **What coin is each of these parts of a dollar equal to?** 1 quarter Put a quarter on top of each $\frac{1}{4}$ dollar. **Two quarters is what part of a dollar?** $\frac{1}{2}$ **Three quarters is what part of a dollar?** $\frac{3}{4}$ **Four quarters is what part of a dollar?** $\frac{4}{4}$, or 1 dollar

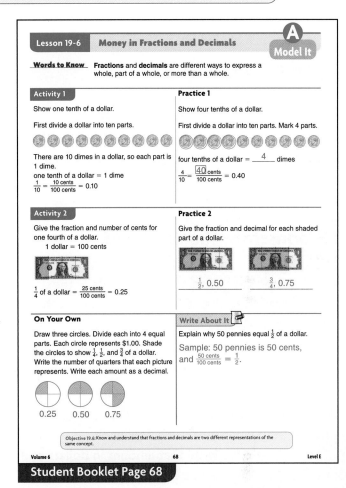

Student Booklet Page 68

Progress Monitoring
Show various numbers of nickels, and ask what fraction shows the same part of a dollar.

Error Analysis
For students having difficulty, have them use play money to model the problems.

Lesson 19-6: Money in Fractions and Decimals

Understand It

Objective 19.6: Know and understand that fractions and decimals are two different representations of the same concept.

Facilitate Student Understanding

Develop Academic Language

Review the meaning of *mixed number*. A value greater than 1 can be written as an improper fraction, which can be converted to a mixed number.

Relate the word *quarter* to its meaning, "one fourth."

Demonstrate the Examples

Example 1 You may want to remind students that a fraction such as $\frac{4}{4}$ is equal to one whole. If the numerator is more than the denominator, then the amount shown is greater than 1.

Review fractions that show more than 1, such as $\frac{5}{4}$, $\frac{6}{4}$, and so on.

Each quarter is what fraction of a dollar? $\frac{1}{4}$ To find the fraction of a dollar shown, you can count by fourths. Stop when you reach one whole. $\frac{1}{4}$, $\frac{2}{4}$, $\frac{3}{4}$, $\frac{4}{4}$ How would we write that amount in dollars and cents? $1.00 Now start at $\frac{4}{4}$ and keep counting. $\frac{5}{4}$ How many wholes are there? 1 How many fractional parts in addition to the whole? $\frac{1}{4}$ What is the amount as a mixed number? $1\frac{1}{4}$ How is one and one-fourth dollars written in dollars and cents? $1.25

Example 2 Each dime is what fraction of a dollar? $\frac{1}{10}$ Have students count by tenths to find the fraction of a dollar. $\frac{10}{10}$ How many wholes are there? 1 How many fractional parts in addition to the whole? 0 Can you write a mixed number? No. Can you write the amount in dollars and cents? Yes: $1.00

On Your Own

Students may write 1 nickel as $\frac{1}{20}$ or $\frac{5}{100}$, and 75 pennies as $\frac{75}{100}$ or $\frac{3}{4}$.

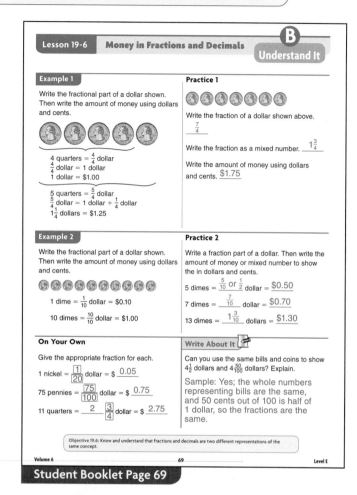

Student Booklet Page 69

Progress Monitoring

Have students come to the board in pairs. Have the first student draw a set of coins, and have the second student give a fraction that shows what part of a dollar is shown. Encourage them to show amounts more than $1.00.

Error Analysis

Remind students that when asked to tell what part of a dollar is shown, that part may be more than 1. Model several examples with coins for students who answer incorrectly. Be sure that they are using the dollar sign and decimal point correctly.

Lesson 19-6 Money in Fractions and Decimals

Objective 19.6: Know and understand that fractions and decimals are two different representations of the same concept.

Observe Student Progress

Computer Tutorial

Some students may benefit from completing a computer tutorial before they attempt the Try It page. A list of the tutorials for each lesson can be found beginning on page x in the front of this book.

Error Analysis

Exercise 1 For students having difficulty writing the correct fraction, have them first model the amount of money shown using quarters or dimes, as appropriate.

Exercise 2 If students answer incorrectly, have them model the fractional part using quarters or dimes. For example, for Exercise 2b have them show 10 dimes to make 1 dollar, move 8 dimes over to one side, and then give the value of the 8 dimes.

Exercise 4 To begin, students can count quarters to make $2.25. Remind students that the amount of money is more than $1.00, so the numerator of the fraction will be greater than the denominator.

Exercise 5 Have students use quarters and skip count until they reach $1.75.

Exercise 6 Guide students to count enough nickels to equal one dollar. Then separate 2 nickels and express that fraction of the total. It can be expressed as $\frac{2 \text{ nickels}}{20 \text{ nickels}}$, $\frac{10 \text{ cents}}{100 \text{ cents}}$, or $\frac{1}{10}$ dollar.

Exercise 7 Suggest students first represent the amount of money for each item with drawings of coins or in cents, and then write the total with a dollar sign and decimal point.

Exercise 8 If students have difficulty deciding which coin to use, ask which coin shows $\frac{1}{10}$ dollar.
Dime

Student Booklet Page 70

Lesson 19-6 Money in Fractions and Decimals — Try It

1. Give the fractional part of a dollar for each amount of money.
   a. $0.25 — $\frac{1}{4}$
   b. $0.50 — $\frac{1}{2}$
   c. $1.25 — $\frac{5}{4}$ or $1\frac{1}{4}$
   d. $0.90 — $\frac{9}{10}$

2. Give the number of cents for each fractional part of a dollar.
   a. $\frac{1}{2}$ — 50 cents
   b. $\frac{8}{10}$ — 80 cents
   c. $\frac{3}{4}$ — 75 cents
   d. $\frac{2}{10}$ — 20 cents

3. Draw a picture to show $\frac{3}{2}$ of a dollar. Then write the amount of money using a dollar sign and a decimal point.
   Sample: [coins]
   $1.50

4. Which fraction of a dollar shows $2.25? Circle the letter of the correct answer.
   (A) $\frac{9}{4}$ B $\frac{4}{9}$
   C $\frac{1}{4}$ D $\frac{2}{25}$

5. The price of a chicken taco is $1.75. If only quarters can be used to pay for the taco, how many quarters are needed?
   7 quarters

6. What fractional part of a dollar is two nickels? Explain.
   Sample: $\frac{2}{20}$, since there are 20 nickels in $1.00.

7. Marissa bought 3 items for her art project. They cost $\frac{5}{4}$ dollar, $\frac{7}{10}$ dollar, and $\frac{1}{2}$ dollar. Express the total amount of money she spent using a dollar sign and a decimal point.
   $2.45

8. Suppose you want to show $\frac{32}{10}$ of a dollar using all the same coins. What coin would you use? How many coins would you show? How much money is that?
   Sample: dime, 32, $3.20

Lesson 19-7: Unit Costs

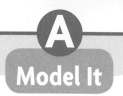

Objective 19.7: Determine the unit cost when given the total cost and number of units.

Teach the Lesson

Materials ☐ Teaching Aid 10 Coins and Bills

Activate Prior Knowledge

Have students review basic division facts and division of decimals by whole numbers, such as 31.2 ÷ 24. **1.3**

Ask students questions such as how much 6 identical items would cost if each one costs $2.40. **$14.40**

Develop Academic Language

Clarify for students that when talking about purchasing several equally priced items, the *total cost* is the amount of money paid for all the items. The total cost can be found by multiplying the cost of a single item by the number of items purchased. The *unit cost*, or cost per item, is found by dividing the total cost by the number of items.

Model the Activities

Activity 1 We need to divide these coins into 5 groups with each group showing the same amount of money. How many half dollars will be in each group? **1** How many dimes will be in each group? **2** How many nickels will be in each group? **1** What is the value of each group of coins? **75¢**

Activity 2 How can you divide 12 dollars between 4 boxes of CD labels? **Divide 12 by 4.** What is the cost of one box? **$3** Have students model this by drawing 4 boxes on a piece of paper and dividing 12 play dollar bills among them.

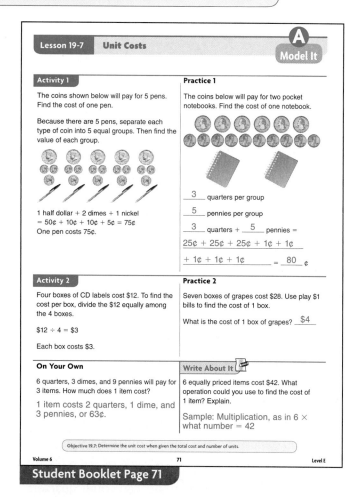

Student Booklet Page 71

Progress Monitoring

If 15 quarters can buy 3 items that cost the same amount, how much does each item cost? **$1.25**

Error Analysis

Check to ensure that students express their answers using appropriate money symbols. For answers that are less than $1, accept either dollar notation ($0.85) or cent notation (85¢).

Volume 6   342   Level E

Lesson 19-7 Unit Costs

Objective 19.7: Determine the unit cost when given the total cost and number of units.

Facilitate Student Understanding

Develop Academic Language

Help students understand the concept of unit cost. Often it is the cost of a single item, as in "Find the unit cost if 4 items cost $32." **$8** However, it can also refer to a unit of measure, as in "If a pound of ground beef costs $4.48, what is the cost per ounce?" **$0.28**

Demonstrate the Examples

Example 1 Point out that when finding unit cost, students are looking to find the cost *per* item. Thus, the total cost is divided by the total number of items purchased. Point out that this is the opposite of finding the total cost.

ENGLISH LEARNERS Remind students that a dozen refers to 12 items; a half dozen refers to 6.

Example 2 Allow students to draw or model the problem. **What is the total cost? $5.10 How many items are there? 6 How do we find the unit cost? Divide $5.10 by 6.** Work through the long division with students. **What is the unit cost? $0.85**

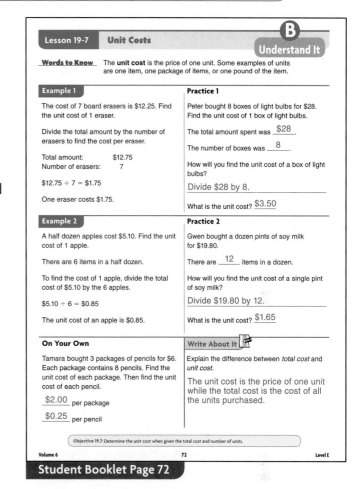

Student Booklet Page 72

Progress Monitoring

What is the unit cost for a dozen items that sell together for $37.80? $3.15

Error Analysis

Remind students to think of finding unit cost as separating a quantity of money in order to put an equal amount with each item. Separation into equal groups calls for division.

Lesson 19-7 Unit Costs

Objective 19.7: Determine the unit cost when given the total cost and number of units.

Observe Student Progress

Computer Tutorial

Some students may benefit from completing a computer tutorial before they attempt the Try It page. A list of the tutorials for each lesson can be found beginning on page x in the front of this book.

Error Analysis

Exercise 1 Suggest that students who are having difficulty use play coins and "deal them out" into 4 groups.

Exercise 2 Check to ensure that students correctly divide by the total number of items in each exercise.

Exercise 3 Point out to students that they need to carefully read each exercise and not assume that the first number given is the divisor. Remind them to always divide total cost by total number of items.

Exercise 5 Remind students that each item is priced as a whole number of dollars. They need to determine which numbers will evenly divide $24.

Exercise 6 Encourage students to look at Amy's answer and think about the reasonableness of it rather than to immediately divide $87 by 3 to find the exact unit cost.

ENGLISH LEARNERS You may need to explain the meaning of *reasonable*.

Student Booklet Page 73

Lesson 19-7 Unit Costs — Try It

1. The total amount of money shown below will buy 4 items that cost the same amount. Circle groups of coins to show the unit cost of each item.

   What is the unit cost? **68¢**

2. Answer each question.
   a. Tanya paid $44.10 for 18 notebooks. What was the cost of each notebook? **$2.45**
   b. 345 counters are distributed equally to 15 students. How many counters did each student receive? **23 counters**
   c. Daryl bought a dozen bagels for $15.60. What was the cost of each bagel? **$1.30**

3. Find the unit cost for each purchase.
   a. a dozen ounces for $4.20 **$0.35**
   b. 65 items for 91 dollars **$1.40**
   c. a half-dozen items for $29.70 **$4.95**
   d. $510 for 30 crates **$17**

4. What is the unit cost if 15 items total $37.50? Circle the letter of the correct answer.
   Ⓐ $2.50   B $562.50
   C $0.40   D $37.50

5. Glen had $24 to spend at the school garage sale. The school did not want to deal with coins, so all prices were in dollars. Could Glen buy 4 of the same thing? 5? 6? 7? Explain.
   Yes; no; yes; no; 4 and 6 divide evenly into $24. 5 and 7 do not.

6. Phil told Amy that he purchased 3 identical items for $87. Amy calculated the unit cost of each item to be $45. Use mental math to decide whether her answer is reasonable. Explain.
   No; $45 is more than half of the total Phil spent. 3 items at $45 each would be much more than the total $87 he spent.

Volume 6: Understand Decimals

Topic 19: Decimals and Money

Topic Summary

Objective: Review the relationship between decimals and money.

Have students complete the student summary page.

You may want to have students work in groups of four with each student analyzing a different choice.

Ask students to share their ideas about each answer choice. Be sure they confirm the correct answer for each problem at the end of the discussion.

Answer Evaluation

1. **A** Students do not understand decimal place value.
   B Students wrote the tenths digit in the numerator and the hundredths digit in the denominator.
   C This choice is correct.
   D Students do not understand decimal place value.

2. **A** Students ordered the decimals from greatest to least.
   B Students did not order the decimals.
   C Students thought 0.33 was smaller than 0.3.
   D This is the correct answer.

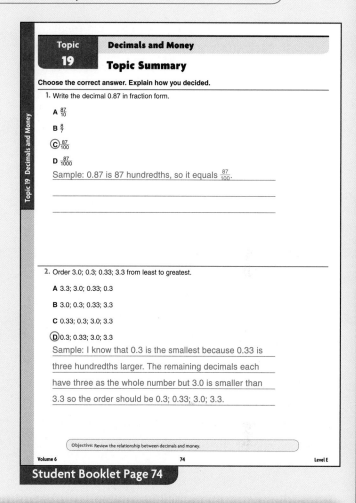

Student Booklet Page 74

 Error Analysis

Exercise 1 It is common for students to write the incorrect denominator when changing a decimal to a fraction. Have students say the decimal aloud so that they hear *hundredths* and write 100 as the denominator.

Exercise 2 Suggest that students use a place-value chart to write each decimal and then compare.

Progress Monitoring

When all assignments for this topic have been completed, assign the corresponding Progress Monitoring page for this topic (Assessment Resources Book, page 19). Be sure students complete the Progress Monitoring page before you administer the final assessment for this volume.

Volume 6
Topic 19
Understand Decimals
Decimals and Money
Mixed Review

Objective: Maintain concepts and skills.

Have students complete the Mixed Review page. Work with each student individually to review results. Identify strengths and weaknesses and correct any misunderstandings.

Error Analysis

Exercise 1 Remind students to round up if the digit to the right of the place to be rounded to is 5 or more. After the indicated place for rounding, all digits to the right should be zeros.

Exercise 2 For students having difficulty, provide Teaching Aid 13 Fraction Models.

Exercise 3 Some students may need play coins to model the amounts of money.

Exercise 4 Some students may need a review of the rules of operations with fractions. Remind students that they need common denominators when adding and subtracting.

Exercise 6 Review the rules for estimation and the meaning of the word *reasonable*.

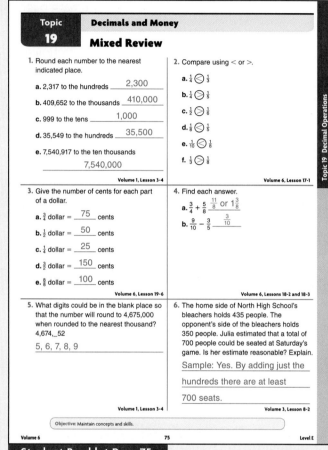

Student Booklet Page 75

Volume 6: Understand Decimals
Topic 20: Decimal Operations and Comparisons
Topic Introduction

Lesson 20-1 extends students' knowledge of adding and subtracting whole numbers to performing those operations with decimals. Lesson 20-2 teaches students to find basic fraction and decimal equivalents; for example, $\frac{50}{100}$ is equivalent to 0.50, or 0.25 is equal to $\frac{1}{4}$. Finally, in Lesson 20-3, students learn to order and compare decimals, fractions, and mixed numbers using number lines.

Lesson	Objective	Student Pages	Teacher Pages	Tutorials
Topic 20 Introduction	20.1 Find basic fraction and decimal equivalents. 20.2 Identify and represent on a number line fractions, decimals, and mixed numbers. 20.3 Compare and order fractions, decimals, and mixed numbers and place them on a number line.	76	348	
20-1 Add and Subtract Decimals	20.1 Add and subtract simple decimals.	77–79	349–351	20a, 20b, 20c
20-2 Decimal and Fractions Equivalents	20.2 Find basic fraction and decimal equivalents.	80–82	352–354	20d
20-3 Use a Number Line	20.3 Identify and represent on a number line fractions, decimals, and mixed numbers.	83–85	355–357	20e, 20f
Topic 20 Summary	Review comparing decimals and fractions.	86	358	
Topic 20 Mixed Review	Maintain concepts and skills.	87	359	

Computer Tutorial

Some students may benefit from completing the computer tutorial before they attempt the Try It page of each lesson. If you are using the electronic components of *Pinpoint Math,* you will find a complete listing of tutorial codes and titles when you access them either online or via CD-ROM.

Volume 6: Understand Decimals

Topic 20: Decimal Operations and Comparisons

Topic Introduction

Objectives: 20.1 Add and subtract simple decimals. **20.2** Find basic fraction and decimal equivalents. **20.3** Identify and represent on a number line fractions, decimals, and mixed numbers.

Materials ☐ Teaching Aid 9 Decimal Models

Distribute Teaching Aid 9. Have students shade 45 squares on a hundredths model. Review how to write a decimal based on the model, and then how to write a fraction based on the model. Discuss why the decimal ends in hundredths and the denominator of the fraction is 100. Repeat using a tenths model and other hundredths models.

Informal Assessment

1. **How many sections is the circle divided into?** 4 **What does that tell you?** The denominator of the fraction. **How many sections are shaded?** 1 **What does that tell you?** The numerator **How can you find an equivalent decimal for $\frac{1}{4}$?** Sample: A quarter of the circle is the same as 0.25.

2. **How do you change a fraction to a decimal?** Sample: Multiply the numerator and denominator by a number that will give me a denominator of 10. **Which fractions will be easier to change to a decimal and why?** $\frac{4}{10}$ and $\frac{8}{100}$ because their denominators are multiples of ten.

3. Have students set up the problems vertically. **You can use the same algorithm you use to add and subtract whole numbers to add and subtract decimals. How do you line up the decimals when you set up the problem vertically?** Line up the decimal points. **Why is this important?** So you add the correct place values.

4. **How many sections are on the number line?** 10 **What fractional part does each section represent?** One tenth **How do you write this as a decimal?** 0.1 If you rewrite each fraction as a number of tenths, you can find them on this number line.

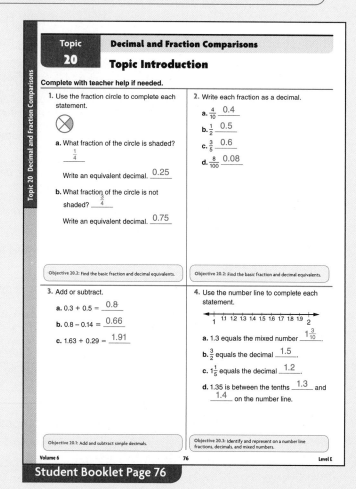

Student Booklet Page 76

Another Way For Exercise 2, students could use decimal models to shade each fraction. This would allow students to count the shaded squares to find the equivalent decimal.

Lesson 20-1 Add and Subtract Decimals

Objective 20.1: Add and subtract simple decimals.

Teach the Lesson

Materials ☐ Teaching Aid 9 Decimal Models

Activate Prior Knowledge

Ask a volunteer to write two money amounts on the board, for example, $3.47 and $16.25. **What does the decimal point do?** It separates the dollars from the cents. **How should I write the problem to add the two amounts?** Line up the decimal points.

Develop Academic Language

Review the place value names for decimals. Point out the similarity between the names *tenths* and *tens*, *hundredths* and *hundreds*. **What is the difference between the words *tens* and *tenths*?** Only the *-th*

ENGLISH LEARNERS The *-th* sound is difficult for some students to distinguish if that sound does not occur in their home languages. You should articulate carefully and point out the difference.

Model the Activities

Activity 1 **What is the first addend?** 60 hundredths or 6 tenths **Can we shade 60 hundredths or 6 tenths of the hundredths grid without counting every square?** Yes; shade 6 columns of 10 squares. **How much will we shade to add 0.65?** $6\frac{1}{2}$ more columns **What is the value of each square?** 0.01 **How many hundredths are shaded in all?** 125 The sum is 125 hundredths. Convert to a mixed number or break the number apart as 100 hundredths and 25 hundredths in order to write 1.25.

Activity 2 Since the decimals are aligned, you can ignore them as you subtract. Subtract as if there were no decimal point. **What is the difference of 4 tenths and 1 tenth?** 3 tenths Guide students to complete the solution.

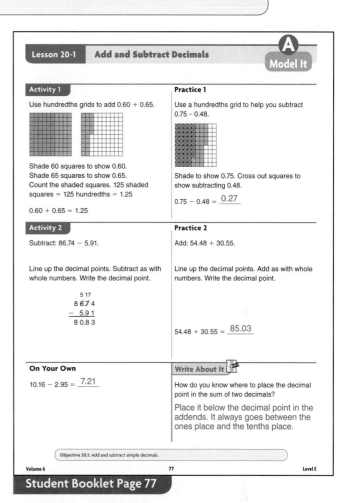

Student Booklet Page 77

Progress Monitoring

Subtract 19.44 from 52.11. 32.67

Error Analysis

Review the places' names. Point out that just as tens should be subtracted from tens and ones from ones, tenths should be subtracted from tenths, and so on. If the places are aligned in the problem, the values and decimal point in the sum or difference will also be aligned. As you guide students to find a solution, use the name of each place: **What is the difference 4 tenths − 1 tenth?** 3 tenths

Lesson 20-1 Add and Subtract Decimals

Objective 20.1: Add and subtract simple decimals.

Facilitate Student Understanding

Activate Prior Knowledge

Ask students how to round 3,561 to the nearest hundred and then to the nearest ten. **To round to the nearest hundred, look at the tens place. The tens place has 6, so round up to 3,600. To round to the nearest ten, look at the ones place. The ones place has 1, so round down to 3,560.**

Develop Academic Language

Discuss the term *estimate* with students. Point out that an estimate is not an exact answer. Remind students that some words, such as *about*, signal that we are reporting an estimate.

Demonstrate the Examples

Example 1 Review rounding to the nearest whole number and review the term *addends*. Help students round each addend to the nearest whole number. **23 and 20** Now find the exact sum. Guide students to add. **Since you rounded to the nearest 1, your answer should be no more than 1 away from your estimate. How close is your answer to your estimate?** The difference is less than 1. **That suggests that your answer is reasonable.**

Example 2 Guide students to estimate. **One number has tenths and hundredths. The other only has tenths.** At the board, align the last digits rather than the decimal points. **What is wrong with this problem?** The decimals and places are not aligned. Have students use a terminal zero to include the same number of decimal places in each number. **What tells us that 14.445 would be the wrong answer?** The difference of the estimate and the answer is about 100, which is too great.

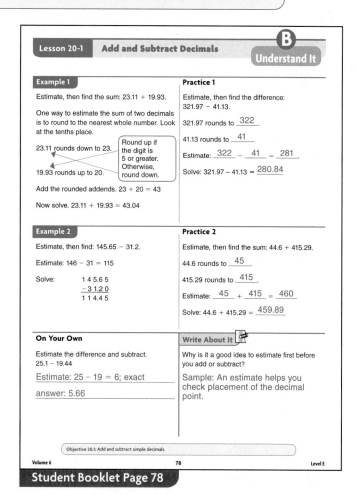

Student Booklet Page 78

Progress Monitoring

Ask students to explain how to estimate, solve, and check 25.6 + 89.06. **Round 25.6 to 26 and round 89.06 to 89. Then find the sum of 26 and 89, which is 115. To add, align the decimals; the sum is 114.66. That is within 1 of 115, so the answer is reasonable.**

Error Analysis

Talk about the value of the digit in each place to help students understand the need to align the decimal points. Discuss terminal zeros on decimal numbers and initial zeros on whole numbers; neither changes the number's value.

Lesson 20-1 Add and Subtract Decimals

Objective 20.1: Add and subtract simple decimals.

Observe Student Progress

Computer Tutorial

Some students may benefit from completing a computer tutorial before they attempt the Try It page. A list of the tutorials for each lesson can be found beginning on page x in the front of this book.

★ Error Analysis

Exercise 2a Ask students how many hundredths are in 69.3, and remind them that they can use a zero to represent the hundredths.

Exercise 4 When students set up the problem vertically, make sure that they align the place values correctly.

Exercise 5 Model Ellie's solution process, and ask students to stop you when you make an error. Add correctly, but count decimal digits in order to place the decimal. Help students explain why the decimal rule for addition and subtraction works; when the decimals are aligned, the place values are aligned.

Exercise 6 Read through the problem with students. Help them understand that together the American cheese and cheddar cheese weigh 5.23 pounds.

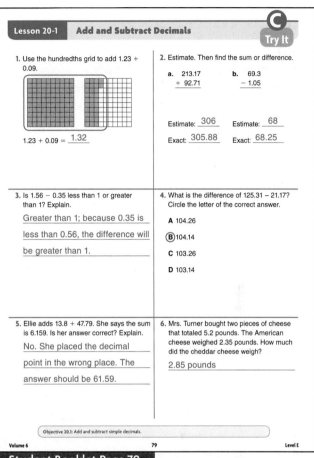

Student Booklet Page 79

Lesson 20-2: Decimal and Fraction Equivalents

Objective 20.2: Find basic fraction and decimal equivalents.

Teach the Lesson

Activate Prior Knowledge

Review reading fractions and decimals through the hundredths place. Have students practice writing equivalent fractions with denominators of 10 or 100. **Four fifths equals how many hundredths?** $\frac{80}{100}$

Develop Academic Language

ENGLISH LEARNERS Write the following on the board: If *halves* are shown, then each part is 1 half or $\frac{1}{2}$. If *fourths* are shown, then each part is 1 fourth or $\frac{1}{4}$.

At the board, have one student draw a regular figure. Have another student divide the figure into halves. Repeat with fourths.

Model the Activities

Activity 1 **What are the first few decimal place values? Tenths, hundredths...** Fractions like $\frac{1}{10}$ or $\frac{2}{100}$ are easy to write as decimals. To write any fraction as a decimal, first we find an equivalent fraction with 10 or 100 as the denominator. **What fraction is shown on the first square?** $\frac{1}{4}$ **What fraction is shown on the next square?** $\frac{25}{100}$ **How are those fractions alike? They name the same amount. How are they different? One has 100 as the denominator. How do we write hundredths in decimal form?** Write the numerator of $\frac{25}{100}$ so it ends in the hundredths place: 0.25. Have students read the equation.

Activity 2 **What do the decimals along the bottom of the number line show? Tenths from 0 to 1** The number line was divided in 10 sections to show tenths. You have to divide those 10 sections into just 5 equal parts to show fifths. **How many sections will be in each fifth? 2 How many parts was it divided in to show fifths along the top? 5 How many tenths are equal to each fifth? Two tenths** **Two tenths equals one fifth.** Have students count by $\frac{2}{10}$s until they reach 0.80, or $\frac{4}{5}$.

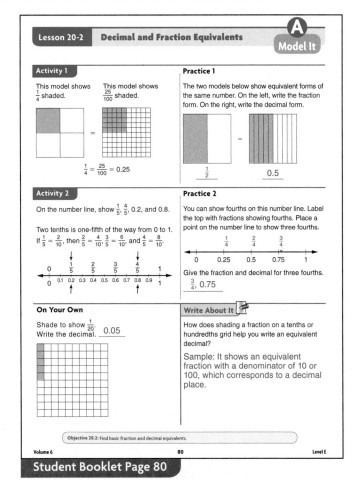

Student Booklet Page 80

Progress Monitoring

How could you use a grid to turn the fraction $\frac{2}{4}$ into an equivalent decimal? Color $\frac{2}{4}$ of a hundredths grid; count the shaded hundredths and write the number to end in the hundredths place. The decimal is 0.50.

Error Analysis

To help students with the number lines, show them how to try to make equal "jumps" as they skip count by the desired fractional unit.

Lesson 20-2 Decimal and Fraction Equivalents

Objective 20.2: Find basic fraction and decimal equivalents.

Facilitate Student Understanding

Develop Academic Language

Review *numerator* and *denominator*. On a grid, the shading represents the numerator and the division of the grid represents the denominator.

Demonstrate the Examples

Example 1 We need to write $\frac{1}{50}$ as a fraction with a denominator of 10, 100, or 1,000. **Why is 100 a good choice?** 100 is the least multiple of 50 that names a decimal place. **How can you change a denominator of 50 to a denominator of 100? Multiply by 2. Do the same thing to the numerator. What is the equivalent fraction in hundredths?** $\frac{2}{100}$ Have students write the decimal. If students are comfortable with this method, you can start teaching them to divide the numerator by the denominator. In that method, emphasize that the fraction bar indicates division.

Example 2 **Read the decimal.** You may want to have students first make a place-value chart for the number. 5 hundredths **How do you find the denominator?** Use hundredths, the name of the last place. **Use the digits you see as the numerator. Remove the leading zeros from the numerator. What number is the numerator?** 5

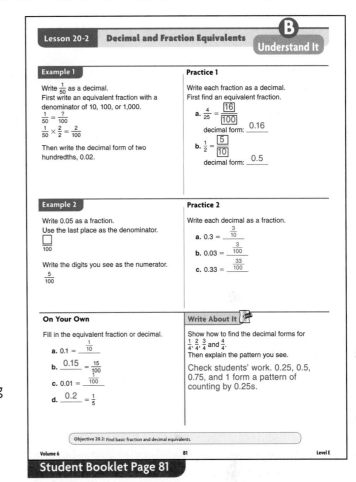

Student Booklet Page 81

Progress Monitoring

Write a decimal equivalent to $\frac{3}{20}$. 0.15 Write 0.30 on the board. **Write a fraction equivalent to this number.** $\frac{30}{100}$ or $\frac{3}{10}$

Error Analysis

Provide tenths and hundredths grids that students can shade to check their work if necessary.

Lesson 20-2: Decimal and Fraction Equivalents

Objective 20.2: Find basic fraction and decimal equivalents.

Observe Student Progress

Computer Tutorial

Some students may benefit from completing a computer tutorial before they attempt the Try It page. A list of the tutorials for each lesson can be found beginning on page x in the front of this book.

Error Analysis

Exercise 1 Help students interpret the fraction as "3 out of every 4" to be shaded, or help them divide the grid to show fourths.

Exercise 2 Help students begin with the decimal form of $\frac{1}{2}$. Then you can provide a few decimals for students to choose from. Last, have students write each decimal as a fraction.

Exercise 3 Have students fill in the fraction and decimal below each picture and then check to see that the equations are true, or have students use the picture to fill in the fractions, and then write each fraction as a decimal.

Exercise 4 Clarify that the question means to find two more numbers that represent the model.

Exercise 5 Students can shade grids, use a number line, or write each fraction as a number of tenths to solve.

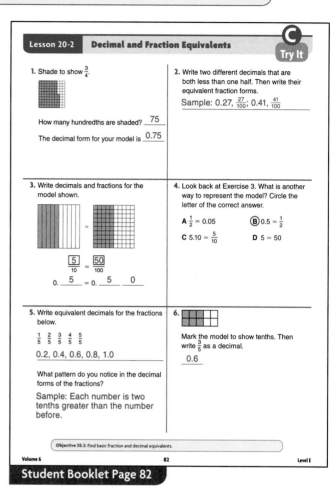

Student Booklet Page 82

Lesson 20-3 Use a Number Line

Objective 20.3: Identify and represent on a number line fractions, decimals, and mixed numbers.

Teach the Lesson

Materials ☐ Teaching Aid 1 Number Lines

Activate Prior Knowledge

Distribute Teaching Aid 1. **Number 3 number lines starting at 0. How did you do the numbering?** If no one suggests numbering by fractional units such as $\frac{1}{2}$, make such a number line.

Develop Academic Language

A number line can show *whole numbers, fractions, mixed numbers,* and *decimals.* Write these terms on the board. Have volunteers come to the board and write one example of each. **Sample: whole number: 2, fraction: $\frac{1}{2}$, mixed number: $2\frac{1}{2}$, decimal: 1.5**

Model the Activities

Activity 1 How many spaces are between 0 and 1 on this number line? **Eight** What do the marks between 0 and 1 represent? **Eighths** How can you locate $\frac{5}{8}$ on the number line? **Start from 0, count 5 spaces to the right.** How can you locate $1\frac{7}{8}$ on the number line? **Start from 1, count 7 spaces to the right.** Point out that $1\frac{7}{8}$ is greater than $\frac{5}{8}$ because $1\frac{7}{8}$ is located to the right of $\frac{5}{8}$.

Activity 2 Help students skip count from 0 to ten hundredths on the number line. What does each tick mark represent? **Two hundredths; 0.02** Where would you start to locate 0.23? **Start at 0.20.** What numbers between 0.20 and 0.30 are marked? **0.22, 0.24, 0.26, 0.28** Between what 2 numbers is 0.23 located? **0.22 and 0.24**

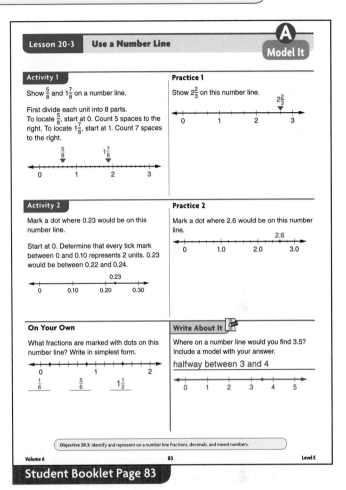

Student Booklet Page 83

⭐ Progress Monitoring

Draw a number line from 0 to 4 on the board. Mark the halves but do not label them. Write a few fractions, mixed numbers, and decimals on the board and have students locate those numbers on the number line.

Error Analysis

Some students may need extra help learning that number lines can have different scales. Have them use Teaching Aid 1 Number Lines to practice making number lines that start at 0 and show counting by 2s, 10s, 100s, halves, thirds, and decimal tenths.

Lesson 20-3 Use a Number Line

Objective 20.3: Identify and represent on a number line fractions, decimals, and mixed numbers.

Facilitate Student Understanding

Develop Academic Language

Discuss the term *benchmark fractions* with students. Explain that learning these common fractions will help them when comparing and ordering fractions.

Demonstrate the Examples

Example 1 How can you locate these benchmark fractions on the number line? **Number by fourths starting at 0.** How can you locate the decimal numbers on the number line? **Number by tenths starting at 0.** Which decimal number will be on the same tick mark as $\frac{1}{2}$? **0.5** The benchmark fractions tell you whether you have placed the decimals in the right places. Between what 2 numbers should $\frac{1}{4}$ be located? **0.2 and 0.3** $\frac{1}{2}$? **0.4 and 0.6** $\frac{3}{4}$? **0.7 and 0.8**

Example 2 If you mark eighths on the number line, then you can also show halves and fourths, since 2 and 4 are factors of 8. Have students draw 7 tick marks between each pair of whole numbers. Which marks represent the halves? **Every fourth mark** Have students label the halves. Which marks represent the fourths? You can help students count by fourths from 0 to 1 as they make jumps along the number line. Help them see that they must pass over one tick mark in each jump in order to make 4 equal jumps as they count.

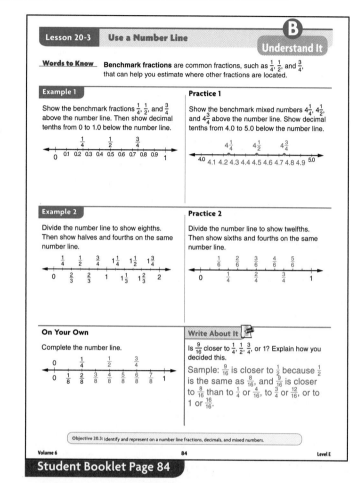

Student Booklet Page 84

Progress Monitoring

Provide students with Teaching Aid 1. **Make a decimal number line. Start at 0 and number by tenths. Now number the same line by fifths starting at zero.** Check students' work.

Error Analysis

Watch for students who are not using consistent intervals when they make number lines.

For Example 2 and Practice 2, students may find it easier to first divide the number line to show the fractional parts with the least denominator, and then to divide those sections.

Volume 6   356   Level E

Lesson 20-3 — Use a Number Line

Objective 20.3: Identify and represent on a number line fractions, decimals, and mixed numbers.

Observe Student Progress

Computer Tutorial

Some students may benefit from completing a computer tutorial before they attempt the Try It page. A list of the tutorials for each lesson can be found beginning on page x in the front of this book.

 Error Analysis

Exercise 1 To get students started, have them write 0 and 1 on the number line and then find the location for $\frac{1}{2}$. Check by having them count spaces between 0 and $\frac{1}{2}$, and between $\frac{1}{2}$ and 1. There should be 6 spaces in each interval. Students may find it easiest to make three separate number lines for twelfths and thirds, twelfths and halves, and twelfths and fourths.

Exercise 2 You can help students convert the mixed numbers to improper fractions, and then count by fourths or halves to locate them. Again, students may find it easiest to make three separate number lines.

Exercise 3 Remind students that 1 equals both $\frac{4}{4}$ and $\frac{2}{2}$. Help students count by halves and fourths.

Exercise 6 Point out that the number will need a place for another digit, and that adding 0 at the end of 0.27 and 0.28 does not change the value of the numbers. Show an example on the board: $0.5 = \frac{5}{10} = \frac{1}{2}$, $0.50 = \frac{50}{100} = \frac{1}{2}$.

Exercise 7 Remind students that a number line is helpful to analyze the numbers. Point out that the fastest runner will have the time that is the lowest number.

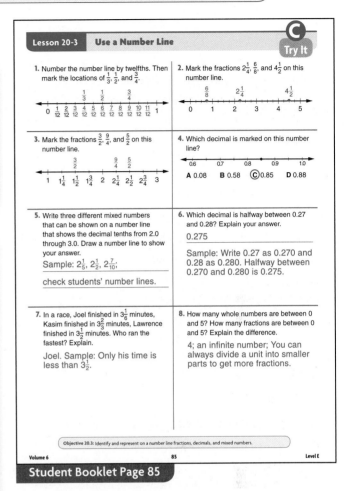

Student Booklet Page 85

Exercise 8 To illustrate the idea of an infinite number of fractions in any interval, draw a large 0–1 number line on the board. Mark $\frac{1}{2}$. Then show a tick mark halfway between 0 and $\frac{1}{2}$ and mark $\frac{1}{4}$. Repeat to get $\frac{1}{8}, \frac{1}{16}, \frac{1}{32}, \frac{1}{64}$. Point out that you can continue forever, getting smaller and smaller fractions.

Volume 6 — Understand Decimals
Topic 20: Decimal Operations and Comparisons

Topic Summary

Objective: Review comparing decimals and fractions.

Have students complete the student summary page. You may want to have students work in groups of four with each student analyzing a different choice. Ask students to share their ideas about each answer choice. Be sure they confirm the correct answer for each problem at the end of the discussion.

Answer Evaluation

1. **A** Students wrote the fraction for the part that is not shaded, and also wrote the decimal incorrectly.

   B Students wrote the fraction for the part that is not shaded, and also wrote the decimal equivalent to $\frac{3}{5}$, not $\frac{2}{5}$.

   C Students used the digits in the numerator and denominator to write the decimal.

   D This choice is correct.

2. **A** Students were unable to locate 0.93.

   B Students were unable to locate 0.723.

   C Students were unable to locate 0.68.

   D This choice is correct.

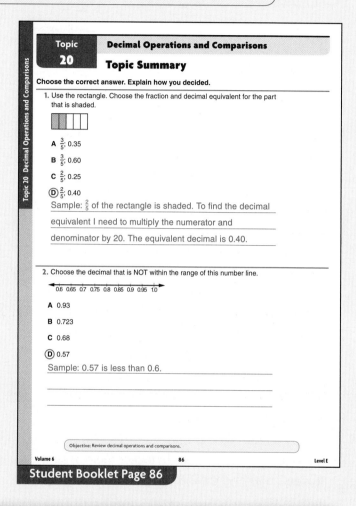

Student Booklet Page 86

 Error Analysis

Exercise 1 Have students count the total number of sections in the rectangle and then count the shaded sections. To convert to a decimal, remind them that they can treat the fraction bar as a division sign or multiply the numerator and denominator by a number that will give them a denominator of 10.

Exercise 2 Have students add zeros to 0.6, 0.7, 0.8, 0.9, and 1.0. The decimals will now look like the multiples of 5 from 60 through 100. Ask questions such as, **If these were whole numbers, where would you find 68?**

Progress Monitoring

When all assignments for this topic have been completed, assign the corresponding Progress Monitoring page for this topic (Assessment Resources Book, page 20). Be sure students complete the Progress Monitoring page before you administer the final assessment for this volume.

Volume 6, Topic 20: Understand Decimals

Decimal Operations and Comparisons

Mixed Review

Objective: Maintain concepts and skills.

Have students complete the Mixed Review page. Work with each student individually to review results. Identify strengths and weaknesses and correct any misunderstandings.

⭐ Error Analysis

Exercise 1 Have students underline the digit in the tens place and then circle the digit to the right in the ones place. If this digit is 5 or greater, students round up. Otherwise, they round down. Remind students that when they are rounding the digit 9 up to 10, as in 1e, they must write the zero part of the 10 where the 9 was and add the 1 part of the 10 to the digit one place to the left.

Exercise 2 Remind students that hundredths have two decimal places. For example, two hundredths equals 0.02, not 0.2.

Exercise 3 Most students will need to copy these problems in vertical form. Check that they are not making errors when copying the numbers.

Exercise 4 Emphasize that students write just one digit at each step. This may help to prevent students writing the product 27 × 8 as 1,656.

Exercise 5 Check whether students are writing the first digit of the quotient above the correct digit of the dividend.

Exercise 6 Encourage students to sketch number lines or use other models to help them make these comparisons.

Student Booklet Page 87

Topic 20 — Decimal and Fraction Comparisons — Mixed Review

1. Round to the nearest ten.
- a. 363 ___360___
- b. 82 ___80___
- c. 1,049 ___1,050___
- d. 619 ___620___
- e. 796 ___800___
- f. 3,508 ___3,510___

Volume 1, Lesson 3-4

2. Write the decimal for each word name.
- a. thirty-five hundredths ___0.35___
- b. eight tenths ___0.8___
- c. four and two hundredths ___4.02___

Volume 6, Lesson 19-2

3. Add or subtract.
- a. 69 − 32 = ___37___
- b. 14 + 56 = ___70___
- c. 182 + 76 = ___258___
- d. 83 − 18 = ___65___
- e. 502 − 172 = ___330___
- f. 648 + 255 = ___903___

Volume 3, Lesson 8-3

4. Multiply.
- a. 3 × 43 = ___129___
- b. 4 × 78 = ___312___
- c. 27 × 8 = ___216___
- d. 81 × 7 = ___567___
- e. 206 × 9 = ___1,854___
- f. 3 × 585 = ___1,755___

Volume 4, Lesson 9-3

5. Write the quotient and remainder.
- a. 76 ÷ 5 = ___15 R1___
- b. 328 ÷ 8 = ___41___
- c. 95 ÷ 7 = ___13 R4___
- d. 4,026 ÷ 3 = ___1,342___

Volume 4, Lesson 10-3

6. Compare. Write >, <, or =.
- a. $\frac{2}{5}$ ◯ $\frac{3}{8}$
- b. 2.5 ◯ 2.55
- c. 0.7 ◯ 0.35
- d. $\frac{50}{100}$ ◯ $\frac{5}{10}$

Volume 6, Lessons 17-1 and 19-3

7. Use estimation to explain why the product of 6 and 47 cannot be 2,442.
Sample: 47 is close to 50, and 6 × 50 is 300. The product must be less than 300.

Volume 4, Lesson 9-2

8. Explain how to change $\frac{3}{20}$ to a decimal.
Sample: Multiply 3 and 20 by 5 to get $\frac{15}{100}$. The decimal is 0.15.

Volume 6, Lesson 20-2

Objective: Maintain concepts and skills.

Volume 6 — 87 — Level E

Teaching Aid Masters

Teaching Aids

1	Number Lines	361
2	Addition Facts Table	362
3	$\frac{1}{4}$-in. Grid Paper	363
4	Centimeter Grid Paper	364
5	Multiplication Facts Table	365
6	Place-Value Charts	366
7	Place-Value Charts	367
8	Place-Value Charts	368
9	Decimal Models	369
10	Coins and Bills	370
11	Fraction Number Lines	371
12	Fraction Strips	372
13	Fraction Models	373
14	Centimeter and Inch Rulers	374
15	First-Quadrant Coordinate Grid	375
16	Four-Quadrant Coordinate Grid	376
17	Inch Grid Paper	377
18	Angles and Triangles	378
19	Area of a Circle	379
20	Complementary and Supplementary Angles	380
21	Dot Paper	381
22	Congruent Shapes	382

Name   Date

Teaching Aid 1 Number Lines

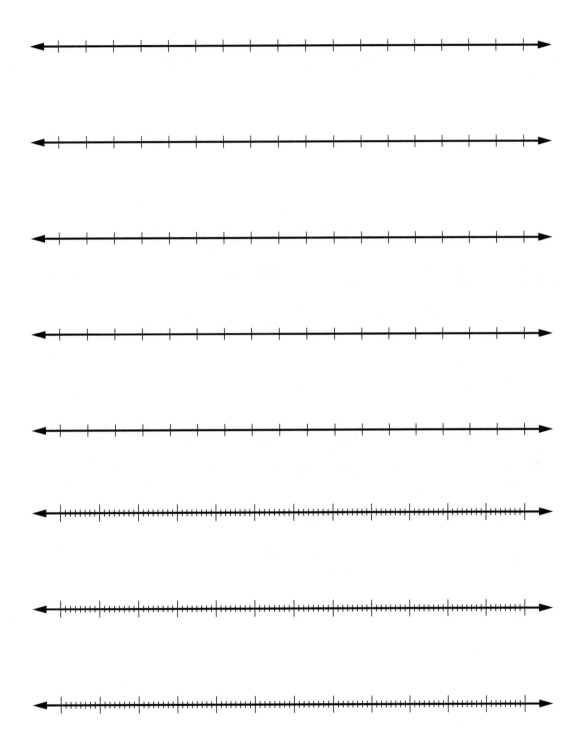

Name   **Date**

Teaching Aid 2 Addition Facts Table

Addition Facts Table

+	0	1	2	3	4	5	6	7	8	9	10
0	0	1	2	3	4	5	6	7	8	9	10
1	1	2	3	4	5	6	7	8	9	10	11
2	2	3	4	5	6	7	8	9	10	11	12
3	3	4	5	6	7	8	9	10	11	12	13
4	4	5	6	7	8	9	10	11	12	13	14
5	5	6	7	8	9	10	11	12	13	14	15
6	6	7	8	9	10	11	12	13	14	15	16
7	7	8	9	10	11	12	13	14	15	16	17
8	8	9	10	11	12	13	14	15	16	17	18
9	9	10	11	12	13	14	15	16	17	18	19
10	10	11	12	13	14	15	16	17	18	19	20

Name **Date**

Teaching Aid 3 $\frac{1}{4}$-in. Grid Paper

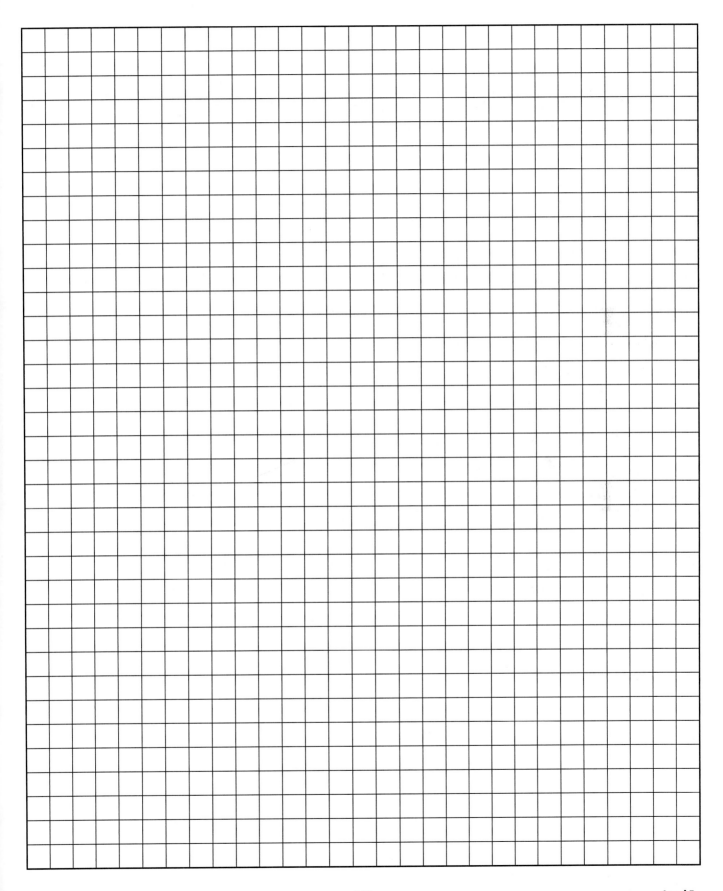

Teaching Aid   Level E

Name	Date

Teaching Aid 4 Centimeter Grid Paper

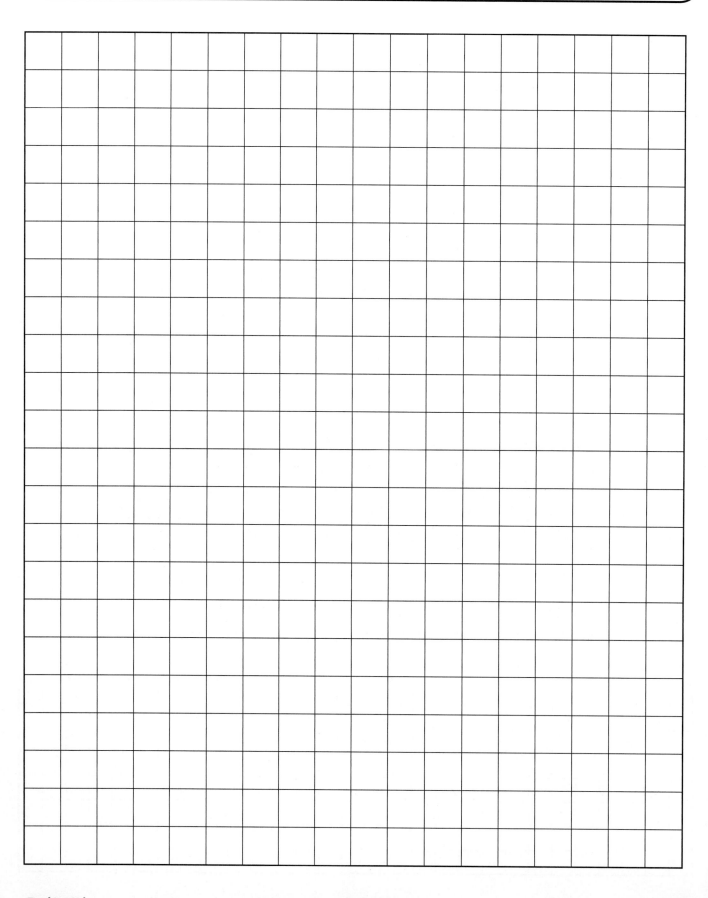

Teaching Aid   364   Level E

Name   **Date**

Teaching Aid 5 Multiplication Facts Table

Multiplication Facts Table

×	0	1	2	3	4	5	6	7	8	9	10
0	0	0	0	0	0	0	0	0	0	0	0
1	0	1	2	3	4	5	6	7	8	9	10
2	0	2	4	6	8	10	12	14	16	18	20
3	0	3	6	9	12	15	18	21	24	27	30
4	0	4	8	12	16	20	24	28	32	36	40
5	0	5	10	15	20	25	30	35	40	45	50
6	0	6	12	18	24	30	36	42	48	54	60
7	0	7	14	21	28	35	42	49	56	63	70
8	0	8	16	24	32	40	48	56	64	72	80
9	0	9	18	27	36	45	54	63	72	81	90
10	0	10	20	30	40	50	60	70	80	90	100

Teaching Aid   Level E

Name Date
Teaching Aid 6 Place-Value Charts

hundreds	tens	ones

hundreds	tens	ones

hundreds	tens	ones

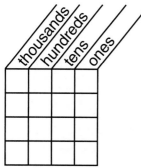 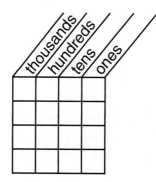 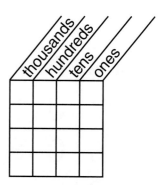

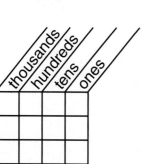 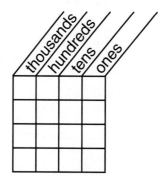 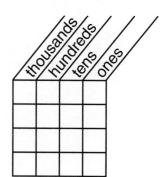

Name Date

Teaching Aid 7 Place-Value Charts

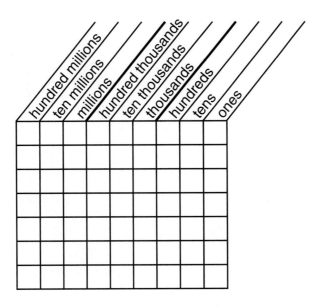

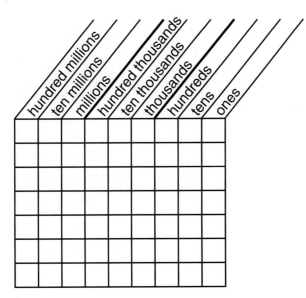

Teaching Aid 367 Level E

Name   **Date**

Teaching Aid 8 Place-Value Charts

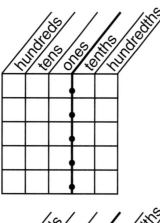 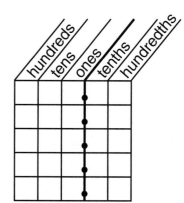

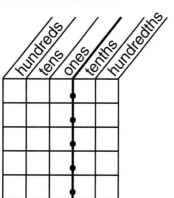 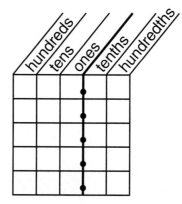

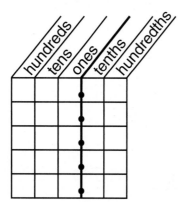 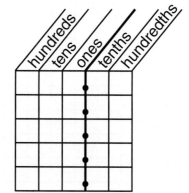

Teaching Aid   368   Level E

Name Date

Teaching Aid 9 Decimal Models

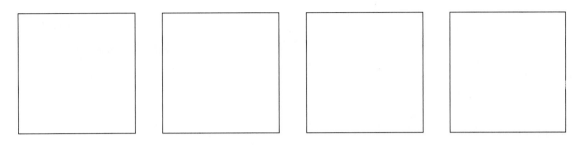

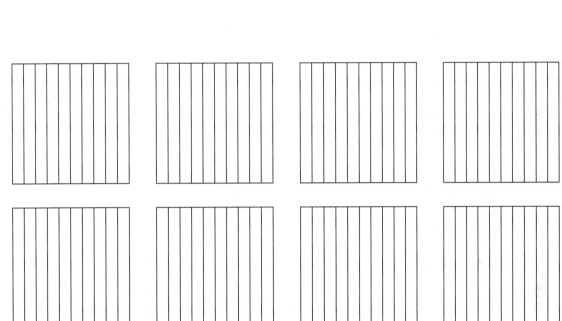

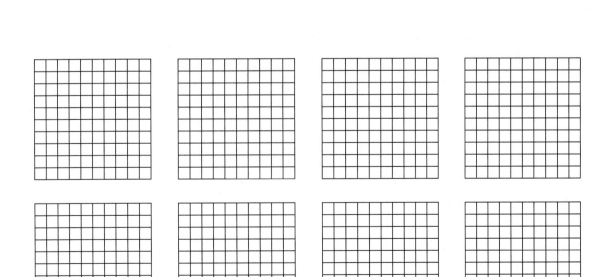

Teaching Aid 369 Level E

Name **Date**

Teaching Aid 10 Coins and Bills

Teaching Aid 370 Level E

Name Date

Teaching Aid 11 Fraction Number Lines

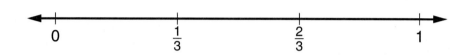

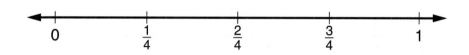

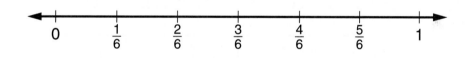

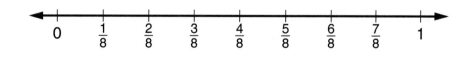

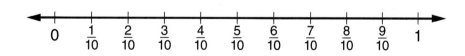

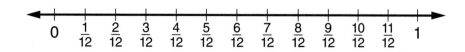

Name Date

Teaching Aid 12 Fraction Strips

1

| 1/2 | 1/2 |

| 1/3 | 1/3 | 1/3 |

| 1/4 | 1/4 | 1/4 | 1/4 |

| 1/5 | 1/5 | 1/5 | 1/5 | 1/5 |

| 1/6 | 1/6 | 1/6 | 1/6 | 1/6 | 1/6 |

| 1/8 | 1/8 | 1/8 | 1/8 | 1/8 | 1/8 | 1/8 | 1/8 |

| 1/10 | 1/10 | 1/10 | 1/10 | 1/10 | 1/10 | 1/10 | 1/10 | 1/10 | 1/10 |

| 1/12 | 1/12 | 1/12 | 1/12 | 1/12 | 1/12 | 1/12 | 1/12 | 1/12 | 1/12 | 1/12 | 1/12 |

Name Date

Teaching Aid 13 Fraction Models

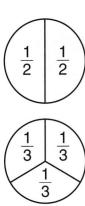
  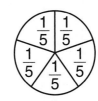 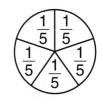
   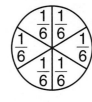 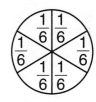

Teaching Aid Level E

Name Date

Teaching Aid 14 Centimeter and Inch Rulers

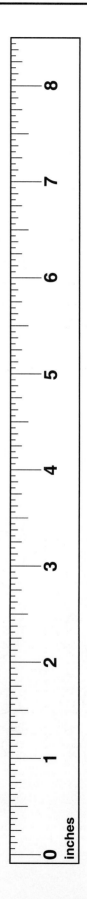

Teaching Aid 374 Level E

Teaching Aid 15 First-Quadrant Coordinate Grid

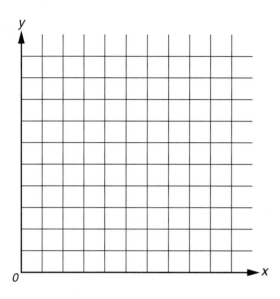

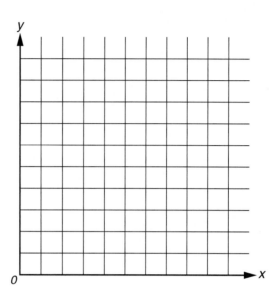

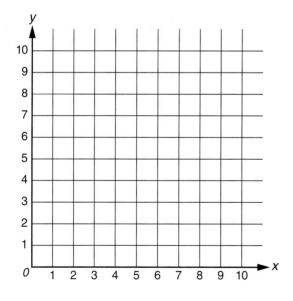

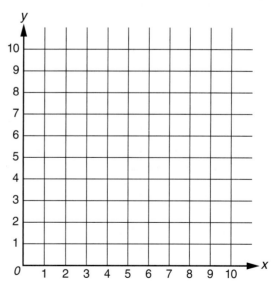

Name   Date

Teaching Aid 16 Four-Quadrant Coordinate Grid

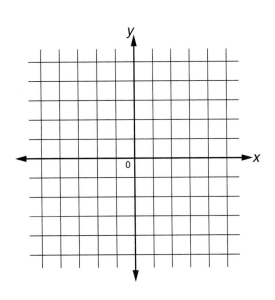
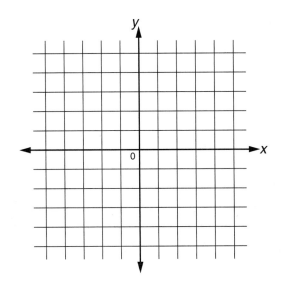
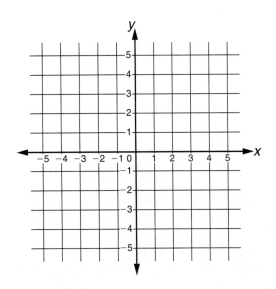
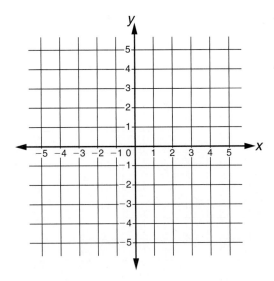

Teaching Aid   376   Level E

Name   **Date**

Teaching Aid 17 Inch Grid Paper

Name Date

Teaching Aid 18 Angles and Triangles

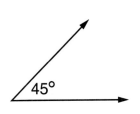
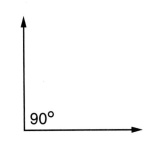
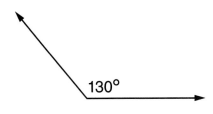

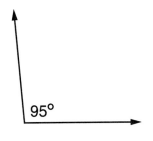
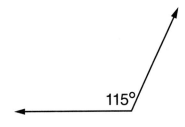
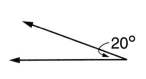

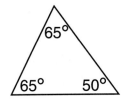
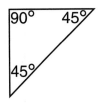
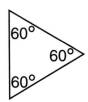

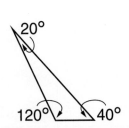
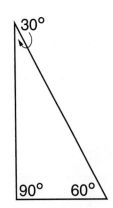
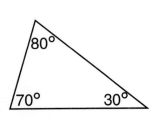

Teaching Aid 378 Level E

Name   Date

Teaching Aid 19 Area of a Circle

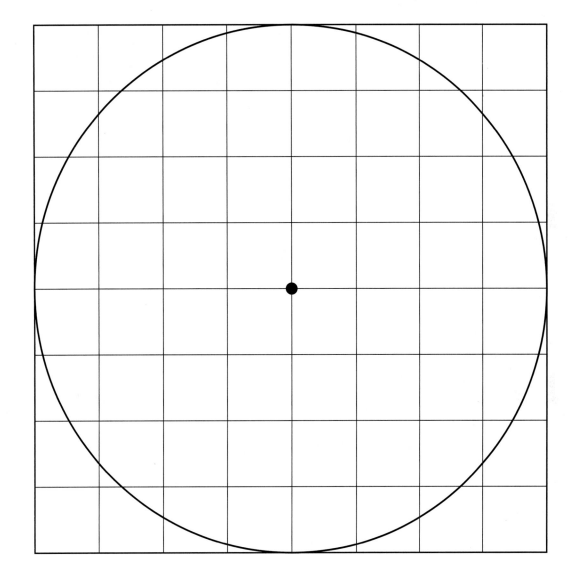

Teaching Aid   379   Level E

Name   Date

Teaching Aid 20 Complementary and Supplementary Angles

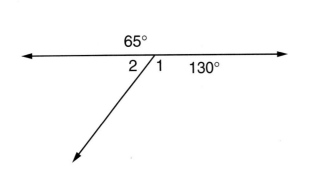

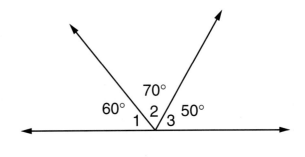

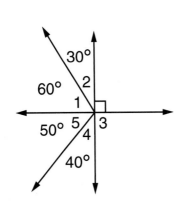

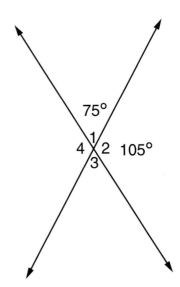

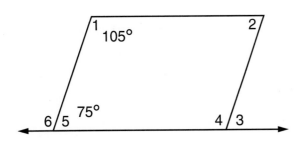

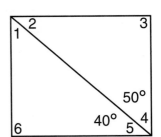

Teaching Aid   Level E

Name **Date**

Teaching Aid 21 Dot Paper

Name   Date

Teaching Aid 22 Congruent Shapes

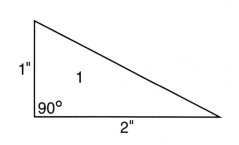
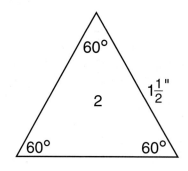
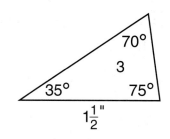

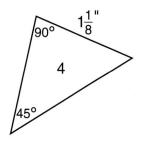
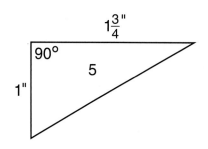
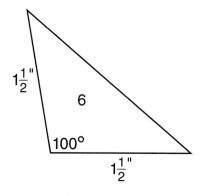

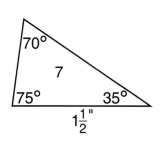
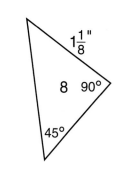
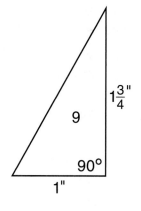

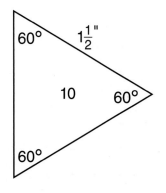
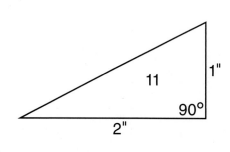
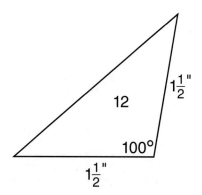

Teaching Aid   382   Level E

Student Action Plan

Name   Date

Student Action Plan Report

Volume 1, Topic 1 Place Value through 100

Component	Objective	Title	Pages	Assign
Assessment Resources		Pre-Test, Volume 1	41–44	
Student Booklet		Topic 1 Introduction	1	
Student Booklet	1.1	Tens and Ones	2–4	
Student Booklet	1.2	Numbers to 100	5–7	
Student Booklet	1.3	Compare and Order Numbers to 100	8–10	
Student Booklet	1.4	Equivalent Forms	11–13	
Student Booklet		Topic 1 Summary	14	
Student Booklet		Mixed Review	15	
Assessment Resources		Progress Monitoring	1	
Computer Tutorial	1.1	1a Modeling Numbers with Base Ten Blocks		
Computer Tutorial	1.1	1b Finding the Value of Coins:		
Computer Tutorial	1.2	1c Working with Whole Numbers, 0 to 100		
Computer Tutorial	1.3	1d Comparing Whole Numbers		
Computer Tutorial	1.3	1e Ordering Whole Numbers, Example A		
Computer Tutorial	1.4	1f Using Equivalent Forms to Represent Numbers, Example A		
Computer Tutorial	1.4	1g Using Equivalent Forms to Represent Numbers, Example B		

Level E

Name　　　　　　　　　　　　　　　　　　　　　　Date

Student Action Plan Report

Volume 1, Topic 2 Place Value through 1,000

Component	Objective	Title	Pages	Assign
Student Booklet		Topic 2 Introduction	16	
Student Booklet	2.1	Numbers to 1,000	17–19	
Student Booklet	2.2	Write Numbers to 1,000	20–22	
Student Booklet	2.3	Compare and Order to 1,000	23–25	
Student Booklet		Topic 2 Summary	26	
Student Booklet		Mixed Review	27	
Assessment Resources		Progress Monitoring	2	
Computer Tutorial	2.1	2a Identifying the Place Value of Each Digit in a Whole Number, Example A		
Computer Tutorial	2.3	2b Comparing Whole Numbers		
Computer Tutorial	2.3	2c Ordering Whole Numbers, Example B		

Name   Date

Student Action Plan Report

Volume 1, Topic 3 Place Value beyond 1,000

Component	Objective	Title	Pages	Assign
Student Booklet		Topic 3 Introduction	28	
Student Booklet	3.1	Place Value to 10,000	29–31	
Student Booklet	3.2	Expanded Notation with Zeros	32–34	
Student Booklet	3.3	Numbers in the Millions	35–37	
Student Booklet	3.4	Round through Millions	38–40	
Student Booklet		Topic 3 Summary	41	
Student Booklet		Mixed Review	42	
Assessment Resources		Progress Monitoring	3	
Assessment Resources		Post-Test, Volume 1	81–84	
Computer Tutorial	3.1	3a Identifying the Place Value of Each Digit in a Whole Number, Example B		
Computer Tutorial	3.2	3b Writing Numbers in Expanded Notation		
Computer Tutorial	3.3	3c Writing Numbers in Word Form		
Computer Tutorial	3.4	3d Rounding		

Name   Date
Student Action Plan Report

Volume 2, Topic 4 Addition and Subtraction Facts

Component	Objective	Title	Pages	Assign
Assessment Resources		Pretest, Volume 2	45–51	
Student Booklet		Topic 4 Introduction	1	
Student Booklet	4.1	Add or Subtract to Solve Problems	2–4	
Student Booklet	4.2	Properties of Addition	5–7	
Student Booklet	4.3	Addition Strategies	8–10	
Student Booklet	4.4	Add Three 1-Digit Numbers	11–13	
Student Booklet	4.5	Relate Addition and Subtraction	14–16	
Student Booklet	4.6	Subtraction Strategies	17–19	
Student Booklet	4.7	Fact Families	20–22	
Student Booklet		Topic 4 Summary	23	
Student Booklet		Mixed Review	24	
Assessment Resources		Progress Monitoring	4	
Computer Tutorial	4.1	4a Choosing the Operation to Solve Addition and Subtraction Problems		
Computer Tutorial	4.2	4b Understanding Compatible Numbers		
Computer Tutorial	4.3	4c Using Addition Fact Strategies		
Computer Tutorial	4.4	4d Finding the Sum of Three Numbers		
Computer Tutorial	4.5	4e Writing Fact Families (Addition and Subtraction)		

Name   Date

Student Action Plan Report

Volume 2, Topic 5 Multiplication Facts

Component	Objective	Title	Pages	Assign
Student Booklet		Topic 5 Introduction	25	
Student Booklet	5.1	Meaning of Multiplication	26–28	
Student Booklet	5.2	Multiply by 2, 5, and 10	29–31	
Student Booklet	5.3	Properties of Multiplication	32–34	
Student Booklet	5.4	Multiplication Strategies	35–37	
Student Booklet	5.5	Basic Multiplication Facts	38–40	
Student Booklet		Topic 5 Summary	41	
Student Booklet		Mixed Review	42	
Assessment Resources		Progress Monitoring	5	
Computer Tutorial	5.1	5a Understanding Multiplication, Example A		
Computer Tutorial	5.1	5b Understanding Multiplication, Example B		
Computer Tutorial	5.3	5c Finding Factors of Numbers, Example A		

Name Date

Student Action Plan Report

Volume 2, Topic 6 Division Facts

Component	Objective	Title	Pages	Assign
Student Booklet		Topic 6 Introduction	43	
Student Booklet	6.1	Meaning of Division	44–46	
Student Booklet	6.2	Properties of Zero and One	47–49	
Student Booklet	6.3	Divide by 2, 3, 4, or 5	50–52	
Student Booklet	6.4	Divide by 6, 7, 8, or 9	53–55	
Student Booklet	6.5	Relate Multiplication and Division	56–58	
Student Booklet		Topic 6 Summary	59	
Student Booklet		Mixed Review	60	
Assessment Resources		Progress Monitoring	6	
Assessment Resources		Posttest, Volume 2	85–90	
Computer Tutorial	6.1	6a Writing Fact Families (Multiplication and Division)		
Computer Tutorial	6.1	6b Using Multiplication to Check Division		
Computer Tutorial	6.3	6c Understanding Patterns in Division		
Computer Tutorial	6.4	6c Understanding Patterns in Division		
Computer Tutorial	6.5	6a Writing Fact Families (Multiplication and Division)		

Name   Date

Student Action Plan Report

Volume 3, Topic 7 Add or Subtract 1- and 2-Digit Numbers

Component	Objective	Title	Pages	Assign
Assessment Resources		Pre-Test, Volume 3	52–54	
Student Booklet		Topic 7 Introduction	1	
Student Booklet	7.1	1-Digit and 2-Digit Numbers	2–4	
Student Booklet	7.2	Add 2-Digit Numbers	5–7	
Student Booklet	7.3	Subtract 2-Digit Numbers	8–10	
Student Booklet		Topic 7 Summary	11	
Student Booklet		Mixed Review	12	
Assessment Resources		Progress Monitoring	7	
Computer Tutorial	7.1	7a Solving Problems Using Addition		
Computer Tutorial	7.2	7b Using the Partial Sums Algorithm		
Computer Tutorial	7.3	7c Using the Same-Change Rule to Subtract		

Level E

Name   Date

Student Action Plan Report

Volume 3, Topic 8 Add or Subtract Multidigit Numbers

Component	Objective	Title	Pages	Assign
Student Booklet		Topic 8 Introduction	13	
Student Booklet	8.1	Add or Subtract Mentally	14–16	
Student Booklet	8.2	Add and Subtract with Estimation	17–19	
Student Booklet	8.3	Add and Subtract 3-Digit Numbers	20–22	
Student Booklet	8.4	Whole Numbers to 10,000	23–25	
Student Booklet	8.5	Add and Subtract Multidigit Numbers	26–28	
Student Booklet		Topic 8 Summary	29	
Student Booklet		Mixed Review	30	
Assessment Resources		Progress Monitoring	8	
Assessment Resources		Post-Test, Volume 3	91–93	
Computer Tutorial	8.1	8a Using the Partial Sums Algorithm		
Computer Tutorial	8.1	8b Using the Distributive Property		
Computer Tutorial	8.1	8c Using the Same-Change Rule to Subtract		
Computer Tutorial	8.2	8d Using the Standard Addition Algorithm, Example A		
Computer Tutorial	8.2	8e Using the Standard Subtraction Algorithm, Example A		
Computer Tutorial	8.3	8d Using the Standard Addition Algorithm, Example A		
Computer Tutorial	8.3	8e Using the Standard Subtraction Algorithm, Example A		
Computer Tutorial	8.4	8d Using the Standard Addition Algorithm, Example A		
Computer Tutorial	8.4	8f Using the Standard Addition Algorithm, Example B		
Computer Tutorial	8.4	8e Using the Standard Subtraction Algorithm, Example A		
Computer Tutorial	8.4	8g Using the Standard Subtraction Algorithm, Example B		
Computer Tutorial	8.5	8d Using the Standard Addition Algorithm, Example A		
Computer Tutorial	8.5	8f Using the Standard Addition Algorithm, Example B		
Computer Tutorial	8.5	8e Using the Standard Subtraction Algorithm, Example A		
Computer Tutorial	8.5	8g Using the Standard Subtraction Algorithm, Example B		

Name   Date

Student Action Plan Report

Volume 4, Topic 9 Use Multiplication to Compute

Component	Objective	Title	Pages	Assign
Assessment Resources		Pre-Test, Volume 4	55–59	
Student Booklet		Topic 9 Introduction	1	
Student Booklet	9.1	Multiply by Multiples of 10	2–4	
Student Booklet	9.2	Estimate Products	5–7	
Student Booklet	9.3	Multiply: Four Digits by One Digit	8–10	
Student Booklet	9.4	Choose a Method and Multiply	11–13	
Student Booklet		Topic 9 Summary	14	
Student Booklet		Mixed Review	15	
Assessment Resources		Progress Monitoring	9	
Computer Tutorial	9.1	9a Using Multiples of 10, 100, and 1,000 to Multiply and Divide		
Computer Tutorial	9.2	9b Using the Partial-Products Method		
Computer Tutorial	9.2	9c Using the Distributive Property		
Computer Tutorial	9.3	9b Using the Partial-Products Method		
Computer Tutorial	9.3	9c Using the Distributive Property		
Computer Tutorial	9.4	9d Choosing a Method to Solve Multiplication and Division Problems		

Name   Date

Student Action Plan Report

Volume 4, Topic 10 Use Division to Compute

Component	Objective	Title	Pages	Assign
Student Booklet		Topic 10 Introduction	16	
Student Booklet	10.1	Divide Multiples of 10	17–19	
Student Booklet	10.2	Estimate Quotients	20–22	
Student Booklet	10.3	Divide by 1-Digit Numbers	23–25	
Student Booklet	10.4	Choose a Method for Division	26–28	
Student Booklet	10.5	Multiplication and Division	29–31	
Student Booklet		Topic 10 Summary	32	
Student Booklet		Mixed Review	33	
Assessment Resources		Progress Monitoring	10	
Computer Tutorial	10.1	10a Using Multiples of 10, 100, and 1,000 to Multiply and Divide		
Computer Tutorial	10.2	10b Estimating Quotients by Rounding Numbers		
Computer Tutorial	10.3	10c Using the Standard Long Division Algorithm, Example A		
Computer Tutorial	10.3	10d Using the Standard Long Division Algorithm, Example B		
Computer Tutorial	10.3	10e Using the Standard Long Division Algorithm, Example C		
Computer Tutorial	10.4	10f Choosing a Method to Solve Multiplication and Division Problems		
Computer Tutorial	10.5	10g Solving Word Problems, Example A		
Computer Tutorial	10.5	10h Solving Word Problems, Example B		
Computer Tutorial	10.5	10i Modeling Division		
Computer Tutorial	10.5	10j Understanding Division		

Name   Date

Student Action Plan Report

Volume 4, Topic 11 Expressions and Equations

Component	Objective	Title	Pages	Assign
Student Booklet		Topic 11 Introduction	34	
Student Booklet	11.1	Write Expressions for Patterns	35–37	
Student Booklet	11.2	Write Expressions	38–40	
Student Booklet	11.3	Write Equations with Unknowns	41–43	
Student Booklet	11.4	Solve Equations with Unknowns	44–46	
Student Booklet		Topic 11 Summary	47	
Student Booklet		Mixed Review	48	
Assessment Resources		Progress Monitoring	11	
Assessment Resources		Posttest, Volume 4	94–98	
Computer Tutorial	11.1	11a Writing and Applying a Rule		
Computer Tutorial	11.3	11b Writing Equations		
Computer Tutorial	11.4	11c Solving Equations, Example A		
Computer Tutorial	11.4	11d Solving Equations, Example B		
Computer Tutorial	11.4	11e Solving Equations, Example C		
Computer Tutorial	11.4	11f Solving Equations, Example D		

Level E

Name Date

Student Action Plan Report

Volume 5, Topic 12 Graphing

Component	Objective	Title	Pages	Assign
Assessment Resources		Pretest, Volume 5	60–73	
Student Booklet		Topic 12 Introduction	1	
Student Booklet	12.1	Compare Data with Graphs	2–4	
Student Booklet	12.2	Record Data	5–7	
Student Booklet	12.3	Mean, Median, and Mode	8–10	
Student Booklet	12.4	Show Data in More Than One Way	11–13	
Student Booklet	12.5	Graph Ordered Pairs	14–16	
Student Booklet		Topic 12 Summary	17	
Student Booklet		Mixed Review	18	
Assessment Resources		Progress Monitoring	12	
Computer Tutorial	12.2	12a Using a Tally Chart to Make a Bar Graph		
Computer Tutorial	12.3	12b Find the Mean, Median, and Mode of a Data Set		
Computer Tutorial	12.4	12a Using a Tally Chart to Make a Bar Graph		
Computer Tutorial	12.4	12c Using a Bar Graph		
Computer Tutorial	12.5	12d Graphing Ordered Pairs		

394 Level E

Name   Date

Student Action Plan Report

Volume 5, Topic 13 Basic Geometric Figures

Component	Objective	Title	Pages	Assign
Student Booklet		Topic 13 Introduction	19	
Student Booklet	13.1	Angles and Lines	20–22	
Student Booklet	13.2	Types of Polygons	23–25	
Student Booklet	13.3	Triangles	26–28	
Student Booklet	13.4	Quadrilaterals	29–31	
Student Booklet	13.5	Circles	32–34	
Student Booklet		Topic 13 Summary	35	
Student Booklet		Mixed Review	36	
Assessment Resources		Progress Monitoring	13	
Computer Tutorial	13.1	13a Identifying and Drawing Parallel Lines		
Computer Tutorial	13.1	13b Identifying and Drawing Perpendicular Lines		
Computer Tutorial	13.1	13c Measuring Angles		
Computer Tutorial	13.1	13d Drawing an Angle		
Computer Tutorial	13.2	13e Classifying Polygons		
Computer Tutorial	13.3	13f Sorting and Classifying Triangles		

Name   Date

Student Action Plan Report

Volume 5, Topic 14 Measurement Conversion

Component	Objective	Title	Pages	Assign
Student Booklet		Topic 14 Introduction	37	
Student Booklet	14.1	U.S. Customary Units	38–40	
Student Booklet	14.2	Basic Metric Prefixes	41–43	
Student Booklet	14.3	Use the Metric System	44–46	
Student Booklet	14.4	Factors in Unit Conversions	47–49	
Student Booklet	14.5	Convert Units within a System	50–52	
Student Booklet		Topic 14 Summary	53	
Student Booklet		Mixed Review	54	
Assessment Resources		Progress Monitoring	14	
Computer Tutorial	14.1	14a Converting Units of Capacity		
Computer Tutorial	14.1	14b Converting Units of Time		
Computer Tutorial	14.1	14c Converting Units of Length		
Computer Tutorial	14.3	14d Using the Metric System to Measure Length		
Computer Tutorial	14.3	14e Using the Metric System to Measure Mass		
Computer Tutorial	14.5	14b Converting Units of Time		
Computer Tutorial	14.5	14c Converting Units of Length		

Name   Date

Student Action Plan Report

Volume 5, Topic 15 Measure Geometric Figures

Component	Objective	Title	Pages	Assign
Student Booklet		Topic 15 Introduction	55	
Student Booklet	15.1	Length	56–58	
Student Booklet	15.2	Perimeter	59–61	
Student Booklet	15.3	Area	62–64	
Student Booklet	15.4	Area of Rectangles	65–67	
Student Booklet	15.5	Volume	68–70	
Student Booklet		Topic 15 Summary	71	
Student Booklet		Mixed Review	72	
Assessment Resources		Progress Monitoring	15	
Assessment Resources		Post-Test, Volume 5	99–109	
Computer Tutorial	15.1	15a Measuring Length to the Nearest Unit		
Computer Tutorial	15.2	15b Finding Perimeter		
Computer Tutorial	15.3	15c Finding Area		
Computer Tutorial	15.5	15d Finding Volume		

Level E

Name   Date

Student Action Plan Report

Volume 6, Topic 16 Meaning of Fractions

Component	Objective	Title	Pages	Assign
Assessment Resources		Pretest, Volume 6	74–80	
Student Booklet		Topic 16 Introduction	1	
Student Booklet	16.1	Basics of Fractions	2–4	
Student Booklet	16.2	Unit Fractions	5–7	
Student Booklet	16.3	Parts and the Whole	8–10	
Student Booklet	16.4	Rename Values Greater Than 1	11–13	
Student Booklet		Topic 16 Summary	14	
Student Booklet		Mixed Review	15	
Assessment Resources		Progress Monitoring	16	
Computer Tutorial	16.1	16a Relating Fractions and Decimals to Models		
Computer Tutorial	16.1	16b Modeling Fractions and Decimals, Example A		
Computer Tutorial	16.2	16c Recognizing and Using Unit Fractions, Example A		
Computer Tutorial	16.2	16d Recognizing and Using Unit Fractions, Example B		
Computer Tutorial	16.2	16e Comparing Unit Fractions		
Computer Tutorial	16.3	16f Modeling a Whole		
Computer Tutorial	16.4	16h Changing Between Mixed Numbers and Improper Fractions		

Name   Date

Student Action Plan Report

Volume 6, Topic 17 Equivalence of Fractions

Component	Objective	Title	Pages	Assign
Student Booklet		Topic 17 Introduction	16	
Student Booklet	17.1	Compare Fractions	17–19	
Student Booklet	17.2	Factor Whole Numbers	20–22	
Student Booklet	17.3	Equivalent Fractions	23–25	
Student Booklet	17.4	Fractions in Lowest Terms	26–28	
Student Booklet	17.5	Interpretations of Fractions	29–31	
Student Booklet		Topic 17 Summary	32	
Student Booklet		Mixed Review	33	
Assessment Resources		Progress Monitoring	17	
Computer Tutorial	17.1	17a Comparing Fractions using the Symbols $<, =, >$, Example A		
Computer Tutorial	17.1	17b Comparing Fractions using the Symbols $<, =, >$, Example B		
Computer Tutorial	17.1	17c Comparing Unit Fractions		
Computer Tutorial	17.1	17d Finding Equivalent Fractions, Example A		
Computer Tutorial	17.1	17e Finding Equivalent Fractions, Example B		
Computer Tutorial	17.1	17f Finding Equivalent Fractions, Example C		
Computer Tutorial	17.2	17g Finding Factors of Numbers, Example B		
Computer Tutorial	17.2	17h Finding the Prime Factorization of a Number		
Computer Tutorial	17.4	17i Simplifying Fractions, Example B		
Computer Tutorial	17.4	17j Simplifying Improper Fractions		

Name Date

Student Action Plan Report

Volume 6, Topic 18 Addition and Subtraction of Fractions

Component	Objective	Title	Pages	Assign
Student Booklet		Topic 18 Introduction	34	
Student Booklet	18.1	Add and Subtract Fractions	35–37	
Student Booklet	18.2	Unlike Denominators: Add	38–40	
Student Booklet	18.3	Unlike Denominators: Subtract	41–43	
Student Booklet	18.4	Add Mixed Numbers	44–46	
Student Booklet	18.5	Subtract Mixed Numbers	47–49	
Student Booklet		Topic 18 Summary	50	
Student Booklet		Mixed Review	51	
Assessment Resources		Progress Monitoring	18	
Computer Tutorial	18.1	18a Adding and Subtracting Fractions with Common Denominators, Example A		
Computer Tutorial	18.1	18b Adding and Subtracting Fractions with Common Denominators, Example B		
Computer Tutorial	18.2	18c Adding Fractions (with Like and Unlike Denominators), Example A		
Computer Tutorial	18.2	18d Adding Fractions (with Like and Unlike Denominators), Example B		
Computer Tutorial	18.3	18e Using Fraction Strips to Subtract Unlike Fractions		
Computer Tutorial	18.4	18f Solving Problems Involving Fractions, Example A		
Computer Tutorial	18.5	18g Subtracting Fractions (with Like and Unlike Denominators)		
Computer Tutorial	18.5	18h Subtracting Fractions with Common Denominators, Example C		

Name Date

Student Action Plan Report

Volume 6, Topic 19 Decimals and Money

Component	Objective	Title	Pages	Assign
Student Booklet		Topic 19 Introduction	52	
Student Booklet	19.1	Understand Money	53–55	
Student Booklet	19.2	Tenths and Hundredths	56–58	
Student Booklet	19.3	Compare and Order Decimals	59–61	
Student Booklet	19.4	Use Coins and Bills	62–64	
Student Booklet	19.5	Decimal Notation for Money	65–67	
Student Booklet	19.6	Money in Fractions and Decimals	68–70	
Student Booklet	19.7	Unit Costs	71–73	
Student Booklet		Topic 19 Summary	74	
Student Booklet		Mixed Review	75	
Assessment Resources		Progress Monitoring	19	
Computer Tutorial	19.1	19a Finding the Value of Coins		
Computer Tutorial	19.2	19b Modeling Fractions and Decimals, Example B		
Computer Tutorial	19.2	19c Representing Money as Decimals and Fractions, Example A		
Computer Tutorial	19.2	19d Representing Money as Decimals and Fractions, Example B		
Computer Tutorial	19.2	19e Representing Money as Decimals and Fractions, Example C		
Computer Tutorial	19.3	19f Comparing Decimals Using the Symbols <, =, >, Example A		
Computer Tutorial	19.3	19g Ordering Decimals		
Computer Tutorial	19.4	19h Using Bills and Coins to Solve Problems, Example A		
Computer Tutorial	19.4	19i Using Bills and Coins to Solve Problems, Example B		
Computer Tutorial	19.5	19a Finding the Value of Coins		
Computer Tutorial	19.6	19c Representing Money as Decimals and Fractions, Example A		
Computer Tutorial	19.6	19d Representing Money as Decimals and Fractions, Example B		
Computer Tutorial	19.6	19e Representing Money as Decimals and Fractions, Example C		
Computer Tutorial	19.7	19j Using Rates to Find the Unit Cost		

Name   Date

Student Action Plan Report

Volume 6, Topic 20 Decimal Operations and Comparisons

Component	Objective	Title	Pages	Assign
Student Booklet		Topic 20 Introduction	76	
Student Booklet	20.1	Add and Subtract Decimals	77–79	
Student Booklet	20.2	Decimal and Fraction Equivalents	80–82	
Student Booklet	20.3	Use a Number Line	83–85	
Student Booklet		Topic 20 Summary	86	
Student Booklet		Mixed Review	87	
Assessment Resources		Progress Monitoring	20	
Assessment Resources		Posttest, Volume 6	110–117	
Computer Tutorial	20.1	20a Adding Decimals, Example A		
Computer Tutorial	20.1	20b Adding Decimals, Example B		
Computer Tutorial	20.1	20c Subtracting Decimals		
Computer Tutorial	20.2	20d Using Models to Find Equivalent Fractions and Decimals		
Computer Tutorial	20.3	20e Graphing Positive Fractions, Mixed Numbers, and Decimals, Example B		

Level E

Volume 1
Words to Know/Glossary

C
compare — Use the symbols < and > to compare two unequal quantities.

D
digits — We use ten digits to make numbers: 0, 1, 2, 3, 4, 5, 6, 7, 8, 9.

G
greatest — Highest

H
hundreds place — The hundreds place is the third place to the left of a number or a decimal point.

I
inequalities — These statements are called inequalities: 3 < 5, 5 > 3.

L
least — Lowest

M
millions period — The millions period includes the millions, ten millions, and hundred millions places.

O
ones period — The ones period includes the ones, tens, and hundreds places.

ones place — The ones place is the first place to the left of a number or a decimal point.

ordering — Writing in order or ordering means to list numbers from greatest to least or from least to greatest.

P
period — a set of three place values.

place — Each digit in a number has a place. Example, in the number 42, the digit 4 is in the tens place, and the digit 2 is in the ones place.

R
rounded — Rounded numbers have values close to the original amount that are often more convenient to use.

T
tens place — The tens place is the second place to the left of a number or a decimal point.

thousands period — The thousands period includes the thousands, ten thousands, and hundred thousands places.

V
value — The value of a digit is determined by the place it is in. Example: In 4,528, the value of the 2 is 2 tens, or 20.

W
writing in order — Writing in order or ordering means to list numbers from greatest to least or from least to greatest.

Volume 2
Words to Know/Glossary

A

addend — A number that is added to another number.

addition — Joining groups or increasing a quantity.

array — A picture that shows a multiplication fact.

associative property of multiplication — You can group factors in any way, and the product will be the same. $3 \times (2 \times 4) = 24$ and $(3 \times 2) \times 4 = 24$.

C

commutative property of multiplication — You can multiply factors in any order, and the product will be the same. $3 \times 8 = 24$ and $8 \times 3 = 24$.

D

dividend — The number you divide.

divisible — A number is divisible by another number when there is no remainder after division.

divisor — The number you divide by.

F

factor — A number that you multiply by another number to equal a product.

M

multiplication — A way to join groups of equal size.

Multiplication Facts Table — The rows and columns in the multiplication facts table can help you find the facts.

N

number line — A line that shows numbers from smallest to greatest.

P

product — The answer to a multiplication problem.

Q

quotient — The answer to a division problem.

R

remainder — The number left over after you divide.

repeated addition — A way to multiply by adding the same number as many times as needed.

S

skip count — Use a counting pattern to multiply.

subtraction — Taking away, comparing, or decreasing a quantity.

Volume 3
Words to Know/Glossary

E
estimate — A number close to the actual answer.

R
regroup — Exchange amounts of equal value to rename a number.

Volume 4
Words to Know/Glossary

C
compatible numbers — Numbers that are close to the actual numbers but easier to work with mentally.

D
dividend — The number you divide.

divisor — The number you divide by.

E
estimate — A number close to the actual answer.

M
mental math — Solving math problems in your head.

multiple — The product of a given number and another whole number.

Q
quotient — The answer to a division problem.

R
rounded — Rounded numbers have values close to the original amount that are often more convenient to use.

Volume 5: Words to Know/Glossary

A

acute angle — An angle that measures less than 90 degrees.

angle — Two rays or segments that meet at a common endpoint.

area — Area is the amount of surface the shape has. It is measured in square units.

C

capacity — The measure of how much something holds.

centi- — A metric prefix meaning one-hundredth.

centimeters — A centimeter (cm) is a measuring unit of length in the metric system. 1 meter = 100 centimeters.

chord — A line segment that connects two points on a circle.

circle — A closed plane figure made up of points that are the same distance from the center.

circumference — The distance around a circle.

congruent — Having the same measure.

cubic units — Units used to measure volume and tell the number of cubes of a given size that are needed to fill a three-dimensional figure.

D

data — Information collected about people or things.

days — A common unit of time measurement.

diameter — A chord that passes through the center of the circle.

E

equilateral triangle — A triangle with all sides congruent and all angles congruent.

F

feet — A unit of length in the U.S. system. 1 foot (ft) = 12 inches.

G

gram — A metric unit for measuring mass.

H

height — The height of a figure is the length of a perpendicular line between the base and the top of the figure.

hexagon — A polygon with 6 sides.

horizontal axis — The horizontal axis on a graph is at the bottom. The numbers go from left to right.

hours — A common unit of time measurement. 24 hours = 1 day.

I

inches — An inch (in.) is a unit of length measurement in the U.S. system. 36 inches = 3 feet = 1 yard.

isosceles triangle — A triangle with exactly two sides congruent and exactly 2 angles congruent.

K

key — The key on a pictograph shows the value of each symbol.

kilo- — Metric prefix meaning one thousand.

kilogram — A kilogram (kg) is metric unit for measuring mass. 1 kg = 1,000 g.

L

length — The measure of how long something is.

line — A set of points that continues without end in both directions.

line plot — A line plot shows one X, or other symbol, for each data item.

line segment — A part of a line between two endpoints.

liter — A metric unit for measuring capacity.

M

mass — The amount of matter in an object.

mean — The sum of the addends in a set of data divided by the number of addends.

measurement system — A measurement system can be based on U.S. units or on metric units.

median — The middle number in an ordered set of data.

meter — A metric unit for measuring length or distance.

mile — A unit of length measurement in the U.S. system. 1 mile = 63,360 inches = 5,280 feet.

milli- — A metric prefix meaning one-thousandth.

milliliter — A metric unit for measuring capacity; 1,000 mL = 1 L.

millimeter — A millimeter (mm) is a measuring unit of length in the metric system. 1 meter = 1,000 millimeters.

minutes — A common unit of time measurement. 60 minutes = 1 hour.

mode — The number that occurs most often in a set of data.

O

obtuse angle — An angle that measures greater than 90 degrees but less than 180 degrees.

octagon — A polygon with 8 sides.

ordered pair — An ordered pair shows the location of a point. The first number is the distance from the x-axis, and the second number is the distance from the y-axis.

ounce — A U.S. unit for measuring weight.

P

parallel lines — Lines that lie in the same plane and do not intersect.

pentagon — A polygon with 5 sides.

perimeter — The distance around all sides of a figure.

perpendicular lines — Lines that intersect to form right triangles.

pi (π) — A ratio of circumference of a circle to its diameter; it is approximately 3.14.

pictograph — A graph that uses symbols or simple pictures to represent quantities.

polygon — A closed figure made up of three or more line segments that meet but do not cross.

pound — A U.S. unit for measuring weight; 16 ounces.

Q

quadrilateral — A 4-sided polygon.

R

radius — A line segment that connects the center of a circle to a point on the circle.

range — The difference between the greatest and least numbers in a set of data.

ray — Part of a line that has one endpoint and continues without end in one direction.

rectangular solid — A three-dimensional figure with 6 rectangular faces.

regular polygon — One whose side measurements are all the same.

right angle — An angle that measures greater than 90 degrees.

right triangle — A triangle with one right angle.

S

seconds — A common unit of time measurement. 60 seconds = 1 minute.

sides — The line segments that make up a polygon.

square units — A unit of area with dimensions 1 unit by 1 unit.

stem-and-leaf plot — A stem-and-leaf plot organizes numbers by their tens digits.

straight angle — An angle measuring 180 degrees.

T

ton — A U.S. unit for measuring weight; 2,000 pounds.

V

vertex (pl: vertices) — The point where two sides of a ray or a polygon meet.

vertical axis — The vertical axis is at the left. The numbers increase from bottom to top.

volume — The number of cubic units needed to fill the space occupied by a solid or three-dimensional figure.

W

weight — The measure of how heavy something is.

X

x-**axis** — On a coordinate grid, the x-axis is horizontal.

Y

yards — A yard (yd) is a unit of length. There are 36 inches in 1 yard.

y-**axis** — On a coordinate grid, the y-axis is vertical.

Volume 6
Words to Know/Glossary

B

benchmark fractions — Common fractions, such as $\frac{1}{4}$, $\frac{1}{2}$, and $\frac{3}{4}$, that can help you estimate where other fractions are located.

C

cancel — Find a common factor of a numerator and denominator and divide by that factor.

cents notation — A way to show the value of money using the ¢ sign.

common denominators — Common denominators are multiples shared by two or more denominators.

D

decimal — A way to express a whole, part of a whole, or more than a whole.

decimal point (.) — Separates the dollars from the cents in dollar notation.

denominator — The bottom number in a fraction. It tells the total number of equal parts.

dime — A coin worth 10¢.

dividend — The number being divided in a division problem.

divisor — The number that divides the dividend.

dollar bill — A bill worth 100¢.

dollar sign ($) — Money can be shown using a dollar sign ($).

E

equivalent — Two number forms that have the same value are equivalent.

equivalent fractions — Two fractions that name the same amount.

F

factor (n) — A number that is multiplied by another to give a product.

factor (v) — To write a number as a product of two or more whole numbers.

fraction — A fraction is a part of a whole or part of a set.

G

greatest common factor — The greatest factor that two numbers share.

H

hundredths place — The second decimal place to the right of the decimal point is the hundredths.

I

improper fraction — A fraction with a numerator greater than its denominator.

in order from least to greatest — To write numbers in order from least to greatest means to make a list with the least number first and each greater number to be right until the greatest number is last.

is greater than (>) — The symbol for *is greater than* is >.

is less than (<) — The symbol for *is less than* is <.

L

least common denominator (LCD) — The least common multiple of the denominator of two or more fractions.

like denominators — Denominators that are the same number.

like fractions — Like fractions have the same denominator. To add or subtraction like fractions, you add the numerators and leave the denominators the same.

lowest terms — A fraction is in lowest terms when the only common factor of the numerator and denominator is 1.

M
mixed number — A number with a whole number part and a fraction part.

N
nickel — A coin worth 5¢.

numerator — The top number in a fraction. It tells the number of parts in the fraction.

P
penny — A coin worth 1¢.

prime number — A whole number greater than 1 that is divisible by only 1 and itself.

Q
quarter — A coin worth 25¢.

quotient — The number that results from dividing.

S
simplest form — A fraction is in simplest form if the numerator and denominator cannot be divided by whole numbers other than 1.

T
tenths place — The first decimal place to the right of the decimal point is tenths.

U
unit cost — The unit cost is the price of one unit.

unit fraction — A unit fraction is one part of a whole that is divided into equal parts.

unlike denominators — Denominators that are not the same number.

W
whole — The whole is equal to one complete set. For the whole, all parts are shaded or included. For example, 8 parts out of 8 parts is the whole, or 1.

Index

A

acute angle, 213–215, SE 5, 21
addition
    of decimals, 349–351, SE 6, 77–79
    estimation with, 127–129, SE 3, 17–19
    facts, See addition facts
    of fractions, 305–307, SE 6, 35–37
    of fractions with unlike denominators, 308–310, SE 6, 38–40
    of mixed numbers, 314–316, SE 6, 44–46
    of multidigit numbers, 136–138, SE 3, 26–28
    of 1-digit numbers to 2-digit numbers, 111–113, SE 3, 2–4
    of 2-digit numbers, 114–116, SE 3, 5–7
    of 3-digit numbers, 130–132, SE 3, 20–22
    of three 1-digit numbers, 57–59, SE 2, 11–13
    strategies, 54–56, SE 2, 8–10
    and subtraction, 60–62, SE 2, 14–16
    using mental arithmetic, 124–126, SE 3, 14–16
    using to solve equations, 184–186, SE 4, 41–43
    whole numbers to 10,000, 133–135, SE 3, 23–25
addition facts, 60–62, SE 2, 14–16
angles, 213–215, SE 5, 20–22
area, 257–259, SE 5, 62–64
    of rectangles, 260–262, SE 5, 65–67
associative property
    of addition, 51–53, SE 2, 5–7
    of multiplication, 79–81, SE 2, 32–34

B

benchmark fractions, 355–357, SE 6, 84
bills, 333–335, SE 6, 62–64

C

capacity, 232–234, SE 5, 39
centigram, 235–237, SE 5, 42
centimeter, 235–243, 251–253, SE 5, 41–44, 47, 56
cents notation, 324–326, SE 6, 53
choosing operations to solve problems, 48–50, SE 2, 2–4
chord, 225–227, SE 5, 33
circles, 225–227, SE 5, 32–34
coins, 333–335, SE 6, 62–64
common denominator, 308–313, SE 6, 38, 41
commutative property
    of addition, 51–53, SE 2, 5–7
    of multiplication, 79–81, SE 2, 32–34
comparing
    data with graphs, 194–196, SE 5, 2–4
    fractions, 286–288, SE 6, 17–19
    fractions and decimals 355–357, SE 6, 83–85
comparing and ordering
    numbers to 100, 9–11, SE 1, 8–10
    numbers to 1,000, 25–27, SE 1, 23–25
    whole numbers and decimals, 330–332, SE 6, 59–61
congruent, 222–224, SE 5, 29
    triangles, 219–221, SE 5, 26
cubic units, 260–262, SE 5, 68
cup, 232–234, SE 5, 39

D

data
    comparing with graphs, 194–196, SE 5, 2–4
    recording, 197–199, SE 5, 5–7
    showing in more than one way, 203–205, SE 5, 11–13
day, 244–246, SE 5, 51
dec-, 216–218, SE 5, 24
decimal point, 324–326, SE 6, 54
decimals
    adding and subtracting, 349–351, SE 6, 77–79
    comparing with fractions, 355–357, SE 6, 83–85
    comparing and ordering with whole numbers, 330–332, SE 6, 59–61
    and fraction equivalents, 352–354, SE 6, 80–82
    and money, 339–341, SE 6, 68–70
    notation for money, 336–338, SE 6, 65–67
denominator, 270–275, 279–281, 286–288, SE 6, 2, 6, 11
    See common denominator; least common denominator; like denominator; unlike denominator
diameter, 225–227, SE 5, 32
dime, 324–326, SE 6, 53, 62
dividend, 92–94, 98–103, SE 2, 44–46, 51–55
divisible, 98–103, SE 2, 51, 54
division
    by 2, 3, 4, or 5, 98–100, SE 2, 50–52
    by 6, 7, 8, or 9, 101–103, SE 2, 53–55
    choosing a method for, 168–170, SE 4, 26–28
    families of facts, 104–106, SE 2, 56–58
    meaning of, 92–94, SE 2, 44–46
    of multidigit numbers by 1-digit numbers, 165–167, SE 4, 23–25
    of multiples of 10, 159–161, SE 4, 17–19
    problems, 171–173, SE 4, 29–31
    to solve equations, 187–189, SE 4, 44–46
divisor, 92–94, SE 2, 44–46
dollar bill, 324–326, SE 6, 53
dollar sign, 324–326, SE 6, 54
doubles, 83, SE 2, 36
drawing angles and lines, 213–215, SE 5, 20–22

E

equations
    using addition and subtraction to solve, 184–186, SE 4, 41–43
    using multiplication and division to solve, 187–189, SE 4, 44–46
equilateral triangles, 219–221, SE 5, 26
equivalent
    decimal and fractions, 352–354, SE 6, 80–82
    forms, 12–14, SE 1, 11–13
    fractions, 292–294, SE 6, 23–25, 41
estimating
    with addition and subtraction, 127–129, SE 3, 17–19
    products, 146–148, SE 4, 5–7
    quotients, 162–164, SE 4, 20–22
expanded form, 22–24, SE 1, 20
expanded notation, 35–37, SE 1, 32–34
expressions
    using unknowns in, 178–180, SE 4, 35–37

F

fact families, See families of facts
factors, 76–78, SE 2, 30, SE 6, 23
    See greatest common factor

factoring whole numbers, 289–291, SE 6, 20–22
families of facts, 66–68, SE 2, 20–22
    division and multiplication, 104–106, SE 2, 56–58
fewer than, 181–183, SE 4, 39
fluid ounce, 232–234, SE 5, 39
foot, 232–234, 241–243, SE 5, 38, 47
fractions
    addition of, 305–310, SE 6, 35–40
    basic units of, 270–272, SE 6, 2–4
    benchmark, 355–357, SE 6, 84
    comparing, 286–288, SE 6, 17–19
    comparing with decimals, 355–357, SE 6, 83–85
    and decimal equivalents, 352–354, SE 6, 80–82
    equivalent, 292–294, SE 6, 23–25
    improper, 279–281, SE 6, 11–13, 29
    interpretations of, 298–300, SE 6, 29–31
    like, 305–310, SE 6, 36
    in lowest terms, 295–297, SE 6, 26–28
    and money, 339–341, SE 6, 68–70
    subtraction of, 305–307, 311–313, SE 6, 35–37, 41–43
    unit, 273–275, SE 6, 5–7

G

gallon, 232–234, SE 5, 39
gram, 235–240, SE 5, 41, 46
graphs
    comparing data with graphs, 194–196, SE 5, 2–4
    graphing ordered pairs, 206–208, SE 5, 14–16
greater than, 181–183, SE 4, 39
greatest common factor, 295–297, SE 6, 27

H

half dollar, 333–335, SE 6, 60
hept-, 216–218, SE 5, 24
hex-, 216–218, SE 5, 24
hexagon, 216–218, 254–256, SE 5, 23, 60
hours, 244–246, SE 5, 51
hundredths, 327–329, SE 6, 56–58

I

improper fractions, 279–281, SE 6, 11–13, 29

inch, 232–234, 241–243, 251–253, SE 5, 38, 47, 56
inequalities, writing, 181–183, SE 4, 38–40
isosceles triangles, 219–221, SE 5, 26

K

kilometer, 235–237, SE 5, 42–43
least to greatest, 330–332, SE 6, 60
least common denominator, 311–319, SE 6, 42, 44

L

least to greatest, 330–332, SE 6, 60
least common denominator, 311–319, SE 6, 42, 44
length, 232–234, 251–253, SE 5, 38, 56–58
like denominators, 308–310, SE 6, 38
like fractions, 305–310, SE 6, 36
line plot, 203–205, SE 5, 11
lines, 213–215, SE 5, 20–22
liter, 235–240, SE 5, 41–42
lowest terms, 295–297, SE 6, 26–28

M

mass, 238–240, SE 5, 46
mean, 200–202, SE 5, 8
mean, median, and mode, 200–202, SE 5, 8–10
measurement system, 244–246, SE 5, 50
median, 200–202, SE 5, 9
mental arithmetic, 124–126, SE 3, 14–16
meter, 235–243, SE 5, 41–47
metric system
    metric prefixes, 235–237, SE 5, 41–43
    metric units, 238–240, SE 5, 44–46
mile, 232–234, SE 5, 38
milliliter, 235–240, SE 5, 42
millimeter, 235–240, SE 5, 43–44
minute, 244–246, SE 5, 51
mixed numbers, 279–281, SE 6, 11–13, 29
    addition of, 314–316, SE 6, 44–46
    subtraction of, 317–319, SE 6, 47–49
Mixed Review
    1: 16, SE 1, 15; 2: 29, SE 1, 27; 3: 45, SE 1, 42; 4: 70, SE 2, 24; 5: 89, SE 2, 42; 6: 108, SE 2, 60; 7: 121, SE 3, 12; 8: 140, SE 3, 30; 9: 156, SE 4, 23; 10: 175,

SE 4, 33; 11: 191, SE 4, 48; 12: 210, SE 5, 18; 13: 229, SE 5, 36; 14: 248, SE 5, 54; 15: 267, SE 5, 84; 16: 283, SE 6, 15; 17: 302, SE 6, 33; 18: 321, SE 6, 51; 19: 346, SE 6, 75; 20: 359, SE 6, 87 mode, 200–202, SE 5, 9
money
    decimal notation for, 336–338, SE 6, 65–67
    in fractions and decimals, 339–341, SE 6, 68–70
    understanding, 324–326, SE 6, 53–55
multiples, 82, SE 2, 35
multiplication
    by 2, 5, and 10, 76–78, SE 2, 29–31
    choosing a method for, 152–154, SE 4, 11–13
    facts, 85–87, SE 2, 38–40
    facts table, 76–78, SE 2, 29
    families of facts, 104–106, SE 2, 56–58
    meaning of, 73–75, SE 2, 26–28
    of multidigit and 1-digit numbers, 149–151, SE 4, 8–10
    multiples of 10, 143–145, SE 4, 2–4
    patterns and properties of, 79–81, SE 2, 32–34
    problems, 171–173, SE 4, 29–31
    to solve equations, 187–189, SE 4, 44–46
    strategies, 82–84, SE 2, 35–37

N

nickel, 324–326, SE 6, 53, 62
numbers to 100, 6–8, SE 1, 5–7
    compare and order, 9–11, SE 1, 8–10
numbers to 1,000, 19–21, SE 1, 17–19
    compare and order, 25–27, SE 1, 23–25
    writing, 22–24, SE 1, 20–22
numbers in the millions, 38–40, SE 1, 35–37
    rounding, 41–43, SE 1, 38–40
numerator, 270–275, 279–281, 286–288, SE 6, 2, 6, 11

O

oct-, 216–218, SE 5, 24
octagon, 216–218, 254–256, SE 5, 23, 60
one, 3–5, 95–97, SE 1, 2–4, SE 2, 47–49
operations
    choosing to solve problems, 48–50, SE 2, 2–4

Index

order from least to greatest, 330–332, SE 6, 60
ordered pairs
    graphing, 206–208, SE 5, 14–16
ounce, 232–234, SE 5, 38

P

parallelogram, 222–224, SE 5, 29
parts and the whole, 276–278, SE 6, 8–10
patterns and properties, 51–53, SE 2, 5–7
penny, 324–326, SE 6, 53, 62
pent-, 216–218, SE 5, 24
pentagon, 216–218, 254–256, SE 5, 23, 60
perimeter, 254–256, SE 5, 59–61
pictograph, 194–196, SE 5, 3
pint, 232–234, SE 5, 39
place value to 10,000, 32–34, SE 1, 29–31
polygons, 216–218, 254–256, SE 5, 23–25, 59–61
pound, 232–234, SE 5, 38
prime number, 295–297, SE 6, 27
products, 76–78, 146–148, SE 2, 30, SE 4, 5–7
properties, 51–53, SE 2, 5–7
    of circles, 225–227, SE 5, 32–34
    of multiplication, 79–81, SE 2, 32–34
    of zero and one, 95–97, SE 2, 47–49
    See associative property; commutative property

Q

quad-, 216–218, SE 5, 24
quadrilaterals, 216–218, 222–224, SE 5, 23–25, 29–31
quart, 232–234, SE 5, 39
quarter, 324–326, SE 6, 53, 62
quotients, 92–94, 98–100, 162–164, SE 2, 44–46, 51–55, SE 4, 20–22

R

radius, 225–227, SE 5, 32–33
range, 194–196, SE 5, 2
ray, 213–215, SE 5, 20
recording data, 197–199, SE 5, 5–7
rectangles, 222–224, 260–262, SE 5, 29, 65–67
regrouping, 114–119, 127–129, SE 3, 5–10, 17–19
regular polygon, 254–256, SE 5, 60
relating addition and subtraction, 60–62, SE 2, 14–16

remainder, 98–100, SE 2, 50–55
repeated addition, 73–75, SE 2, 26
right angle, 213–215, SE 5, 21
right triangle, 219–221, SE 5, 27
rounding through millions, 41–43, SE 1, 38–40

S

second, 244–246, SE 5, 51
showing data in more than one way, 203–205, SE 5, 11–13
simplest form, 305–310, SE 6, 36
skip count, 73–75, SE 2, 27
square units, 257–262, SE 5, 62, 65
standard form, 22–24, SE 1, 20
strategies
    of addition, 54–56, SE 2, 8–10
    of multiplication, 82–84, SE 2, 35–37
    of subtraction, 63–65, SE 2, 17–19
stem-and-leaf plot, 197–199, SE 5, 6
subtraction
    and addition, 60–62, SE 2, 14–16
    of decimals, 349–351, SE 6, 77–79
    estimation with, 127–129, SE 3, 17–19
    of fractions, 305–307, SE 6, 35–37
    of fractions with unlike denominators, 311–313, SE 6, 41–43
    of mixed numbers, 317–319, SE 6, 47–49
    of multidigit numbers, 136–138, SE 3, 26–28
    strategies, 63–65, SE 2, 17–19
    of 2-digit numbers, 117–119, SE 3, 8–10
    of 3-digit numbers, 130–132, SE 3, 20–22
    using mental arithmetic, 124–126, SE 3, 14–16
    using to solve equations, 184–186, SE 4, 41–43
    whole numbers to 10,000, 133–135, SE 3, 23–25
symbols (<, =, >), 9–11, 25–27, 330–332, SE 1, 8–10, 23–25, SE 6, 59

T

tally marks, 12–14, SE 1, 12
tens, 3–5, SE 1, 2–4
tens and ones, 3–5, SE 1, 2–4
tenths, 327–329, SE 6, 56–58
ton, 232–234, SE 5, 38

Topic Summary
    1: 15, SE 1, 14; 2: 28, SE 1, 26; 3: 44, SE 1, 41; 4: 69, SE 2, 23; 5: 88, SE 2, 41; 6: 107, SE 2, 59; 7: 120, SE 3, 11; 8: 139, SE 3, 29; 9: 155, SE 4, 22; 10: 174, SE 4, 32; 11: 190, SE 4, 47; 12: 209, SE 5, 17; 13: 228, SE 5, 35; 14: 247, SE 5, 53; 15: 266, SE 5, 83; 16: 282, SE 6, 14; 17: 301, SE 6, 32; 18: 320, SE 6, 50; 19: 345, SE 6, 74; 20: 358, SE 6, 86
trapezoids, 222–224, SE 5, 30
tri-, 216–218, SE 5, 24
triangles, 219–221, SE 5, 23, 26–28

U

unit conversion
    factors and divisors in, 241–243, SE 5, 47–49
    within a system, 244–246, SE 5, 50–52
unit cost, 342–344, SE 6, 71–73
unit fractions, 273–275, SE 6, 5–7
units of U.S. measure, 232–234, SE 5, 38–40
unknowns, 178–180, SE 4, 35–37
unlike denominators
    addition of fractions with, 308–310, SE 6, 38–40
    subtraction of fractions with, 311–313, SE 6, 41–43
using coins and bills, 333–335, SE 6, 62–64

V

vertex, 206–208, 213–215, SE 5, 15, 20
volume, 263–265, SE 5, 68–70
weight, 232–234, SE 5, 38
whole, 276–278, SE 6, 8–10
whole numbers
    comparing and ordering with decimals, 330–332, SE 6, 59–61
    factoring, 289–291, SE 6, 20–22

W

word form, 22–24, SE 1, 20
writing
    equations and inequalities, 181–183, SE 4, 38–40
    numbers to 1,000, 22–24, SE 1, 20–22

Y

yard, 232–234, 241–243, SE 5, 38, 47

Z

zero, 35–37, 95–97, SE 1, 32–34, SE 2, 47–49